SHOOTING THE FRONT EASTERN OPERATIONS
VOLUME ONE

WITHOUT FLYERS, NO TANNENBERG

Aviation on the Eastern Front of 1914—
Evolution of a Critical Role For Modern Warfare

BY TERRENCE J. FINNEGAN,
HELMUT JÄGER & CARL J. BOBROW

DAWDLE PUBLISHING, LLC

Durham, New Hampshire

ISBN - 978-1-955072-02-1

Shooting the Front Eastern Operations Volume 1: Without Flyers, No Tannenberg

By Terrence J. Finnegan, Carl J. Bobrow & Helmut Jäger

terrencefinnegan.com, shootingthefront.com

Published by Dawdle Publishing, LLC
Durham, NH 03824 dawdlepublishing.com

This book was edited by Elaine Mintzer and designed by Bridget Finnegan. bridgetfinnegan.com

All of the images contained within this book are public domain unless otherwise marked.

Front Cover:
Modified painting. Original by Hans Rudolf Schulze.

Back Cover: *Battle of Tannenberg* by Hugo Vogel (1917).
Generals depicted are identified below.

DEDICATION

This book is dedicated to my late wife Susan, my life companion for 40 years. Susan was a pioneer—breaking new ground for the USAF proving a female could do a great job every time. She served her nation, lived life to the fullest and in her last years was a disabled veteran. Susan's courage and creativity always inspired me through the decades and I remain devoted to her memory.

—Terry Finnegan

SR-71 photo on a mission over the Pacific Theater taken by Susan Finnegan from a KC-135 in 1979.

TABLE OF CONTENTS

ACKNOWLEDGMENTS

In 2013, Carl Bobrow asked me to consider writing *Shooting the Front* for the Eastern Front.

He did so, as he was familiar with my prior work. I agreed to take on the project, because I wanted to apply the same, analytical- military scholarship of my early-related work, to the Eastern Front— an area I had not previously explored. The Eastern Front was probably the most important theater of the entire war; if the Germans had lost at Tannenberg, they would not have had the fortitude and determination to proceed with the rest of the war.

I want to thank Carl for the detailed Russian data he's provided and his extreme attention to content and detail, that has informed the book, greatly. As well as for "keeping the faith" throughout the writing and rewriting of the text. Helmut Jäger's work ethic has been herculean. He not only contributed to the content and thought process but also provided invaluable insights into original sources that required translation from the German. Helmut served as writer, editor, proofreader, and translator from the German. Both Helmut and Carl are my co-authors.

Over the years, the three of us consulted with a number of noted scholars on the subject of World War I military aviation. We were fortunate to have these experts advise us, as the field is very limited. They generously invested their time in our evolving manuscript, replete with typos, improper translations, and misinterpreted facts. They provided us with insights and meaningful input that allowed us to more carefully and thoroughly craft the text. I want to thank the following contributors, for all that you have done for this book over the years: Dr. James Davilla, Dr. James Streckfuss, Aaron Weaver, Steve Suddaby, Mike Hanlon, Sal Compagno, Dr. Gregory Vitarbo, Andrew Smoot, Dr. Von Hardesty, Jon Barrett, Dr. Dieter Gröschel, James Pratt, Kurt Schwarz, Sebastian Rosenboom, Jon Guttman, Dr. Andrzej Olejko, George Fuderer, Dr. Jenifer Van Vleck, Dr. Lance Bronnenkant, Dr. Hannes Täger and Dr. Sarah J. Tisch-Small.

In recent years, Carl made a breakthrough discovery in Russian aviation research, by accessing Russian works in progress as well as acquiring files that were buried in Moscow for over 100 years. Marat Khairulin provided us with photographs and Russian documents and invested countless hours in correspondence with Carl on a variety of subjects—which included the history of Russian aviation from inception to the revolution.

The time that I've spent on this book has been bittersweet. Susan, my wife of 40 years, was dying of Alzheimer's for most of these years and my personal life was consumed as a caregiver. She passed away on the morning of 28 September 2019. My caring for Susan, overlapped with the writing and coordination of source and substance, with Carl and Helmut. Both of them became my core support, as Susan's world came to an end.

This book and all of the work I've done on it, is dedicated to Susan's memory—a reminder that our lives are only fulfilling if we make them so, with the ones we love.

I had the pleasure of working with Elaine Mintzer, a dedicated editor whose attention to detail on the proper use of the English language was awe inspiring. Elaine's proofreading and editing the manuscript resulted in the best oration on what this amazing chapter of military history could have provided. In many ways, this book is dedicated to Elaine for helping all of us reach the excellence that we all required and expected.

We would like to thank our readers in advance, for taking interest in a subject we are passionate about. Our main objectives, were to document a period in time and the multiple military aviation breakthroughs that changed the scope of intelligence. We also wanted to pay tribute to those who shaped aviation during the early stages of the Great War, who carved a path for future generations of aviators and intelligence experts.

PREFACE

Though I'd written two books before (*Shooting the Front, Allied Aerial Reconnaissance in the First World War* and *'A Delicate Affair' On the Western Front*), a number of factors made the writing of this third book, a radically different experience. I was older, my late wife was ill, the subject matter was far more complex to construct a coherent historical chronicle (as the data and sources were sparse) and I had two co-authors. In summary, I was negotiating new and more arduous terrain in attempting to craft a cogent and coherent narrative.

My early and extensive work in government intelligence, fueled my passion to keep going back to intelligence issues and breakthroughs, during the Great War. I was compelled to continue to pursue areas of study related to aviation's history and its impact, that hadn't been fully explored. Carl's knowledge of aircraft and access to all-things Russian as well as his academic stature and skills, and Helmut's life experience—flying German planes, working with NATO, as well as his academic studies and skills, were essential to this effort. This book was a thoroughly collaborative effort.

Our collective task, was to take all the data available, prune it for relevance, then try to make sense of it. The only logical way to present the data, was in chronological order. The book only covers the period from August 1914 to April 1915. We chose to cover the battles in detail for a number of reasons. They intersect with all of the data on aviation and they provide a historical record of events, that have been documented before in original sources but never written about in detail. Our access to new sources in both German and Russian, allowed us to introduce newly recorded history.

We chose to provide an extensive history of Russian aviation intelligence in Chapter 2, because it offered new facts and insights into Russia's early efforts, that we felt added value. Carl had access to a unique body of documents and sources, that were deemed to be important in expanding the scope of the book.

Though the focal point throughout our work was on the data, we felt an imperative to present to the reader, an intermittent chronicle of history's open issues, ambiguous outcomes, and controversial conclusions. Though we had our own individual viewpoints, we saved most of them for the conclusions section. In the "final thoughts" section of the conclusion, we chose to briefly share various insights, theories, and global implications. It was integral and vital to the narrative, to articulate some of the more poignant issues surrounding the events we've carefully documented. The "final thoughts" are a collection of all of the authors' perceptions. Our hope is that both the data and our reflections, stimulate further thought, discussion, and research, for generations to come.

INTRODUCTION

The Great War-Eastern Front

The time period between the Napoleonic era and 1914 experienced profound developments in warfare strategy. The fortress legacy initiated by Vauban in the 18th Century, no longer served as a functional means to win wars. Modern technological advances had shaped the strategic and logistical initiatives on the battlefield. These advances were reflected in the campaigns on the Eastern Front—an area so vast that it dwarfed the landscape of the Western Front.

As Russia's Northwest Front entered Germany's *Ostpreussen* [East Prussia] region, Germany fortified their campaign through aviation intelligence. This enabled them to track the Russian campaign and report to commanders as frequently as sorties could be generated. German commanders regularly accepted aerial reconnaissance reports and quickly maneuvered their forces against known enemy positions.[1] Germany's "home field advantage" proved critical in validating intelligence. Not only were they informed of terrain covered, but leadership benefited from a communications network that enabled a near-real-time delivery of aviator observations.

Intelligence collection was successfully supported by a well-established wireless intercept operation. Logistic and maintenance support to aeroplanes sustained the initial fleet, tasked to monitor any Russian advance. In contrast, Russia lacked an adequate infrastructure as they invaded Germany. Their aviators had neither knowledge of the region nor an effective communication network to serve advancing Russian armies. They also employed a failed logistic and maintenance element to service both airframes and rotary engines. Operating without aerial reconnaissance to validate decisions, Russian aviators were still tasked as message carriers. In summary, Russian commanders failed to place value on aviation in the campaign.

Research and Resources

Organizing the story of aviation's role on the Eastern Front, was a significant challenge. The primary sources for research were not aviation centric. Prevailing secondary sources barely touched on the roots of aviation on the Eastern Front. Primary published history on aviation's role in the Great War, for the most part focused on the Western Front. Secondary sources reflected an intellectual bias for analyzing and recording the history of the West, due to the inaccessibility (over the past century) of archives in Eastern Europe and Russia. Soviet period histories and sources were suspect and tended to reflect a politically-tainted view. Limited source materials, constrained knowledge of the Eastern Front, and biased published history records, all leaned towards western interpretations. In recent years, a bounty of previously unknown German and Russian records surfaced, uncovering a body of primary source material never before published in the West. Russian archives such as the *Rossiiskii gosudarstvennyi voenno-istoricheskii arkhiv (RGVIA)* [Russian State Military History Archive] have become a goldmine of rare and important documents on Russian aviation experience.

Researching Germany's aviation role required review of the Reichsarchiv's monumental 14-volume history, *Der Weltkrieg 1914 – 1918,* published during the interwar years. Critical to our analysis were two volumes of *Kriegswissenschaftliche Abteilung der Luftwaffe. 3. Sonderband: Mobilmachung, Aufmarsch und erster Einsatz der deutschen Luftstreitkräfte im August 1914* (published in Berlin in 1939) was well known and has been accessible to scholars for some time. Research on the subject of German aviation in 1914 was made available in the extremely rare *8. Sonderband: Die Luftstreitkräfte im Zusammenhang mit den Operationen des Jahres 1914* (completed in 1943 but never published). *8. Sonderband* proved to be the most useful source in describing *Fliegertruppe's* early operations, generated from detailed aviator reports.

Shooting the Front Eastern Operations also benefited from a resource that became available at the National Archives and Records Administration (NARA) in College Park, MD. In the 1920's the US Army, in coordination with the Reichswehr, agreed to share planning and operations documents. German army records also covered *8. Armee* in 1914. Not only were the original German records made available, we also discovered a translated English version that was originally written and reviewed by the staff of the US Army War College. Key to the analysis, was the *8. Armee War Diary,* which provided daily accounts of battle related events. Actual *Fliegertruppe* aviation reports were published and subsequently included in the chronological *8. Armee* accounts and this data is integral to our analysis.

The Book's Organization

Shooting the Front Eastern Operations, Volume I, addresses aviation's role in the Russian invasion of *Ostpreussen*[1] and is divided into five parts. In Part I the roots of aviation in Germany and Russia are discussed. Germany's *Preussisch-Deutsche Fliegertruppe* [Prussian-German Air Force] and Russia's *Rossiyskiy Imperatorskiy Voyenno-Vozdushnyy Flot (RIVVF)* [Russian Imperial Military Air Fleet] share similar accounts of how aviation served military requirements. Part II, "A Change in Command," explores the initial accomplishments of aviation in the opening days of the war. German *8. Armee's* commander *Generaloberst* v. Prittwitz and Russian Northwest Front commander *General ot kavalerii* Zhilinsky both demonstrated limited knowledge of aviation's potential. Prittwitz's myopic application of reconnaissance resulted in his inability to gauge enemy intentions and his disastrous decision to withdraw from *Ostpreussen.* He is replaced after three weeks. Zhilinsky's failures resulted in Russia's catastrophic defeat at Tannenberg. He manages to hang on as overall commander, until being replaced in early 1915.

After documenting the antecedents to Tannenberg, Part III is devoted to the actual battle of Tannenberg, an archetypal success on the modern battlefield—where German commanders effectively employed aviators and eventually grew to depend upon them as assets. In contrast, the archetypal antecedents of failure were fueled by Russia's military leaders, who had disdain for aviation's potential. Tannenberg was a Russian catastrophe; early on in the war it carved the landscape of the war's future.

In Part IV, The Masurian Lakes Campaign, we discuss how an exhausted aviation force needed to sustain operations tempo. The Russian retreat out of *Ostpreussen* was skillfully accomplished by *General ot kavalerii* v. Rennenkampf. The eventual battle of the Masurian Lakes required consistent and timely reconnaissance to correctly posture against the enemy. Part V, the final section, summarizes the acceptance by both sides of aviation's critical role in fighting a modern-day battle. Commanders' decision-making required detailed information on rear echelons and the Russians finally achieved that milestone by operationally fielding the Il'ya Muromets. In spite of early acquisition of French aeroplanes the Russians initially lacked the organizational skills and infrastructure to successfully implement and utilize their aerial fleet.

The ultimate goal of *Shooting the Front Eastern Operations* is to invite the reader into the world of aviation and how the aerial reconnaissance mission it commanded evolved tactically and strategically. In recounting the role that the Eastern Front played in history, we have told the story of how a complexity of innovations ushered in a new era of intelligence that transformed the shape of warfare for generations to come.

1 Ostpreussen will be used throughout instead of East Prussia to provide a more definitive understanding of the German identity during the campaign.

RUSSIAN RANKS

General Feldmarshal	Gen.FM	General of the Army
General ot infanterii	Gen.inf	General of Infantry
General ot kavalerii	Gen.kav	General of Cavalry
General ot artillerii	Gen.art	General of Artillery
General-leytenant	Genleyt	Lieutenant General[1]
General-maior	Genmai	Major General[2]
Polkovnik	Polk.	Colonel
Podpolkovnik	Podpol	Lieutenant Colonel
Voiskovoi starshina	Vstar	Cossack Lieutenant Colonel
Kapitan	Kapt.	Major
Yesaul or Esaul	Es.	Cossack Major
Rotmistr	Rotm.	Cavalry Major
Shtabs-kapitan	Sh.Kap.	Staff Captain
Shtabs Rotmistr	Sh.Rotm.	Cavalry Staff Captain
Podyesaul	Pod'es.	Cossack Staff Captain
Poruchik	Por.	Lieutenant
Sotnik	Sot.	Cossack Lieutenant
Podporuchik	Podpor.	Second Lieutenant
Kornet	Korn.	Cavalry 2nd Lieutenant
Khorunji or Khorunzhii	Khor.	Cossack 2nd Lieutenant
Praporshchik	Prap.	Ensign of Reserve
Starshiy Unter-Ofitser	St.Uo.	Senior Non-Commissioned Officer

GERMAN RANKS

Generalfeldmarschall	FM	Field Marshal or General of the Army
Generaloberst	GO	Colonel General
General der Infanterie	GdI	General of Infantry
General der Kavallerie	GdK	General of Cavalry
General der Artillerie	GdA	General of Artillery
Generalleutnant	Genltn	Lieutenant General3
Generalmajor	Genmaj	Major General4
Oberst	Obst	Colonel
Oberstleutnant	Obstltn	Lieutenant Colonel
Major	Maj	Major
Hauptmann	Hptm	Captain
Rittmeister	Rittm	Cavalry Captain
Oberleutnant	Obltn	1st Lieutenant
Leutnant	Ltn	Lieutenant

1 Equivalent to Major General in US Army rank.
2 Equivalent to Brigadier General in US Army rank.
3 Equivalent to Major General in US Army rank.
4 Equivalent to Brigadier General in US Army rank.

PART I

AVIATION PRIOR TO THE "UNKNOWN WAR"

CHAPTER I

Preussisch-Deutsche Fliegertruppe

A Vision for Aviation – At a Cost

Germany's introduction to military aviation reflected the whims of Kaiser Wilhelm II and his family. Their dominance in shaping Prussian military culture influenced aviation development. Their ultimate goal was to fortify military operations against Germany's enemies. *Fliegertruppe* [Air Service] were subsequently instituted and put in place to augment the *Preussisch-Deutsche* federal army ranks. German aviation initially focused on airships.

Kaiser Wilhelm II symbolized Germany's interest in aviation as a tool for military applications.

Ferdinand Graf von Zeppelin was the lead advocate for airships and his advocacy was reinforced by the Berlin Military Administration, the Prusso-German War Minister, and the Kaiser. Critical funding for airships far surpassed investments in aeroplane development, as Germany's initial vision was influenced by Zeppelin's massive-rigid airship designs. Perceptions began to change when Orville Wright (in 1909) personally demonstrated to the Germans the full breadth of an aeroplane's capabilities.[1] In 1910, *Generaloberst* Helmuth von Moltke's *Chef des Grossen Generalstabs* [Chief of the Grand General Staff], view also shifted in favor of aeroplanes. German military leaders changed their focus and rigorously monitored France's early initiatives in aeroplane development. The incentive to raise public awareness of this credible threat resulted in the implementation of a media-based public outreach campaign. Kaiser Wilhelm's brother, Prince Heinrich, was chosen to lead the movement to inspire German investment in aeroplanes. In 1912 the movement was

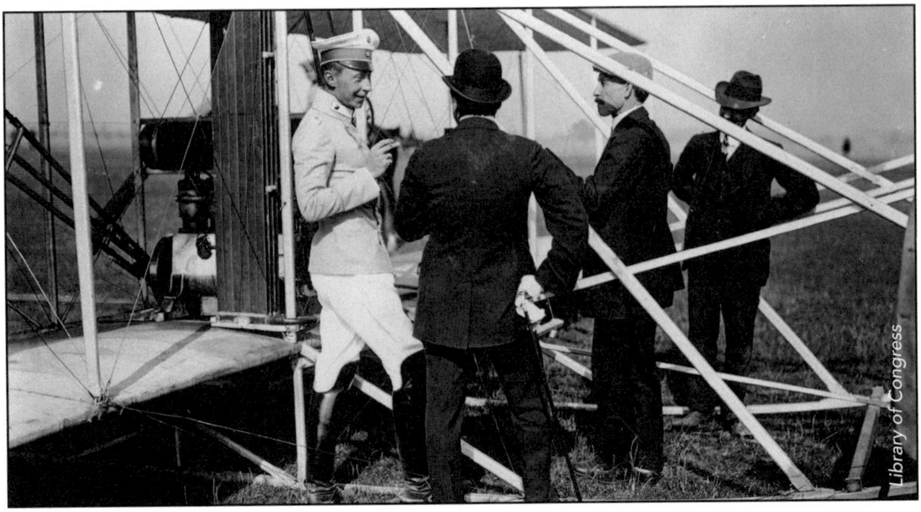

On 2 October 1909, Crown Prince Wilhelm of Germany (on the left) became the first member of royalty to fly in an aeroplane. Orville Wright (wearing light cap facing the prince) flew Wilhelm on a fifteen-minute flight. Wright established a record later the same day, by flying his aeroplane to an altitude of 488 meters.

launched. Citizens provided a solution to defense budget limitations and created a private sponsorship program. "*Nationalflugspende für das Militärflugwesen*" became the cornerstone of the initiative culminating in the successful launch of Germany's aeroplane industry.[2]

German aviation associations (totally comprised of civilian membership) procured over six million Marks to provide pilot training. Royal advocacy provided financial support for training, as well as funding for airframes and motors. Designs complied with rigorous military requirements for a rugged, dependable, two-seat aerial reconnaissance aeroplane. At the same time, the German military kept a close watch on France's aviation developments. The 1913 Paris Air Show provided the Germans with valuable insights. French aeroplanes had the capacity to utilize armaments such as machine guns. German aeroplanes initially lacked this capacity. It became clear to the Germans that they needed to invest more resources into research and development, and their subsequent advances enabled them to set records.[3]

Viktor Stoeffler set the aeroplane distance record on 14 October 1913 by flying 2253 kilometers in a single day in an Aviatik "Arrow" biplane. The Germans set two world records for aviation just before the war began. On 11 July 1914, Reinhold Boehm estab-

Generaloberst Helmuth Johannes Ludwig Graf von Moltke (also known as Moltke the Younger) served as Chef des Generalstabes der deutschen Armee at the start of the war.

Oberstleutnant Erich Ludendorff, head of the "Aufmarschabteilung" [Mobilization and Deployment Department] of Der Grosse Generalstab, emerged as visionary for Fliegertruppe.

Kaiser Wilhelm's brother, Prince Heinrich (left), led the movement to inspire Germans to invest in aeroplanes. This spurred a private sponsorship program, "Nationalflugspende für das Militärflugwesen." He is seen speaking with Edmund Rumpler, director of Rumpler's aviation firm.

EARLY GERMAN AVIATORS

Viktor Stoeffler (right) set the aeroplane distance record on 14 October 1913, by flying 2253 kilometers in a single day, in an Aviatik "Arrow" biplane.

Reinhold Boehm flew an Albatros biplane on 11 July 1914, and set a world record for nonstop flight, of twenty-four hours twelve minutes, covering an estimated 2173 kilometers—a record that was not broken until thirteen years later.

Heinrich Oelerich set a world altitude record of over 7620 meters in a Deutsche Flugzeug-Werke (DFW) aeroplane on 14 July 1914.

lished a world record for a non-stop flight of twenty-four hours, twelve minutes (flying a German-made Albatros biplane) and this record remained unbroken until thirteen years later. Three days later, Heinrich Oelerich set a world altitude record of over 7620 meters in a *Deutsche Flugzeug Werke (DFW)* biplane.[4]

Along with Ludendorff, Hauptmann Hermann von der Lieth-Thomsen was a leader in formulating Generalstab vision, aviation's role. Thomsen flew on Zeppelin Z-V over Ostpreussen, at the beginning of the war.

These records served to rebrand Germany as a leader in military aviation development and innovation.

Matt Bowden, in his thorough examination of *Fliegertruppe's* commitment to the Western Front of 1914, provided an in-depth analysis of the evolving role of German aerial reconnaissance, prior to the onset of war. The work of General Erich Ludendorff, *The General Staff and its Problems,* was integral to his analysis. Chapter I, Volume I, covered *GO* v. Moltke's *Generalstab* August 1912 analysis of aeroplanes versus airships, which highlighted the era of French dominance in utilizing aeroplanes to support their armies.[5] Moltke's focus on this "new and important weapon," contained five proposals:

1. Mobile field reconnaissance flights and an aeroplane park for the army headquarters.
2. One mobile field reconnaissance flight for each army corps, including the reserve corps.
3. One mobile field reconnaissance flight for the cavalry corps.
4. Fortress reconnaissance flights for the important fortresses.
5. Artillery squadrons for the army corps.[6]

DÖBERITZ FLUGPLATZ AERODROME

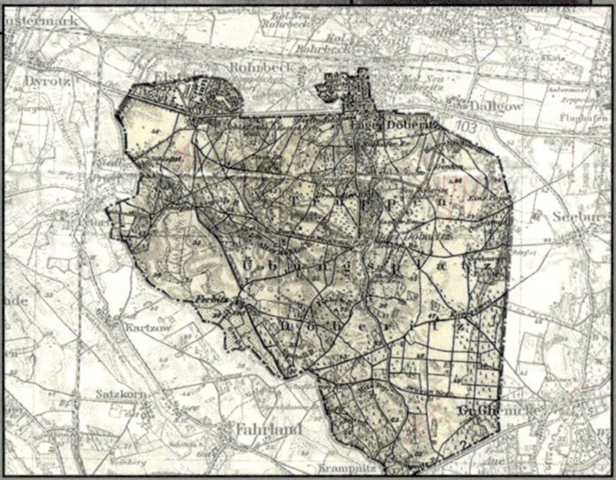

Döberitz Flugplatz (aerodrome) near Berlin, was established in 1901 for aeronautics (airship/balloon) training. In 1910, the Provisional Aviation School was set up and subsequently became the primary school for Fliegertruppe. A contemporary map shows Döberitz.

Döberitz has the distinction of being the birthplace of Fliegertruppe training. Four future aviators enjoy a respite from the training experience.

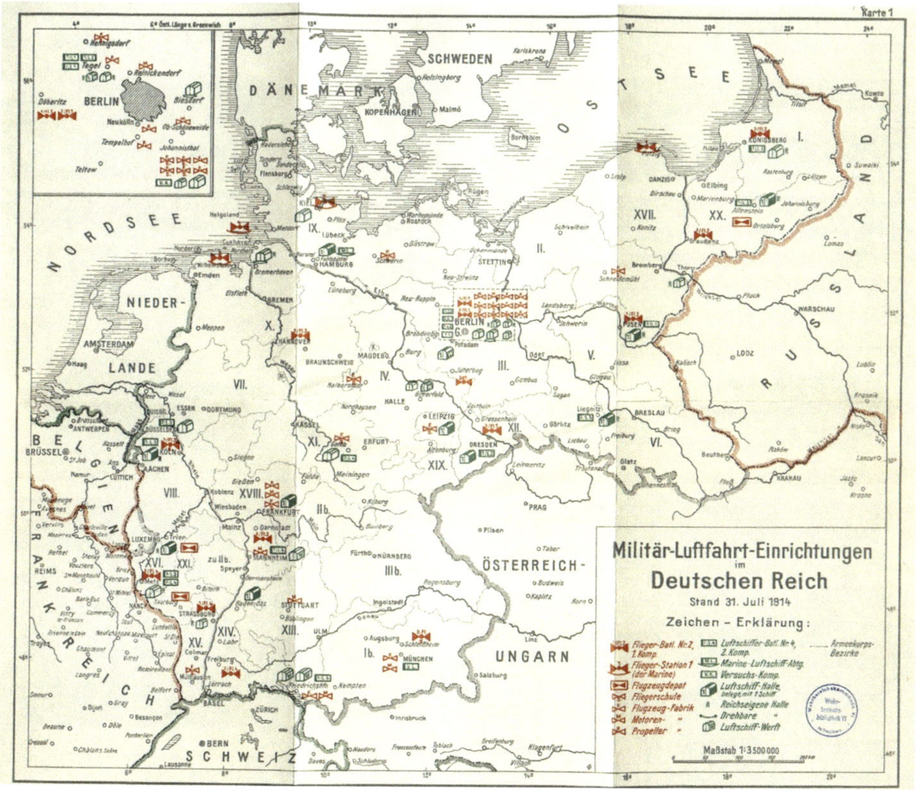

Germany's Generalstab perceived the Eastern Front as a holding action. The majority of Fliegertruppe aerodromes and facilities supported the Western Front prior to the war. Source: Karte 1, Anlagenband zu Die Militärluftfahrt bis zum Beginn des Weltkriegs 1914, Berlin 1941. Larger map available at shootingthefront.com

Preussisch-Deutsche Fliegertruppe aviator training began as a military initiative but initially relied on civilian expertise and assets. German aeroplane manufacturers had established their own facilities, which subsequently set the standard for military training requirements that served the entire industry. *Oberstleutnant* Erich Ludendorff, head of the important Mobilization and Deployment Department [*Aufmarschabteilung*] of the Grand General Staff [*Der Grosse Generalstab*], and his assistant, *Hauptmann* Hermann Thomsen, emerged as leaders in *Preussisch-Deutsche Fliegertruppe's* prewar thinking on aviation. At Air Station Döberitz (birthplace for *Preussisch-Deutsche Fliegertruppe*) both Ludendorff and Thomsen experienced the new aeroplanes first-hand through orientation rides. In doing so they deepened their understanding of how aviation could facilitate military planning and execution.

PRE WAR ORGANIZATION (AS OF 1 OCTOBER 1913)

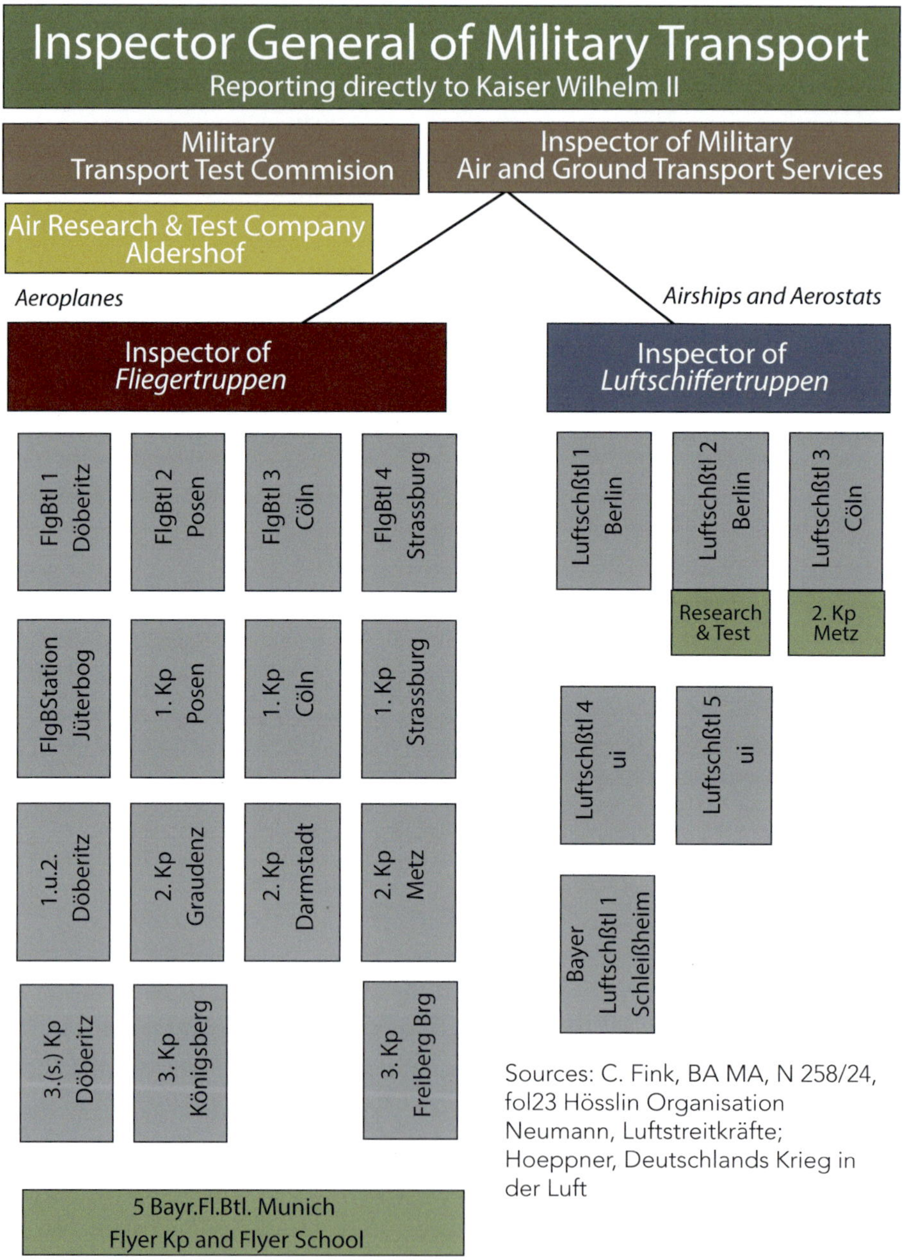

Sources: C. Fink, BA MA, N 258/24, fol23 Hösslin Organisation Neumann, Luftstreitkräfte; Hoeppner, Deutschlands Krieg in der Luft

Fliegertruppe in 1913. Outlines aviation as a military transport resource reporting directly to Kaiser Wilhelm II. Both aeroplanes and airships were organized through the Generalinspektion des Militärverkehrswesens (Inspector General of Military Transport).

Germany's Festungen (Fortress) network was designed to maintain a defensive bulwark against a possible Russian invasion. The Russian Fortress network played a major role in allocation of RIVVF flying resources.

Their collective advocacy helped fuel the argument for both airship and aeroplane armament priorities, as well as for deployment of more aeroplanes to complement the long-favored Zeppelin airship fleet.[7]

Der Grosse Generalstab Views the Russian Threat

Once the war began, Germans focused on mobilizing forces on two simultaneous fronts. France was their first priority. France's defeat in 1871 remained the catalyst for driving the German military towards employing state-of-the-art weaponry. Russia's visible military presence on the *Ostpreussen* border, was of equal concern.

Despite Russia's vast armies, Germans regarded operations on the Eastern Front as a holding action. Their priorities rested with defeating France first, before any efforts against Russian mobilization could take hold. Once victory over France was secured, *Oberste Heeresleitung's (OHL)* [Army High Command] strategy was to move east for the final assault on Russia.

OHL needed to take into account the role *Festungen* [Fortresses] played in fighting Russia. Prior to 1914, Europe still held fortresses, but their emphasis was on conducting and maintaining the offensive. Fortresses applied and utilized Vauban's (the 18th century military genius) tradition of achieving military dominance through strategic positioning. The aging strategic *Festungen* network in Germany was designed to maintain a defensive bulwark against a possible Russian invasion. All fortress commanders were subordinate to the *8. Armee* commander. The most important fortress in *Ostpreussen* was Königsberg, supported by rail and sea.

Though many *Generalstab* officers in *OHL* understood the criteria to succeed on the Western Front, they lacked a depth of understanding when they looked east towards Russia. German pre-war intelligence had limited knowledge of the Russian war plan. How quickly could the Russians advance into Germany? Alternatively, did the Russians intend to concentrate forces against Austro-Hungarian forces, serving in Galicia? Early on, pre-war German estimates of Russia's capabilities led to a specific strategy. A piecemeal application of forces (attacking and defeating one Russian army after another) was considered the best possible course of action. Therefore, *OHL's* planning priorities were to position seven *Armeen* against the west, while maintaining *8. Armee* to fend off the Russians to the east. Once war commenced, both the slow pace of Russia's deployment and slow advance to the battleground, confirmed the wisdom of the Germans' initial military strategy.

The continued development of *OHL's* defense against the Russians, necessitated extensive military intelligence work. The German army *Nachrichtendienst* [Intelligence Service] was established to organize military intelligence. *Nachrichtendienst* provided staff to assist operations in planning. The vital wireless intercepts were accomplished by *Nachrichtenwesen* [Signal Troops].[8] *Hauptmann* Walter Nicolai—an intelligence professional with great

ZEPPELINS

The Zeppelin Z-IV [LZ 16 (production number)] was the primary airship over the Eastern Front until decommissioned in the fall of 1916. Z-IV's notoriety prior to the war, was achieved on 3 April 1913, when it inadvertently landed near Lunéville, France. The French military released Z-IV back to Germany the next day, after an extensive examination of the technology that made Zeppelin a world leader in airship design.

depth in understanding the Russian threat, possessed an expertise that enabled him to effectively manage counterintelligence operations against Russian agents in *Ostpreussen*. In April 1913 Nicolai operated from *Gouvernement Festung* Königsberg, expanding the *Nachrichtendienst* mission to cover both intelligence gathering and counter-espionage. *Nachrichtendienst* acquired information that outlined Russian plans for invading *Ostpreussen*, utilizing two armies to attack from Vilna and Warsaw military districts.[9] *Hauptmann* Nicolai was later chosen and promoted to *Major*, to lead *Abteilung IIIb*, the *Generalstab's* most significant military intelligence department, operating counterintelligence against France and Russia. He served in that capacity throughout the entire war and was later promoted to *Oberstleutnant*.[10]

When war began, *OHL's GO* v. Moltke clarified his primary strategy against the Russians, in emphasizing his priorities on the Eastern Front. Germany had to defeat Russia as well as defend their weaker ally (the Austro-Hungarian *kaiserlich und königlich (k.u.k.)* ["imperial and royal"] *Armee*.[11] Moltke envisioned a complete destruction of the threat posed by the Russian "steamroller." In a message to the *8. Armee* commander, *Generaloberst* Maxmilian von Prittwitz und Gaffron, he wrote, "If the Russians approach, never choose the defensive, only offensive, offensive, offensive!"[12] The first days of the war proved Moltke's strategy effective in shaping the outcome of the critical battle of Tannenberg.

Fliegertruppe in the Ostpreussen Region

As hostilities unfolded, *Fliegertruppe* was comprised of aviation units with a commander and six aeroplanes (each with an aircrew consisting of a pilot and aerial observer), supported by maintenance and other necessary personnel. *Fliegertruppe* in the *Ostpreussen* region, was comprised of four *Feldfliegerabteilungen* [aviation units] with six aeroplanes each. One *Feldfliegerabteilung (FFA)* was assigned to *8. Armee* and the remaining three were each assigned to the first line *Armee-Korps* with *8. Armee's* solitary reserve corps (*I. Reserve-Korps*, operating without a *Feldfliegerabteilung*). Four *Festungsfliegerabteilungen* [fortress aviation units] with four aeroplanes each, served the German Eastern Front region [Königsberg, Graudenz, Boyen (Lötzen), Thorn].[13]

Generalstab's efforts in operationalizing the aeroplane were impacted by prevailing views on airships. Though prior to the war, early

LEUTNANT CARL FINK

Leutnant Carl Fink led Fliegertruppe efforts with aerial photography.

Fink was an early graduate of Döberitz. He learned to fly on a Farman-type aeroplane.

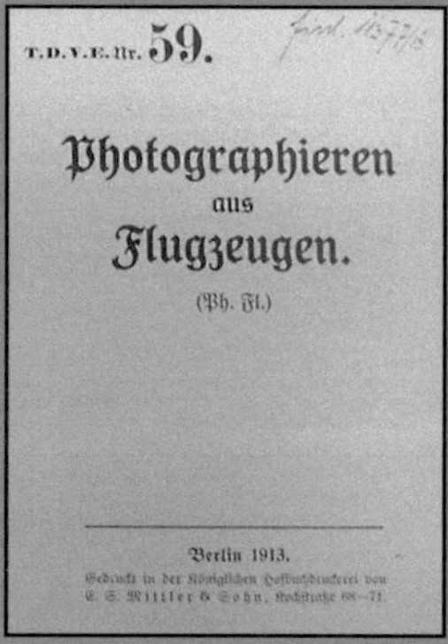

T.D.V.E.Nr. 59.

Photographieren aus Flugzeugen.

(Ph. Fl.)

Berlin 1913.
Gedruckt in der Königlichen Hofbuchdruckerei von E. S. Mittler & Sohn, Kochstraße 68–71.

Fink's manual on aerial photography Photographieren aus Flugzeugen (published in 1913) covered essentials such as the need for an easy-to-handle aerial camera, optical principles relating to aerial photographic practice, and priorities for observation and aerial photographs.

Thorn maneuvers demonstrated airships had difficulty defending themselves from hostile artillery or bombs, and captive balloons were at risk from artillery fire.[14] *8. Armee* was supported by two Zeppelin airships—Königsberg's Zeppelin Z-IV and Posen's Zeppelin Z-V. A third dirigible, Schütte-Lanz-built Airship S.L.2, initially assigned to *8.* Armee, was diverted to the south to assist *k.u.k. Armee-Oberkommando (A.O.K.)* with additional strategic reconnaissance over *Galicia*.[15]

Hauptmann Walter Nicolai, led Germany's military counterintelligence effort on the Russians, prior to the war, as Nachrichtendienst des Generalstabes (Secret Service of the General Staff). Operating from Festung Königsberg, Nicolai's Nachrichtendienst acquired information outlining Russian plans for invading Ostpreussen.

A single aeroplane park at Dirschau near Danzig, serving *8. Armee's Fliegertruppe*, provided logistic support and maintenance repairs to all aviation units. *Feldfliegerabteilung* typically maintained a ground staff of approximately 120 men for maintenance and repair. Reserve equipment, spare parts, workshops, and other necessary logistic and maintenance functions, were supported by the aeroplane park. Aeroplanes were flown to new operating locations. When flight was impossible, aeroplanes were disassembled and directed to a new location by railroad or trucks. Similarly, supporting staff, war consumables, aeroplane shelters, equipment and material utilized for front line technical and logistic support, also employed trucks for transport.[16]

Aerial Photography's First Applications

In the early stages of the war, aerial reconnaissance centered on observer sightings. Aerial photography was introduced over time. Information dissemination focused on the written word, as it appeared in the *Meldungen* [field reports] of aviators. Aircrews either landed in

ZEISS PISTOLENKAMMER

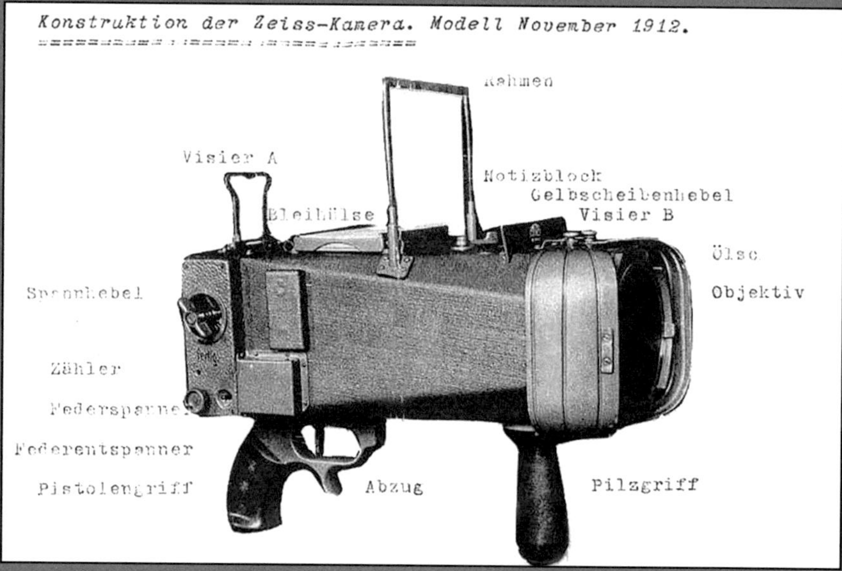

The first German aerial camera was the Zeiss "Pistolenkammer" [Pistol Camera] developed in November 1912.

close proximity to subordinate units, or dropped the *Flieger-Meldung* message in a container with visible streamers attached. The German aviator, *Leutnant* Andre Hug, recalled that an aerial observer dropped a message in a sack, attached to "a large colored ribbon of red, white, and black, of a length six to seven meters. The sack was then thrown out of the plane with a message attached; one could easily observe from the ground, that a colored projectile was coming down."[17]

As the war progressed, aerial photography evolved and became the foundation of aerial reconnaissance. The key to Germany's progress in this new phase of intelligence, was a *Leutnant* with a passion for photography. *Ltn* Carl Fink enlisted in the Prussian Army in 1907 and received his commission through the 165. *Infanterie Regiment*. His depth of knowledge of photography was both artistic and practical. In 1911, *Ltn* Fink volunteered for military aviation service and became a successful pilot in the first Prussian military aviation school at Döberitz. At this moment in time, most military staff failed to consider photography as a pivotal addition to aerial reconnaissance. *Ltn* Fink disagreed with their bias and subsequently became an advocate for a more aggressive application of aerial photography to map restitution and intelligence.[18]

ETRICH TAUBE DESIGN DRAWINGS

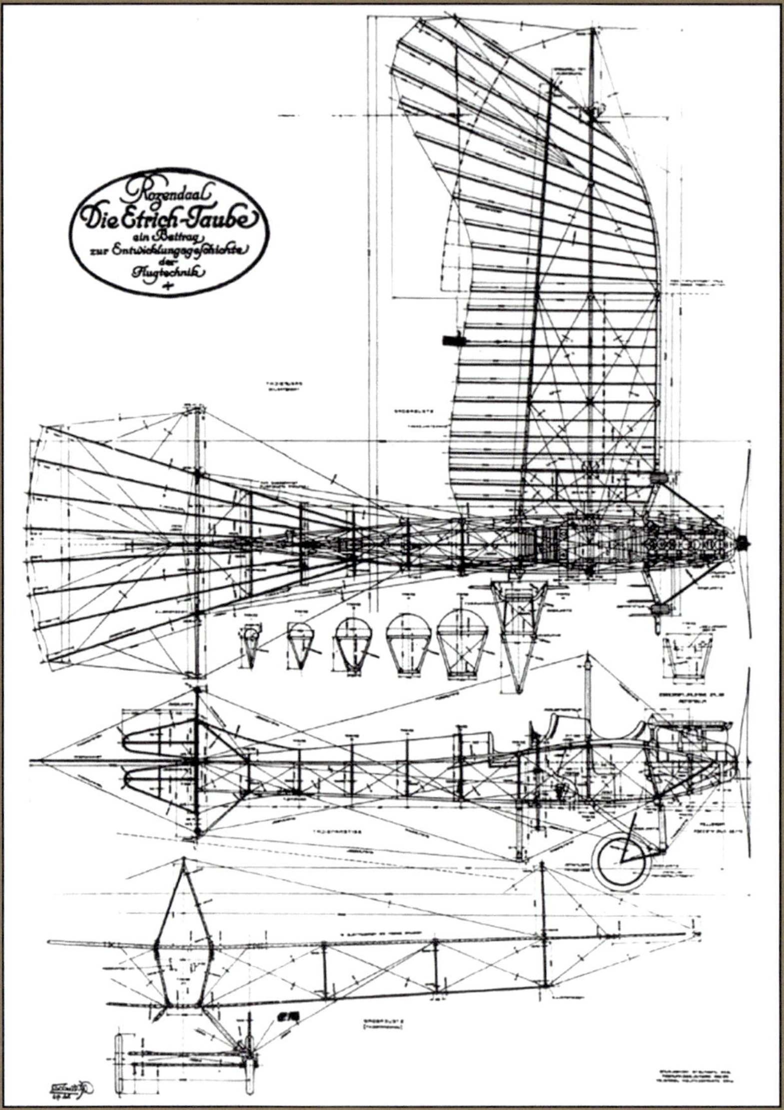

He authored *Fotografie aus Flugzeugen* [Photography from Aeroplanes], which provided a brief introduction to the most effective methods for shooting photographs from aeroplanes. His innovative guidelines were a vital addition to aviation training at Döberitz. At the 1911 Kaiser Manöver (held at Mecklenburg in September), Fink demonstrated initiative by using a camera and taking aerial photographs of enemy targets while flying sorties in Albatros and Rumpler Taube aeroplanes. Upon landing, Fink proceeded to develop his photos inside a tent he designed to serve as a mobile photographic laboratory.[19]

In 1912 Fink was charged with training Döberitz observers on the intricacies of aerial photography. The following year Fink authored a more detailed instruction manual, entitled *"Photographieren aus Flugzeugen" (Ph.Fl)*.

Key points were:

a. A uniform aerial camera was required that was robust and easy to handle. This prevented mistakes from being made.
b. Optical principles should be as easy to understand within the reporting system, as principles of ballistics are, in the standard firing instructions written for infantry and artillery.
c. Instructions should always include any and all details that should be observed, to ensure all photographs are of use to the military. Photographic reports should be included.
d. Handling of aerial photographs in the photographic laboratory should be standardized, and designed with a specified military purpose in mind.
e. Supplies should always be accessible, to insure successful operations.[20]

Ltn Fink's advocacy for a more effective aerial camera, resulted in his collaborating with Alexander Ernemann (owner of a photography firm in Dresden) to design a prototype. Together, they applied the basic principles of Theodor Scheimpflug, a well-known Austrian geodesist. Early on, Scheimpflug developed a systematic method and apparatus for correcting perspective distortions of aerial photographs, shot at an oblique angle. Fink adapted Scheimpflug's theory and created a process that allowed oblique aerial photographs to be accurately mensurated. His work culminated in the creation of a camera that elevated and surpassed all prior efforts, designed to assist in military operations. Of equal importance at this early stage, Fink and Ernemann developed tools to create and generate accurate maps from aerial photographs.[21]

At this time, Zeiss (the world's leading firm for developing cameras) began collaborating with Fink, to develop a new state-of-the art aerial camera. They produced a hand-held *Pistolenkammer* with a 25 cm focal length that created photographic images on 9x12 cm plates. Aerial photographs acquired from the Fink and Zeiss camera prototype, convinced skeptics that aerial photography provided far greater detail than traditional hand-drawn sketches by aerial observers.[22]

German aerial photographic inventory finally took shape, as the prospect became a clear reality. The initial distribution of the *Pistolenkammer* went to *Festungsfliegerabteilungen (FestFA)*. German military staffs perceived aerial photographs as having their greatest value in defense of fortresses. The initial aerial photographic inventory comprised two 9x12 cm *Pistolenkammer* with a 25 cm focal length objective. The mobile *Feldfliegerabteilungen (FFA)* were equipped, soon afterwards, with a trunk for a photographic laboratory, one fold-up darkroom tent, and materials for 14 days of operations.[23]

When the war commenced *Ltn* Fink was mobilized and proceeded with his field unit, *FFA 7*, to the Western Front.[24] The following November he, and other outstanding air crew members, were selected for *Brieftauben Abteilung Ostende (BAO)*, a unit created for strategic aerial bombardment against high value targets such as London. In March of 1915, the German command transferred *BAO* to the Eastern Front. *BAO* was hosted on a railroad train and *Ltn* Fink was ordered to establish a mobile photographic lab, photographic assessment facilities, and power supply on railroad cars. Within a short period of time he was given three railroad cars, equipped with all the necessary equipment. His expertise transformed and served as the foundation for the development and organization of all aerial photography for *Fliegertruppe*.[25]

Germany's evolving commitment to aerial photographic reconnaissance in the opening stages of the war, varied with the particular aircrew flying over the evolving battlefield. *Alte Adler* [Old Eagles] veteran Andre Hug, recalled aerial photography training as "very poor before the war." Despite *Fliegertruppe's* trained and combat-ready aviators, Hug stated that only one or two aerial observers in his unit possessed a true interest in aerial photography. In turn, they developed their own photographic techniques. "It was a question of personal interest," said Hug, who reported that *Fliegertruppe* had (at the onset) no training standards for reconnaissance photography. Initially, aerial observers were neither issued aerial cameras nor field glasses. But by October 1914, Hug reported his *Feldfliegerabteilung* had been provided with a "suitable camera for his unit."[26]

Though *Fliegertruppe* suffered from the lack of adequate photographic equipment during the first weeks of the war, this never impeded their early intelligence initiatives. Visual reconnaissance proved to be effective. As the war progressed, the various initiatives taken by the *Feldfliegerabteilung* commanders and aerial observers ignited further breakthroughs in cultivating Germany's knowledge of the art.

GERMAN AEROPLANES

German Aeroplane Inventory, fielded in August 1914

Type	Engine	Takeoff Weight	Speed	Climbing Perform.
Albatros-Doppeldecker 1912 (Doppeltaube)	100 PS Mercedes	720 kg	90 km/h	—
Albatros-Doppeldecker 1913	100 PS Mercedes	950 kg	100 km/h	800 m in 8 bis 10 Minuten
Albatros-Doppeldecker 1914	100 PS Mercedes	900 kg	105—108 km/h	800 m in 5 bis 8 Minuten
L. V. G.-Doppeldecker 1913/14	100 PS Mercedes	1100 kg	100 km/h	800 m in 9 bis 12 Minuten
Aviatik-Doppeldecker 1913/14	100 PS Mercedes bzw. 100 PS Argus	970 kg	100 km/h	800 m in 11 bis 12 Minuten
D. F. W.-Doppeldecker 1913/14 (Mars-Doppeldecker)	100 PS Mercedes	970 kg	100—120 km/h	800 m in 10 Minuten
Euler-Doppeldecker 1913/14	100 PS Mercedes	600 kg	90—100 km/h	—
A. E. G.-Doppeldecker 1913/14	100 PS Argus	970 kg	105 km/h	800 m in 13 Minuten
Otto-Doppeldecker 1913	100 PS Mercedes oder Argus	1045 kg	100 km/h	800 m in 9 bis 12 Minuten
Rumpler-Taube 1913	100 PS Mercedes oder Argus	870 kg	90—100 km/h	800 m in 15 Minuten
Gotha-Taube 1913/14	100 PS Mercedes	1010 kg	90—100 km/h	800 m in 12 Minuten
Jeannin-Taube 1913/14	100 PS Mercedes oder Argus	1035 kg	100—110 km/h	800 m in 13 Minuten

Payload 200 kg (pilot, observer, equipment). Fuel Reserve 4 hrs flight. Data taken from acceptance protocols, different data may result from different production series.

Germany published a comprehensive list of aeroplanes that were operating in August 1914. English translation of column headings and footnotes is provided. Source: 3. Sonderband

GERMAN AEROPLANES

Luft-Verkehrs-Gesellschaft (L.V.G.) – Doppeldecker 1913/14 (also known as L.V.G. B.I) – assigned to Feldfliegerabteilung 16 – subordinate to 8. Armee

Rumpler-Taube 1913 – assigned to Feldfliegerabteilung 14 - subordinate to I. Armee-Korps.

Rumpler-Taube 1913 – assigned to Feldfliegerabteilung 14 - subordinate to I. Armee-Korps.

AUGUST 21, 1914. FLIGHT

AIRCRAFT "MADE IN GERMANY"

WHICH MAY BE EMPLOYED AGAINST THE ALLIES.

ALTHOUGH up to the present little information of a reliable character has been allowed to transpire regarding the movements and actual work done by aircraft in the present war, there can be little doubt that the air fleets of allies as well as enemies are and will be playing an important part. In fact it seems already that the efficiency or otherwise of the aerial equipment of the contesting powers will in no small measure influence the final issue. As even approximately accurate figures as to the actual number of machines are naturally not obtainable, we have thought it of considerable interest here to describe briefly the different types of machines either already in use by the German Government or immediately available. From these and a table of dimensions and general characteristics of the machines our readers will be able to form a good idea how the various types of aeroplanes produced in Germany compare with our own. In order to facilitate identification all the machines have been given reference numbers in the table, the illustrations, and the descriptions, which will enable our readers to refer quickly and with the minimum of trouble, the description and particulars in the table corresponding to a given illustration or *vice versa*.

1. The Ago-Arrow Biplane

is, as the name implies, of the type deriving its name from the peculiar backward slope of its main planes. When viewed from in front the upper plane is straight,

1. The Ago-Arrow biplane.

whilst the lower one possesses in addition to its backward slope, a comparatively great dihedral angle. The *fuselage*, which is of rectangular section, tapers toward both ends, the streamline form being further improved by a turtle back in which openings have been cut out for the engine

AEROPLANES "MADE IN GERMANY."

Reference No.	Make of Machine.	Where Built.	Type.	No. of Seats.	Span (in ft.). ft.	in.	Length (in ft.). ft.	in.	Area (in sq. ft.).	Weight. Empty. lbs.	Weight. Loaded. lbs.	Fuel Capacity (hours).	Speed (m.p.h.).	Radius of Action.* miles	Climbing (3,281 ft. in – mins.).	Engine.	Type.	No. of Cyls.	h.p.
1	Ago ...	... Johannisthal ..	... Arrow T.B. ...	2	45	6	26	2	450	—	—	4	68	136	—	A.†	Vert.	4	150
2	Ago ...	... Johannisthal ...	... Racing Mono...	1	29	6	26	3	162	—	—	—	84	—	8	A.	Vert.	4	150
3	Albatros	... Johannisthal ...	... T.B. ...	2	47	5	26	2	403	1360	1850	4	68	136	8	A.†	Vert.	4	100
4	Albatros	... Johannisthal ...	... Taube ...	2	50	0	34	0	380	1320	1790	3	64	96	—	M.	Vert.	6	100
5	A.E.G.	... Hennigsdorf (Berlin) ...	T.B. ...	2	54	0	35	6	530	—	—	—	62	—	—	N.A.G.	Vert.	4	100
6	A.E.G.	... Hennigsdorf (Berlin) ...	Later type ...	2	48	0	32	0	440	1540	2080	4	70	140	—	B.	Vert.	6	110
7	Aviatik...	... Mulhouse ...	... T.B. ...	2	45	10	26	2	500	1520	2040	4	62	124	15	A.†	Vert.	4	100
8	Aviatik	... Mulhouse ...	... Later type ...	2	35	0	24	0	390	900	—	—	—	—	—	O.	R.	9	114
9	D.F.W.	... Leipzig ...	... Taube ...	2	46	0	37	6	340	1320	1840	4	68	136	—	M.	Vert.	6	100
10	D.F.W.	... Leipzig ...	... Arrow T.B. ...	2	56	0	29	6	500	1540	2060	4	65	130	—	M.	Vert.	6	100
11	D.F.W.	... Leipzig ...	... Crescent wings	2	46	0	27	6	430	1425	1945	4	74	148	5	M.	Vert.	6	100
12	D.F.W.	... Leipzig ...	... Arrow type ...	2	44	0	25	0	450	1500	2240	6	85	255	5½	A.D.	Vert.	6	120
13	Etrich ...	... Liebau ...	... Taube ...	2	47	0	32	3	320	1420	1940	4	59	118	—	A.†	Vert.	4	100
14	Fokker	... Schwerin ...	... Mono. ...	2	36	0	24	0	240	1445	2180	4	71	142	—	A.	Vert.	4	100
15	Goedecker	... Mainz ...	... Taube ...	2	46	0	36	0	400	1100	1620	4	60	120	—	M.	Vert.	6	100
16	Gotha ...	... Gotha ...	... Taube ...	2	47	0	33	6	370	—	—	—	62	—	—	A.†	Vert.	4	100
17	Gotha ...	... Gotha ...	... Taube ...	2	46	0	28	0	300	—	—	—	74	—	—	M.	Vert.	6	100
18	Halberstadt ...	—	Taube ...	2	48	0	30	0	340	1496	2016	4	92	184	—	M.	Vert.	6	100
19	Hansa	—	Taube ...	2	46	0	28	0	308	—	—	—	72	—	—	M.	Vert.	6	100
20	Harlan ...	... Johannisthal ...	... Arrow Taube ...	2	47	0	34	9	300	1250	1770	4	62	124	15	A.†	Vert.	4	100
21	Jatho ...	... Hannover ...	... Steel Taube ...	2	47	6	39	6	320	1060	1730	7	78	273	—	M.	Vert.	6	100
22	Jeannin	... Johannisthal ...	... Steel Taube ...	2	42	0	30	0	250	1320	1880	4	75	150	—	A.	Vert.	6	120
23	Kondor	... Essen ...	... Taube ...	2	46	0	32	2	330	—	—	—	62	—	—	A.†	Vert.	4	100
24	L.F.G....	... Reinickendorf ...	... Arrow T.B. ...	2	40	6	26	6	350	—	—	—	62	—	—	A.†	Vert.	4	100
25	L.V.G.	... Johannisthal ...	... T.B. ...	2	46	0	29	6	550	—	—	—	62	—	—	A.†	Vert.	4	100
26	L.V.G.	... Johannisthal ...	... T.B. ...	2	50	0	31	0	480	1670	2190	4	65	130	—	M.	Vert.	6	100
27	Otto ...	.. Munich ...	... P.B. ...	2	49	0	34	6	430	—	—	—	68	—	—	A.†	Vert.	4	100
28	Roland (L.F.G.)	Reinickendorf ...	.. Arrow T.B. ...	2	42	6	26	3	430	1380	2150	5	62	155	—	M.	Vert.	6	100
29	Rumpler	... Johannisthal ...	... Taube ...	2	43	6	32	3	340	880	1400	4	62	124	—	A.†	Vert.	4	100
30	Rumpler	... Johannisthal ...	... Taube ...	2	46	0	27	0	300	1320	1840	4	74	148	—	M.	Vert.	6	100
31	Rumpler	... Johannisthal ...	... T.B. ...	2	42	6	—		450	—	—	—	—	—	—	M.	Vert.	6	100
32	Schwade ...	—	P.B. ...	2	60	0	—		500	1012	1532	4	60	120	—	St.	R.	7	80
33	Sommer	... Berlin ...	... Arrow T.B. ...	2	33	0	26	0	350	—	—	—	—	—	—	G.	R.	9	100
34	Union ...	... Berlin ...	... Arrow T.B. ...	2	33	0	23	0	350	1210	1845	5	68	170	—	A.D.	Vert.	6	120

T.B. = Tractor biplane. Mono. = Monoplane. P.B. = Propeller or "Pusher" biplane. A. = Argus. B. = Benz. A.D. = Austro-Daimler. G. = Gnome. M. = Mercedes. O. = Oberursel. St. = Stahlhertz. Vert. = Vertical. R. = Rotary.

* The radius of action given applies to machines fitted with standard tanks and carrying an observer in addition to the pilot. By substituting larger tanks for the observer the radius of action may be greatly increased.

† Also supplied with 100 h.p. 6-cyl. Mercedes.

British aeroplane journal, "Flight Aero Weekly" listed known German aeroplanes in their August 21, 1914 edition.

Jeannin "Stahltaube" was the most used Taube version. The name "Stahltaube" (steel dove) indicated the use of steel struts for the fuselage. Jeannin "Stahltaube" was assigned to Feldfliegerabteilung 14 (subordinate to I. Armee-Korps), Feldfliegerabteilung 15 (subordinate to XX. Armee-Korps) and Königsberg's Festungfliegerabteilung (FestFA) 5.

Albatros-Doppeldecker 1914 (Albatros B.I) – assigned to Feldfliegerabteilung 17 - subordinate to XVII. Armee-Korps.

Gotha-Taube 1913/14 – assigned to Festungfliegerabteilung (FestFA) 5 – subordinate to Festung Königsberg

CHAPTER II

Rossiyskiy Imperatorskiy Voyenno-Vozdushnyy Flot

Russia's Military Air Fleet is Established

The creation of a balloon observation unit on 7 February 1885 launched the Russian military's implementation of aerial platforms. In 1890 the Military Department established a *Uchebnyy vozdukhoplavatel'nyy park (UVP)* [Aeronautical Training Park] commanded by *Poruchik* [Lieutenant] Alexander Matveevich Kovanko. The UVP operated near St. Petersburg at the Volkovo Field Artillery Range. Kovanko was charged with commanding a balloon observation unit during the 1905 Russo-Japanese War.[1]

Tsar Nicholas II and his entourage witness a balloon assent at the Aeronautical Training Park - Gatchina on 19 June 1904 (o.s.). Displayed on the balloon is the flag of the Aeronautic Department

Russia's commitment to aviation fully emerged in 1908. *Russkiy invalid* (a Russian veteran journal representing the views of the Russian War Ministry) indicated that approximately 300 military aeroplanes were initially necessary to provide aerial reconnaissance support for the Russian army.[2] Two years later, Russia's primary aviation military force, *Rossiyskiy Imperatorskiy Voyenno-Vozdushnyy Flot (RIVVF)* [Imperial Russian Military Air Fleet], was established. Three years prior to the war's outbreak, the Russian War Council clarified their commitment, "The fundamental task of the air service is close and long-range reconnaissance for military purposes."[3] One senior Russian staff officer further defined their evolving point of view: "The aeroplanes in their current state (probably, a long-lasting one) can serve for short-range reconnaissance only, but their main purpose will ultimately be to provide communication

Russo-Japanese War in 1905 Russian balloon deployment during the Battle of Liao Yang, Manchuria [Russo-Japanese War, 1905].

support (between troops and fortresses, airships and command staffs). They are the fleet's messenger vessels."[4]

Russian aviation found its greatest advocate in Tsar Nicholas II's cousin, Grand Duke Alexander Mikhailovich. On 11 February 1910 (Gregorian), the Grand Duke, in his capacity as chair of *Osobyy komitet po usileniyu voyennogo flota na dobrovol'nyye pozhertvovaniya* [Special Committee for Strengthening the Navy with Voluntary Donations] announced Russia's new vision for military aviation:

> Now all countries have entered the era of aviation, the success of which is almost impossible to account for. Controlled aircraft undoubtedly represent a new formidable weapon of

Grand Duke Alexander Mikhailovich, cousin of Tsar Nicholas II, was the driving force behind Russia establishing civilian and military aviation. He created the organization known as Otdel Vozdushnogo Flota (OVF) [the Department of the Air Force].

struggle in the near future ... In Russia, aviation is only just beginning to take its first steps. We must hurry, so as not to be wounded again, due to the incompleteness of our weapons. There can be no victory in future wars without an air fleet.[5]

Alexander Matveevich Kovanko, commander of the Aeronautical Training Park from 1885-1909 and head of the Officer Aeronautical School - Gatchina in 1910-1916.

Grand Duke Alexander (also chairman of a non-profit organization known as *Osobom komitete po usileniyu flota* [Special Committee for Strengthening of the Air Fleet]) took the initiative within Russian military and civilian circles, to establish a quasi-public organization known as *Otdel Vozdushnogo Flota (OVF)* [the Department of the Air Force]. Tsar Nicholas II approved this initiative to provide support for officers and private individuals—giving them the opportunity to receive flight training. The organization also established logistical supplies for aviation support to both army and navy air units. Alexander worked with *General ot kavalerii* [General of Cavalry] Baron Aleksandr Wilhelm Andreas von Kaul'bars (OVF chairman) on establishing a military air arm and emphasized the need for it to be built with foreign technology.[6]

In May of 1910, the Russian aviation equivalent to Germany's Döberitz, was formed outside of St. Petersburg *Ofitserskaya Vozdukhoplavatel'naya shkola*—Gatchina [Officer Aeronautical School]. *Polkovnik* Sergey Alekseyevich Ul'yanin, was commander. A second facility, *Sevastopol'skoy aviatsionnoy shkole* [Sevastopol Aviation

On 25 July 1909, French aviator Louis Blériot piloted his Blériot aeroplane on a 36-minute flight across the English Channel. This historic achievement spurred aeroplane manufacturing and development throughout Europe.

School,] was established near Sevastopol in 1911 to facilitate training year-round.[7]

General ot kavalerii Baron Aleksandr Wilhelm Andreas von Kaul'bars (OVF chairman) worked with Grand Duke Alexander on establishing a military air arm that became the foundation of RIVVF.

In 1911, the Aviation Department had enough men and machines to establish experimental flight units to determine which aeroplanes were best suited for Russian military operations. The chief of the Engineering Command requested that aeroplanes participate in Petersburg Military District field exercises. The experience proved invaluable and demonstrated aviation's powerful potential, particularly with regard to short- and long-range reconnaissance. Maneuvers demonstrated that pilots required specific aeronautical charts (maps) to support intelligence. Aeroplanes were identified as an effective means to communicate with troops isolated from the main force. In addition, aeroplanes could now be used to combat aerostats, the prevailing aerial platform of the time. A significant observation that held sway in the coming war, was that soldiers should camouflage themselves against aerial reconnaissance. Lastly, the aeroplane's propensity for breakdowns necessitated repair depots be established.[8]

General-Adjutant Georgi Antonovich Skalon, commander of Warsaw Military District, cabled Grand Duke Alexander, stating, "Our brave pilots should be proud of their courageous and skillful flying in support of our army." Skalon added, "It is extremely desirable to provide aeroplanes to each corps." The Tsar (upon review) wrote in the margin, "In my opinion, we should give airships up as hopeless."[9]

When Louis Blériot flew across the English Channel in 1909, the

POLKOVNIK SERGEY ALEKSEYEVICH UL'YANIN

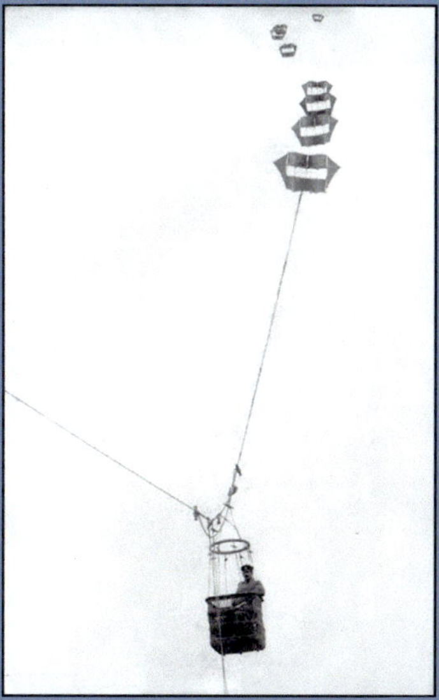

Ul'yanin was initially trained in 1895 as a vozdukhoplavatel [balloon pilot]. One of his first aeronautic achievements was the design of a man-lifting kite that sent an observer aloft faster than a balloon. Ul'yanin is in the observation basket.

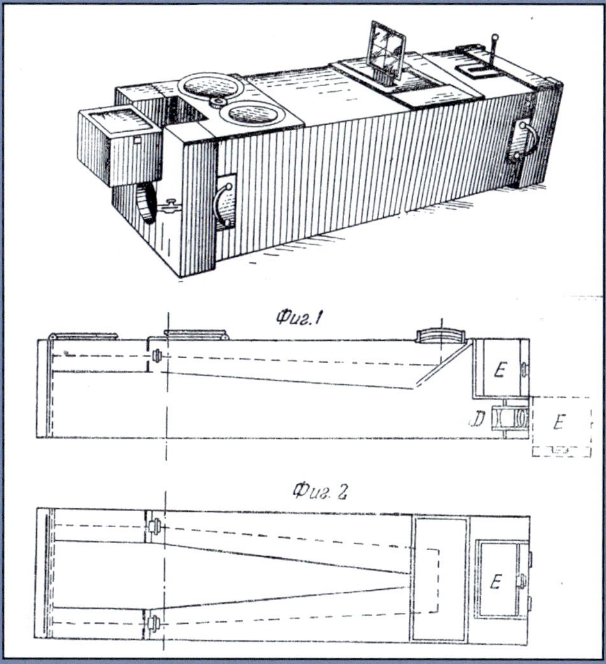

Ul'yanin developed his first aerial camera to work from the man-lifting kite or balloon. He personally employed the camera as a vozdukhoplavatel during the Russo-Japanese War (1904-1905).

In May 1911, Podpolkovnik [Lieutenant Colonel] Ul'yanin commanded Ofitserskaya Vozdukhoplavatel'naya Shkola (OVS) [Officer Aeronautical School] at Gatchina. Minister of War Sukhomlinov (dark coat in center), General ot kavalerii Baron von Kaul'bars (right of Sukhomlinov) and Ul'yanin (right of Kaul'bars) pose with military pilots in front of a Tellier Mono Type Militaire aeroplane. Civilian Sergei Isaevich Utochkin (second from left) was one of Russia's greatest prewar aviators who also trained RIVVF pilots prior to the war.

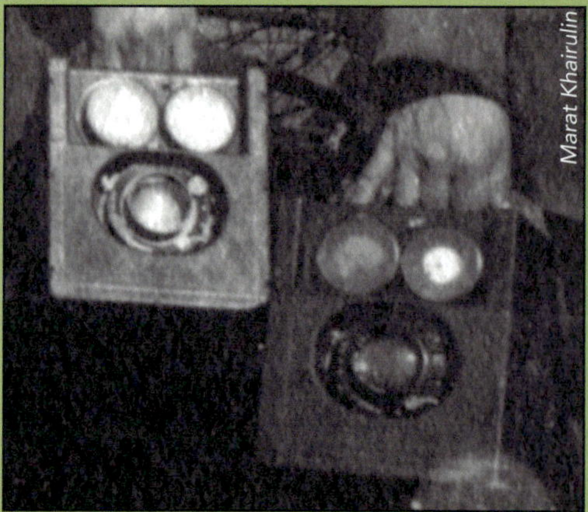

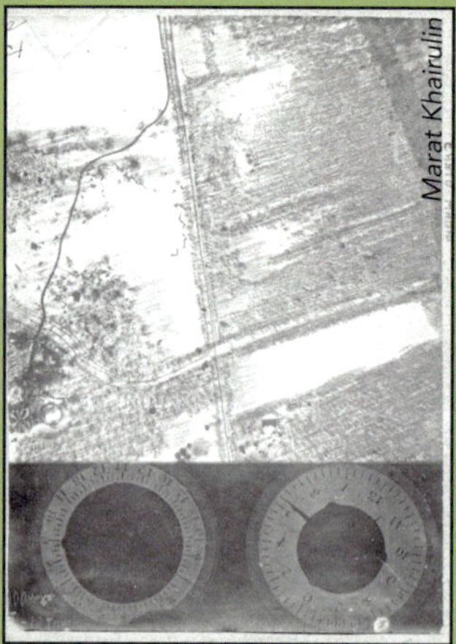

In 1912, Ul'yanin design his most famous aerial camera which served as the mainstay of RIVVF aerial cameras. It was the most unique of all aerial cameras employed in the war. The 13 x 18 cm plate not only imaged the target but also the readings from an altimeter and chronometer.

excitement generated from his achievement prompted some Russian officers to foresee important reconnaissance potential: "Aircraft heralded the end of cavalry, that a dwarf who could see will conquer a blind giant.'"[10] Minister of War Sukhomlinov gained greater confidence in aviation after learning of Blériot's flight. After visiting the Sevastopol Aviation School in October 1911, Sukhomlinov became convinced that Russia's aviation should be pursued with urgency and determination.[11] Sukhomlinov's Engineering Command provided a plan that specified ten aviation units be created and established within the aeronautical (balloon) units.[12]

On 12 August 1912, Minister of War Sukhomlinov signed Decree No. 397 transferring aeronautics and aviation divisions from the Engineering Command to the General Staff, headed by *General-maior* Mikhail Ivanovich Shishkevich. This newly established aviation unit provided both organization and training based on Engineering Command's "Plan of Army Air Service" and issued a new "General Plan of Army Aeronautics and Aviation." The plan specified that aviation units be divided into four categories: *Armeysky Aviaotryad* [Army Detachments], *Korpusnoy Aviatsiony* [Corps Detachments], *Krepostnyye Aviatsionnyye* [Fortress Detachments] and *Otryad Osobogo Naznacheniya* [Special Purpose Detachment]. *Generál-kvartirméyster* [Quartermaster-General] now commanded *Armeysky Aviaotryad*. Their mission was to conduct long-range reconnaissance, destroy the enemy's aeroplanes, and execute aerial bombardment.[13]

In the fall of 1912, *Vozdukhoplavatel'naia Chast' Glavnogo Upravleniia General'nogo Shtaba (GUGSh)* [Aeronautics Unit of the Main Administration of the General Staff] was created under *Stavka* [Supreme Military Headquarters] to supervise Russia's military aeronautical affairs.[14] *GUGSh*'s central function was to ensure that Russia's aviation supported Russian army echelons. Russian military districts now served as administration centers.[15] A Russian army corps in 1914 represented "a perfect formation of all service branches in the required mutual tactical relationship to conduct independent activities at war," and tactical aviation was to be fully dedicated to that requirement.[16]

On 20 September 1912, Minister of War Sukhomlinov signed off on Russia's Order No. 508 outlining the military aviation architecture. The follow-

AEROPLANES

Neiuport IV aeroplanes were imported from France and built under license by Dux, PRTV/Shchetinin, and Antara. In 1914 this aeroplane served in 17 Korpusnoy Aviatsionnyy Otryad (KAO).

Farman HF.16 aeroplanes were imported from France - built under license by Dux, Lebedev, and Shchetinin. About 300 HF.16s were built in Russia.

AERONAUTICAL TRAINING

One of the first dedicated aviation centers in Europe was the Gatchina Aviation School established near St. Petersburg in 1885. In 1910, Gatchina became the Officer Aeronautical School and subsequently in 1914, was integrated into the Gatchina Military Aviation School.

Russian military pilot trainees and aeroplanes on review at Kulikovo Field Sevastopol Aviation Officer School, 30 June 1911.

The Sevastopol Aviation School's Kulikovo airfield was active in promoting Russia's military aviation. The more optimal climate (vs. Gatchina) allowed for year-round training. One of the first Il'ya Muromets pilots, Shtabs-kapitan Rudnev, is shown flying a Farman IV over the hangers.

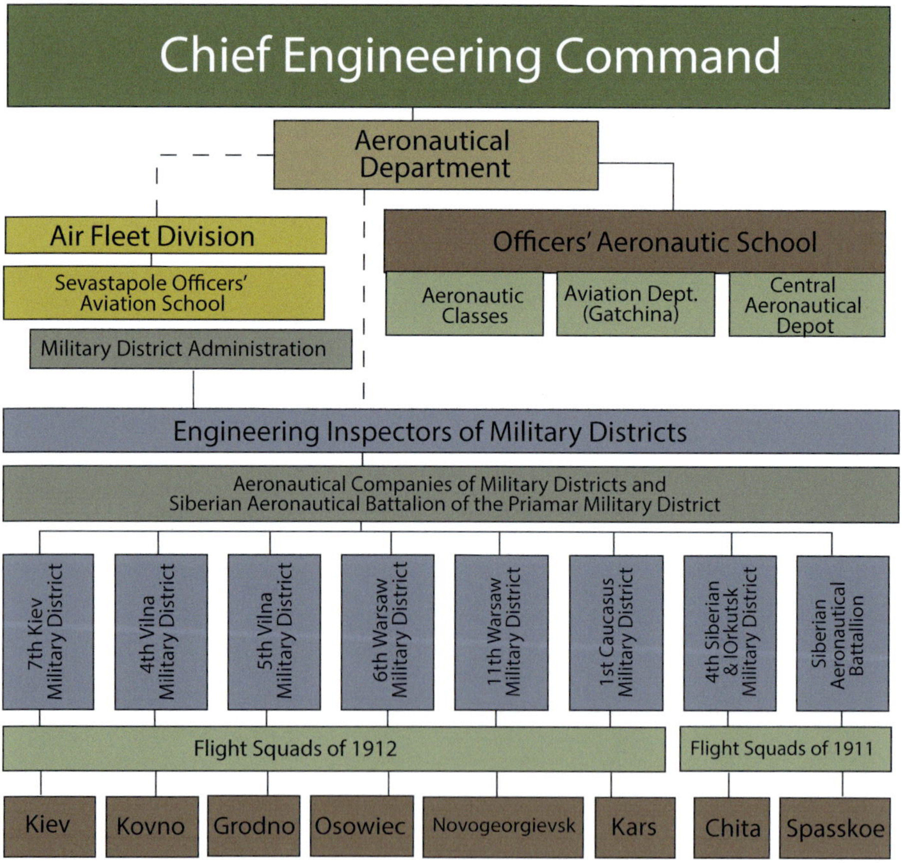

The first organizational plan for the RIVVF created in 1912, "General Plan of Army Aeronautics and Aviation." The plan detailed the various divisions and aviation units. Source: Carl J. Bobrow

RIVVF aviators at Gatchina's Officer Aeronautical School in front of a Farman VII. Poruchik Nikolay Nikolaevich Danilevskiy (standing) as well as Poruchik Vyacheslav Grigorievich Baranov (2nd from R) are two RIVVF pilots who participated in the August 1912 Krasnoye Selo (St. Petersburg) maneuvers.

ing year, the Russian "General Plan" envisioned a concept of military aviation that included the abbreviated nomenclature that subsequently became the standard for *RIVVF* in the war. The Russian *Generál-kvartirméyster* of the army headquarters, was charged with establishing and operating ten *Armeysky Aviaotryad (AAO)* [Army Aviation Detachment] for warfare missions. These included long-range reconnaissance, destroying enemy aeroplanes, and bombing facilities. Chiefs of corps' headquarters established and operated 37 *Korpusnoy Aviatsionnyy Otryad (KAO)* [Corps Aviation Detachment], providing short range reconnaissance. In turn, fortress headquarters chiefs were to have *Krepostnyye Aviatsionnyye Otryad (KrAO)* [Fortress Aviation Detachment], provide short-range reconnaissance to protect the vital fortress area.[17] Each *KrAO* was to have eight aeroplanes with eight more in reserve. *KrAO* were to have a total of 64 airframes for active service and sixty-four more in reserve. By July 1914, the Russians established five *KrAO* at Kovno, Grodno, Osovets, Novo Georgievsk (present-day Modlin), and Brest-Litovsk.[18] Russian protocol for aerial reconnaissance also assigned the airship for army and front support. However, Russia's small airship inventory failed in providing long-range aerial reconnaissance. The Il'ya Muromets eventually served as the strategic aerial reconnaissance resource but was not operationally effective until February 1915.[19] *RIVVF* also had 46 captive balloons attached to 11 balloon units mostly assigned around fortresses.[20]

On 19 October 1912, the Aeronautical unit presented to Minister of War Sukhomlinov, the "General Plan of Army Aeronautics and Aviation," covered the strategy for training flight personnel.[21] Sukhomlinov received counter proposals to the evolving aviation architecture from *General-Inzhener* [Engineer General] Nikolay Fomich Alexandrov, Chief of Engineering Command. "Aeronautics in the army is being developed intensely and will be developed dramatically," said Alexandrov. "I am much concerned that we will not succeed in the existing conditions and will compromise a new technical direction for Russian aeronautics."[22] On 13 February 1914, the Russian War Ministry approved Tables of Organization and Equipment for Military Aviation for *Aviatsionny Otryad (AO)* [Aviation Department]. *Korpusnoy Aviatsionny Otryad (KAO)* [Corps Aviation Detachment] and *Polevoy Aviatsionnyy Otryad (PAO)* [Field Aviation Detachment] were subordinated to *AO*.[23]

Russia placed its first order for aeroplanes from France in January 1912, acquiring Maurice Farman MF.7 and Blériot XI. The order was changed to Farman MF.11 and Nieuport IV following a military aeroplane competition in France.[24] Russia's initial cadre of Farman MF.11s and Nieuport IVs

General-maior Mikhail Ivanovich Shishkevich, chief of the Aeronautical Unit of the General Staff Main Directorate, led implementation of "General Plan of Army Aeronautics and Aviation" that created Armeysky Aviaotryad [Army Aviation Detachments], Korpusnoy Aviatsionniy [Corps Aviation Detachments], Krepostnyye Aviatsionnyye [Fortress Aviation Detachments] and Otryad Osobogo Naznacheniya [Special Purpose Aviation Detachment].

1913 INTERNATIONAL COMPETITION

In September 1913, the Russian military held an open international competition at Korpusnoy Airfield (St. Petersburg) to determine the initial suite of military aeroplanes for RIVVF. Competing aeroplanes were required to fly with a full load for at least 3 hours at 80 km/h or more. R-BVZ/Sikorsky S-10 won the competition. The aeroplane did not serve on the Northwest Front.

Morane Saulnier-G at the 1913 Russian Military Trial at Korpusnoy Airfield (St. Petersburg) achieving fourth place in the competition.

RIVVF flew both Deperdussin machines that were built in serial production by Lebedev in Russia, as well as being imported from France. The Deperdussins primarily flew in support of the Southwest Front.

provided aerial reconnaissance to a depth of 60 kilometers—suitable for Russian corps' planning and operations. Airships were necessary for strategical reconnaissance at distances exceeding 60 kilometers.[25]

Russia's aviation industry lacked necessary capital to purchase key components of additional motors and propellers from France and were unable to produce and maintain an inventory of spare parts. Once war commenced, the Main Directorate of the General Staff was given an additional budget of 8.9 million rubles to support aviation development and he subsequently generated an order for 281 aeroplanes to *PRTV*, *Dux*, and *R-BVZ*.[26]

When Russia's Technical Committee of the Main Military and Technical Directorate met in March 1914, they decided to order 48 "Deperdussin," 42 "Morane-Saulnier Type," as well as 42 indigenously produced "Sikorsky-XI" aeroplanes. That summer, aviation units were fielded across the nation.[27] When Russia mobilized at the start of the war, they had approximately 244 available aeroplanes. *KAO* units arrived at forward landing grounds in accordance with the plan for deployment of armies. *KrAO* remained with their fortresses. Aviation company and subordinate corps detachments were in place when war was declared.[28]

Innovative Thinking for Aviation's Employment

A year prior to the Great War, Russians were well aware of French aviation support to ground operations. Subsequently, they initiated Russian aerial training exercises with artillery to improve targeting. Once the aerial reconnaissance sortie was completed, the Russian aviator returned to the artillery battery and dropped a message bag attached to a banner cloth. Artillery horsemen retrieved and delivered the message bag, with artillery targeting information, to the battery. Aerial orientation was also provided to artillery observers, which familiarized them with how reconnaissance supported artillery operations.[29] Prior to the war, aeroplanes were evaluated as potential platforms for firing a machine gun. In 1913, *Sh. Kap.* Victor Rodionovich Poplavko first tested the concept of firing a machine gun while in flight. His tests showed spent bullet cartridges put the aeroplane's motor at risk. In March 1914, Ul'yanin's staff including Poplavko, conducted tests to ascertain the effectiveness of firing a machine gun while an aeroplane was in flight—targets included tethered balloons and "dummy" troop targets.[30]

Minister of War Sukhomlinov and staff discussed mobile battlefield operations in 1912. Sukhomlinov proposed the Russian army employ a corps-size raiding force to conduct deep strikes into *Ostpreussen*. The Russian strike force was designed to keep Germany's armies off balance, while Russia's main thrust to the west concentrated on defeating Austro-Hungary's *k.u.k. Armee* to the south. Sukhomlinov envisioned using mobile infantry in horse-drawn wagons to follow and support the cavalry spearheads. The innovative concept was abandoned, as staff more carefully considered the threat to ground operations by aerial reconnaissance. Aerial detection was deemed to be the greatest threat to any and all surprise attacks.[31]

Shtabs-kapitan [Staff Captain] Pyotr Nikolayevich Nesterov, 11th *KAO* commander, was regarded as one of the most innovative *RIVVF* aviators prior to the war. He earned the title of *Voennyi Letchik* [Military Pilot]. Nesterov's counterpart, and an aviator of equal stature, *Podyesaul* [Cossack Staff Captain] Vyacheslav Mikhailovich Tkachev, summed up Nesterov's achievements: "As an active, energetic officer and *KAO* commander, Nesterov continually pushed the envelope of aeroplane design, performance, and tactics. A firm proponent of theoretical and technical

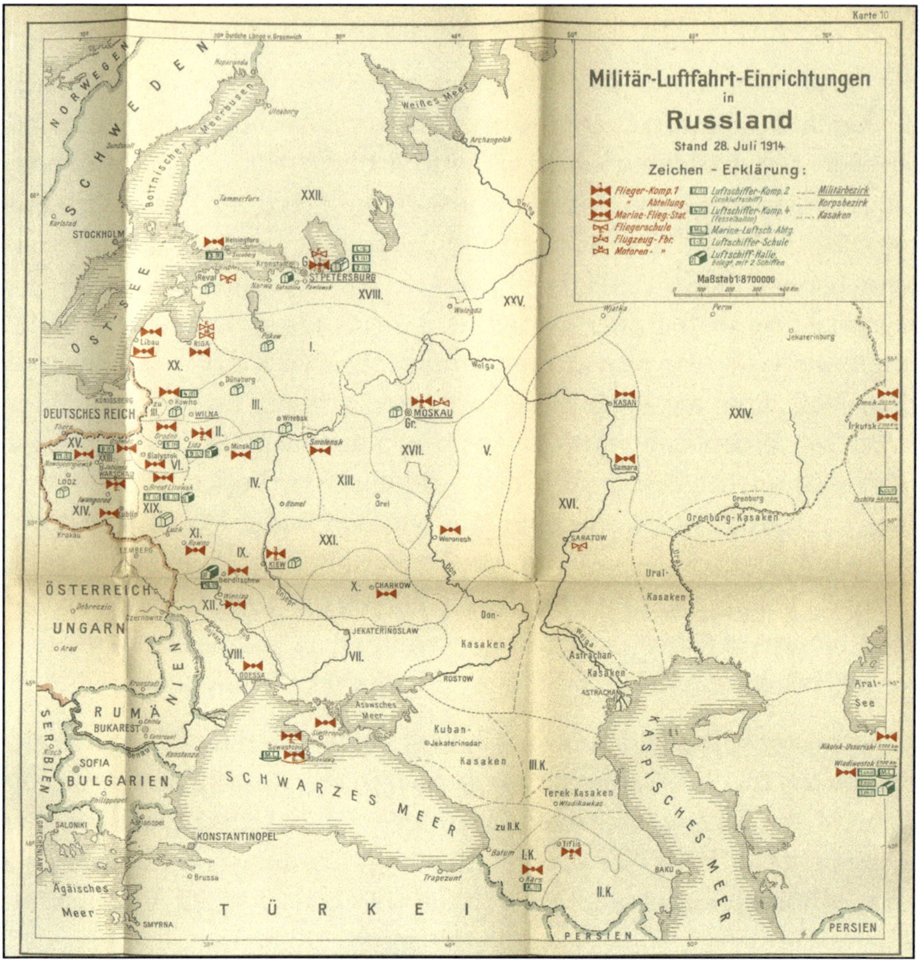

When Russia mobilized at the start of the war, they had approximately 244 available aeroplanes. German intelligence compiled known locations. This particular map was published in 1941. Source: Karte 1, Anlagenband zu Die Militärluftfahrt bis zum Beginn des Weltkriegs 1914, Berlin 1941.

training in aviation, Nesterov had little patience for those who regarded flying as instinct, spirit, or adventure." Nesterov and Tkachev flew in support of the Southwest Front when war commenced.[32]

Executing Russian Plans

Despite the clear threat facing Russia from Germany at the onset of World War I, Russia's military priorities remained focused on Austria-Hungary and the Galician region. Germany's *Ostpreussen* region represented a more formidable challenge. France

VOENNYI LETCHIK [MILITARY PILOTS]

Voennyi Letchik [Military Pilot] badges worn on the chest to honor distinguished RIVVF graduates from Gatchina and Sevastopol training schools. The first RIVVF aviators honored as Voennyi Letchik were instrumental in advancing Russia's aviation with their flying skills, technological application to aeroplane roles, and missions, as well as heroic performance over the battlefield. Source: Smithsonian Institution National Air and Space Museum

Voennyi Letchik [Military Pilot] Shtabs-kapitan [Staff Captain] Pyotr Nikolayevich Nesterov, 11th KAO commander, was looked upon as one of the most innovative RIVVF aviators, prior to the war. Nesterov (left) and his mechanic Gennady Mikhaylovich Nelidov, pose in front of the Nieuport IV that he successfully flew for 3 hours from Kiev to Odessa on 1 March 1914, followed by an 8-hour sortie from Kiev to Gatchina on 11 May 1914.

Voennyi Letchik, Podyesaul [Cossack Staff Captain] Vyacheslav Mikhailovich Tkachev was one of the first pilots trained at Sevastopol Aviation School. On 12 October 1913, he flew a record 2,400 km. flight from Kiev to Odessa to Kerch to Taman to Yekaterinodar, in a Nieuport IV. Tkachev commanded 20th KAO in 1914.

had already spent decades trying to position Russia's main forces to attack Germany via the *Ostpreussen* region. As events unfolded in the summer of 1914, Russian military leadership (*Stavka*), perceived the Northwest Front as secondary to the Southwest front.[33]

As Russia mobilized its forces, they assessed the unfolding German commitment to forces at both Western and Eastern fronts, in an effort to succeed at regional objectives. *8. Armee* was the only German army at the time, facing Russia in *Ostpreussen*. It was comprised of 11 infantry divisions. The Russian Northwest Front allocated 29 divisions. Battalion strength was based on the respective size of forces committed. At the time, they estimated approximately 480 Russian battalions against 130 comparable German battalions. The number of artillery pieces in place, indicated Russia held an advantage with 5,800 Russian guns against Germany's 774 artillery pieces. The Russians possessed confidence at the start of the war. On 10 August, *General-leytenant* Nikolai Nikolaevich Yanushkevitch, *Stavka* chief of staff, notified *Gen.kav* Yakov Grigorievich Zhilinsky that initial reconnaissance had determined that *8. Armee* was made up of only four infantry corps in *Ostpreussen*, augmented by a few reserve divisions.[34]

Grand Duke Nikolai Nikolaevich, Russian commander-in-chief, worked with and through *Stavka*. *Stavka's* senior role in conducting war against Germany and the *k.u.k. Armee*, was limited. Their main responsibility rested in where Russian units were to be assigned. Baranovichi was selected as *Stavka*'s operating location because of its proximity to the center of the Eastern Front. This location provided access to three railway lines, enabling more effective communication with both Northwest and Southwest Fronts. Grand Duke Nikolai traveled via train and met up with his two senior Russian front commanders, Northwest Front's *Gen.kav* Zhilinsky and Southwest Front's *General ot artillerii* Nikolai Iudovich Ivanov.[35] War Minister Sukhomlinov and the War Ministry remained in Petrograd to serve administrative and logistic requirements at the fronts.[36]

When war commenced, *Stavka* concluded that Russian aviation efforts required leadership at each front: *Zaveduyushchego organizatsiyey aviatsionnogo dela v armiyakh YugoZapadnogo fronta* [Head of Organization of Aviation Affairs for Armies of the Southwestern Front] and *Zaveduyushchego organizatsiyey aviatsionnogo dela v armiyakh Severo-Zapadnogo fronta* [Head of Organization of Aviation Affairs for Armies of the Northwestern Front]. Grand Duke Alexander was subsequently appointed to the Southwest Front position and *Gen.kav* Baron Kaul'bars took charge of the Northwest Front. On 14 August 1914, Grand Duke Nikolai stated that their function was to serve "the purpose of making aviation and aeronautic affairs in the combat armies, more orderly." Grand Duke Alexander assumed authority over Kaul'bars. Russian military commanders considered Alexander and Kaul'bars as vehicles for further understanding aviation's possibilities, not as agents for handling *RIVVF* resources.[37]

In their memoirs, *RIVVF* pilots recalled that nearly all military commanders were unfamiliar with aviation capabilities in 1914.[38] Despite Russian army maneuvers (including *RIVVF* pilots and staff and several discussions on integrating aeroplanes with ground force operations), Russian commanders and staff lacked a unified point of view

ADDING WEAPONRY

Marat Khairulin

Voennyi Letchik, Shtabs-kapitan Victor Rodionovich Poplavko conducted Russia's first series of flying tests (1913) with an aeroplane (Farman HF.15), firing a Maxim machine gun. Poplavko, wearing helmet and goggles, demonstrates to Grand Duke Alexander (right), the gun test features. Poplavko commanded 21st KAO, supporting Second Army staff in the Northwest Front invasion of Ostpreussen.

Marat Khairulin

The second compatibility test for the machine gun, was conducted in March 1914. Poplavko (forward on the machine gun) and future Il'ya Muromets pilot Voennyi Letchik, Podporuchik Pankratiev (seated in pilot's compartment) about ready to take off.

on aviation tactics as it applied to modern warfare.[39] The pre-war *Provisions on Field Administration of Armed Forces* also failed to include details regarding the administration of aviation at the front.[40]

(left to right) Tsar Nicholas II, General-adyutant Vladimir Borisovich Frederiks and Grand Duke Nikolai Nikolaevich Commander-in-Chief Imperial Russian Army, in discussion, at Stavka's headquarters at Baronovichi.

When *KAOs* entered the first weeks at the front, they were more than challenged in their ability to acquire vital supplies and maintenance support. Russian army commanders were responsible for aviation assets and it quickly became clear to them that sustaining and supporting an aerial reconnaissance force required far more attention than commanding traditional cavalry.

Providing maintenance and parts for the Russian inventory of French-designed aeroplanes overwhelmed available transport and logistical maintenance. In particular, maintaining the Gnome rotary engine (a unique component of the French-designed aeroplane) often required maintenance crews cannibalize inoperative aeroplanes for their spare parts. At first, *KAOs* deployed whatever spare inventory they had in stock, that could easily be transported in conjunction with respective aeroplanes. But in the coming weeks, as the battle front moved, priorities shifted as *RIVVF* assets frequently lost out to the more critical demand for men and army supplies. Russian commanders were forced to cease relying on *RIVVF*, as aeroplanes became scarce.[41]

In the first two weeks, demand for increased missions pushed requirements beyond the service life of the aeroplanes. This complicated

IGOR IVANOVICH SIKORSKY

Carl J. Bobrow

Igor Ivanovich Sikorsky worked for R-BVZ as their chief designer. The Il'ya Muromets was his masterpiece.

Marat Khairulin

Baron Kaul'bars and Sikorsky on the front balcony of the Russkiy vityaz at the 1913 Russian Military Trial, Korpusnoy Aerodrome, St. Petersburg.

The Bee's Home Magazine Page

What Dame Fashion is Offering

A Fur-Trimmed Cloak and an Evening Gown—Fully Described by Olivette

Little Bobbie's :: Pa ::

By WILLIAM F. KIRK

A Biplane That Will Carry Ten Passengers

A New and Terrible War Machine, Reviewed by Garrett P. Serviss

By GARRETT P. SERVISS

Sikorsky's Grand as reported in an Omaha, Nebraska newspaper as "A New Terrible War Machine."

The first Muromtsy (factory No.107) was completed in October 1913. Both cabane and nose platform are visible. The four engines are German-built Argus engines.

The second Muromtsy (factory No.128) built by R-BVZ set a record, with a 2,600-kilometer round trip flight from St. Petersburg to Kiev. The flight, by Sikorsky and crew, took place on 12 July 1914.

R-BVZ mechanic, Vladimir Sergeyevich Panasyuk, and Sikorsky, pose in front the of the Il'ya Muromets (factory No.128). Panasyuk assisted Sikorsky in all of his Russian aeroplane projects.

KAO's function and support. In an effort to limit the potential for catastrophic loss of aviation resources, Grand Duke Nikolai issued Order No. 6 on 12 August 1914. It declared that aviation should be used only in cases of "actual necessity."[42] Russian aviator Tkachev commented later in life, that unfortunately, no one was able to determine the limits of such necessity.[43]

Russian Military Intelligence Struggles

Following the Russo-Japanese War, attempts were made to improve Russia's capability, including instituting and maintaining an effective military intelligence organization. In 1908, Military Districts established intelligence elements to monitor adversaries across borders. In 1911, Minister of War Sukhomlinov created the first modern counter-intelligence service. It was based upon *General-Leytenant* Fedor Fedorovich Palitsyn's success in laying the foundation for a secret intelligence service. *Polk.* Nikolay Avgustovich Monkevits led the 5th (Intelligence) Department of the 1st Quartermaster-General, charged with the responsibility of gathering intelligence. Sections were subsequently organized based upon geographical areas of study.[44]

Russian military intelligence suffered a major setback when *Stavka* disbanded the Intelligence Department of *GUGSh*. Experienced officers were transferred into non-intelligence related functions in various lower army staffs. Their positions were assumed by less experienced functionaries. *Polk.* Monkevits, Russia's military intelligence chief, was transferred to a Russian field army as a member of staff. *Stavka*'s calculation that the war was not going to last long, proved fatal. It was not until 1916, that *Stavka* took steps to reform the military intelligence structure to better serve the Eastern Front.[45]

Despite the self-inflicted wounds of realigning military intelligence, Russia possessed an impressive espionage program prior to the war. The plans of Germany's eastern fortresses and rail networks were in Russian hands. Intelligence on German and Austro-Hungarian fortresses (most notably the Przemyśl *Festung*), allowed staff from the Warsaw Military District to prepare plans and print thousands of copies. The documents concerning Przemyśl, proved invaluable in Russia's success in capturing that strategic fort. *General der Infanterie* Hermann Karl von François was alarmed when he and his staff examined Russian documents, as they included detailed plans of *Gouvernement Festung* Königsberg "accurate to the smallest detail. Only by espionage, could the Russians have gained this valuable material."[46]

Sikorsky's Miracle

Russia's singular triumph in aerial dominance proved to be Igor Ivanovich Sikorsky's extraordinary prewar four-engine aeroplane, *Russkiy Vityaz* (also known as the "Grand"). The "Grand" evolved into an even larger and more ambitious project, the Il'ya Muromets. Sikorsky demonstrated that his large aircraft had a practical purpose, both as a long-range transport for commercial use, as well as possessing military applications. The first Muromtsy (factory No.107) was completed in October 1913, followed by type b *"Beh"* with similar design features. In 1914, Sikorsky and his team (a dedicated group of followers from the Kiev Polytechnic Institute) built and flew Il'ya Muromets (factory No.128), a second-generation design. On 30 June 1914, Sikorsky took off from Komendantsky Field near St. Petersburg for a 2,600-kilometer round trip flight to Kiev. The success of this flight, unambiguously demonstrated the viability of his design and the prospects for long-distance transport by aeroplane. Detailed news of the flight reached an astonished audience throughout Europe as well as the United States.[47]

One of the obstacles in Russia's aviation production, was in designing and operating suitable engines. This was a major problem for Il'ya Muromets from the onset. The most efficient engine for Il'ya Muromets was the Argus engine from Germany. Prior to the war, Argus engines were ordered; a few arrived and were installed. When war was declared, the contract for German-built Argus engine was immediately terminated and engine substitutes such as French Salmson water-cooled radial were tried, but proved unsuitable. Designing quality engines for Russia's aeroplane fleet (including Il'ya Muromets) remained a major challenge throughout the war.

The Imperial Russian Navy immediately showed interest

in Il'ya Muromets. After flying in *Russkiy Vityaz, Kapitan 2-go ranga* Boris Petrovich Dudorov, an officer of the Baltic Fleet, envisioned a long-range aerial reconnaissance platform. Il'ya Muromets No.107 became the property of the Russian Baltic Fleet and testing commenced soon afterwards. When the war began, Il'ya Muromets No.107 was stranded for repairs at Moonsund Archipelago, the Russian navy base north of the Gulf of Riga. *Stars.Lient* Lavrov expressed concern that Sikorsky's state-of-the-art technology was vulnerable to capture and exploitation. On 3 August 1914, a false alarm that a German destroyer was enroute resulted in Lavrov setting fire to Il'ya Muromets No.107. This destroyed the aeroplane and terminated Russian navy interest in the Il'ya Muromets.[48]

The Belated Utilization of Aerial Photography

Russia had a theoretical advantage as a leader in early aerial camera development. *Polk.* Sergey Alekseyevich Ul'yanin and *Podpolkovnik* Vladimir Filippovich Potte designed aerial cameras that became the mainstay of *RIVVF* aerial photography throughout the war. In 1908, *Polk.* Ul'yanin first developed a 43cm focal length camera that provided aerial photographic reconnaissance coverage from kites and airships. Russian aeronauts received training on aerial photography, as well as wireless technology. In 1912, Ul'yanin then created a handheld 25 cm aerial plate camera that included two magazines holding six 13x18 cm glass plates each. The Ul'yanin aerial camera was unique, as it provided a three-in-one photograph on a 13x18 cm glass plate, showing terrain coverage and the altitude of the aeroplane at the time of exposure. Ul'yanin also conceived the world's first portable photo laboratory, a two-wheeled device with a canvas-covered dark room, ideal for maneuvering armies. Neither Ul'yanin's aerial camera nor photo laboratory were in stock at the war's commencement.[49]

In addition, in the summer of 1912, *Podpol* Potte developed (at Gatchina airfield) the world's first semi-automatic aerial film camera that contained a rotating single shutter. It was known as the "Potte" 21 cm aerial camera.[50] The operator simply had to press the trigger and keep track of the coverage, using a stopwatch to

GEORGY IVANOVICH LAVROV

Voennyi Letchik, Starshi Leytenant [Senior Lieutenant] Georgy Ivanovich Lavrov, one of the first Russian naval pilots. In 1915, Lavrov flew Il'ya Muromets aeroplanes.

Lavrov flight tested Il'ya Muromets (factory No.107) for the Russian Baltic Fleet. Standing on the right float is Sikorsky and R-BVZ mechanic Panasyuk, on the left float. No. 107 was destroyed by Lavrov at the opening of the war, as a result of his concern that an approaching German ship was en route to capture the aeroplane.

determine extent of coverage. The 9 kg film camera produced 50 frames of 13 x 18 cm format aerial photographs using a 30-meter film roll. This process was analogous to a continuous stream of photographs on a tape. Shutter operation and film advancement were carried out through winding spiral springs, using just enough tension to move the 30-meter film base for 50 exposures. Potte's aerial cameras provided surveillance strip coverage at the scale of 100 meters to one cm of the target area.[51]

Another prominent camera designer at the time, was *Por.* Vladimir Fedorovich Gelgar, who was among the first 23 officers to earn *Voennyi Letchik* from the *Sevastopol'skaya ofitserskaya shkola aviatsii* [Sevastopol Officer Aviation School]. Gelgar was a Russian photographic inventor and advocate for military aviation applications. He was recognized in 1911 for inventing the "*Razvedchik*" [Scout], a camera that employed a telephoto system. He developed a photographic system that acquired consecutive cinematic images. The film was automatically parachuted to a processing crew on the ground. Each image included: time of photo, arrow of compass, flight altitude, and angle of aeroplane at time of shooting. Photographic material and subsequent prints were developed by Gelgar's mobile laboratory within 10 minutes.[52] Further innovations included night photography that effectively utilized magnesium flares with parachutes. Lastly, he designed high-explosive aerial bombs with a spherical body that were developed and utilized during the Balkan Wars. Despite his significant contribution to Russian aerial photography efforts in the military, Gelgar's career prospects dissipated as war commenced. In 1915, he was listed as one of the *Poruchiks* of a reserve horse artillery battalion stationed in Orel—an insignificant post. His last known assignment was in an aeronautical battalion in Petrograd.[53] He was innovative and ahead of his time, though the prevailing military structure failed to acknowledge his contribution, relegating him and his work to historical obscurity and insignificance.

CAMERAS

Marat Khairulin

Potte aerial camera provided surveillance strip coverage using a 30-meter cartridge film base for 50 exposures. The bulb was used to advance the clockwork driven film.

Затвор находился внутри аппарата и действовал при нажатии на грушу. Скорость работы затвора (моментально и с выдержкой) устанавливалась на «вентиле экспозиций», укрепленном на трубке, идущей от груши.

Прицел-видоискатель служил для визирования на снимаемый предмет, а компас отмечал направление (азимут) на объект съемки. Аппарат давал возможность фотографировать с обычным ладшафтным объективом и через приставку-бинокль. Конструкция камеры обеспечивала обработку экспонированной пластинки непосредственно в фотоаппарате. Готовое изображение получалось автоматически через несколько минут на полном свету, причем рука снимающего не касалась чувствительных пластин.

Рис. 32. Фотоаппарат «Разведчик» В. Ф. Гельгара для крупномасштабной съемки

Съемка производилась так. После установки камеры на штативе оба окуляра бинокля ставились на деление + 6, затем бинокль крепился к передней доске. На приставке к нижней части камеры помещался бак для проявления. Коробка кассеты передвигалась вперед до появления номера пластинки; вентиль ставился на требуемое время выдержки, а поворотом рычага *5* пластинка устанавливалась в фокальную плоскость. Затем через прицел и мушку наводили в центральную часть (участок) предмета, освобождали стрелку компаса, записывали азимут и сжатием груши затвора производили экспонирование. Для получения панорам местности предлагалось

"Razvedchik" [Scout] camera invented in 1911 by Voennyi Letchik, Poruchik Vladimir Fedorovich Gelgar that used a telephoto enlargement system.

PART II

A CHANGE IN COMMAND

CHAPTER III

Observing the Opening Salvos

COMMAND THINKING AT THE START

Fear of the unknown took hold on both sides of the Eastern Front. The formidable Russian empire, with the largest standing army, presented a serious risk to Germany. Russia commenced mobilizing their army. They calculated the campaign required fifteen days to mobilize forces against Germany. Fifteen days into the war, only a third of Russia's military might was in place to conduct the offensive. This forced Russian generals to go to war and conduct a successful campaign against Germany, before their full capabilities were firmly in place.[1]

General ot kavalerii Yakov Grigorievich Zhilinsky, commander of Northwest Front.

8. Armee commander, *GO* v. Prittwitz, faced Russian Northwest Front commander, *General ot kavalerii* Yakov Grigorievich Zhilinsky, (former chief of staff of the Imperial Russian Army).[2] The Russian command made a critical decision in splitting their Northwest Front's advancing armies into two subordinate thrusts—*General ot kavalerii* Paul Georg Karl Edler von Rennenkampf's First [Niemen] Army and *General ot kavalerii* Aleksandr Vasilyevich Samsonov's Second [Narev] Army. This provided *8. Armee* with a fateful opportunity to conduct a campaign, fighting one Russian army at a time. *Gen. kav* Rennenkampf and *Gen.kav* Samsonov were veteran division commanders from the Russo-Japanese war, while their adversary, *GO* v. Prittwitz, had no combat experience commanding a division or an *Armee-Korps*.[3]

Zhilinsky's lack of experience in fighting a major campaign, was clear from the beginning. He ordered divisions to transfer from the corps into fortress garrisons to protect the flanks and rear of his

Northwest Front. His orientation reflected the prior century's strategic posture. This fateful decision reduced both the infantry and cavalry strength of the Russian Second Army by one third. Two-fifth's of *Gen.kav* Samsonov's artillery was also diverted to fortress garrisons.[4] This arcane view of military strategy also affected *RIVVF* deployments. A Farman HF.16 from Fortress Kovno *KrAO* was assigned to support Rennenkampf in Russia's initial advance into *Ostpreussen*. The Fortress Kovno commander instead chose to keep the aeroplanes at the fortress. This decision was made to support his own defense; the aeroplanes did not support the campaign.[5] In contrast, Prittwitz fully supported his campaign by ordering infantry and artillery, normally maintained at the *Festung*, to be deployed to the field units.

When war commenced, Russia possessed a strategic knowledge of Germany's eastern provinces; however, Russian intelligence lacked the critical details necessary for commanders to formulate strategy. The possibility of Germany playing for time (using inferior forces within *Ostpreussen*'s natural defensive positions), was recognized by prewar Russian intelligence. This interpretation was derived from analyzing Germany's peacetime military exercises. *Stavka* had no clear picture as to the number of army corps that the Germans planned to deploy to the eastern *Ostpreussen* border, in an effort to counter any Russian offensive. Staff were more than wary that Germany was engaged in generating a disinformation campaign to cover up its actual intent. Russia's objectives were to attack *Festung* Königsberg with the First Army from the Niemen River and the Second Army from the Narev River to the southwest.[6]

Post-war German military historians analyzed lessons learned by examining initial Russian war plans and conceptual operations for *Ostpreussen*. They realized that once the German threat in *Ostpreussen* was eliminated, a second maneuver could easily be deployed to invade Germany from Silesia, then advance towards Berlin. After the war began, *Stavka* discovered that the Germans sent more units to the Western Front. This change in the enemies' force posture, influenced Stavka to reduce the number of Russian armies opposite *Ostpreussen*. *General ot infanterii* Platon Alekseevich Letschitzki's Ninth Army moved south from *Ostpreussen* and *General ot infanterii* Alexey Ermolaevich Evert's Tenth Army left Warsaw, to fight *k.u.k. Armee* forces.[7]

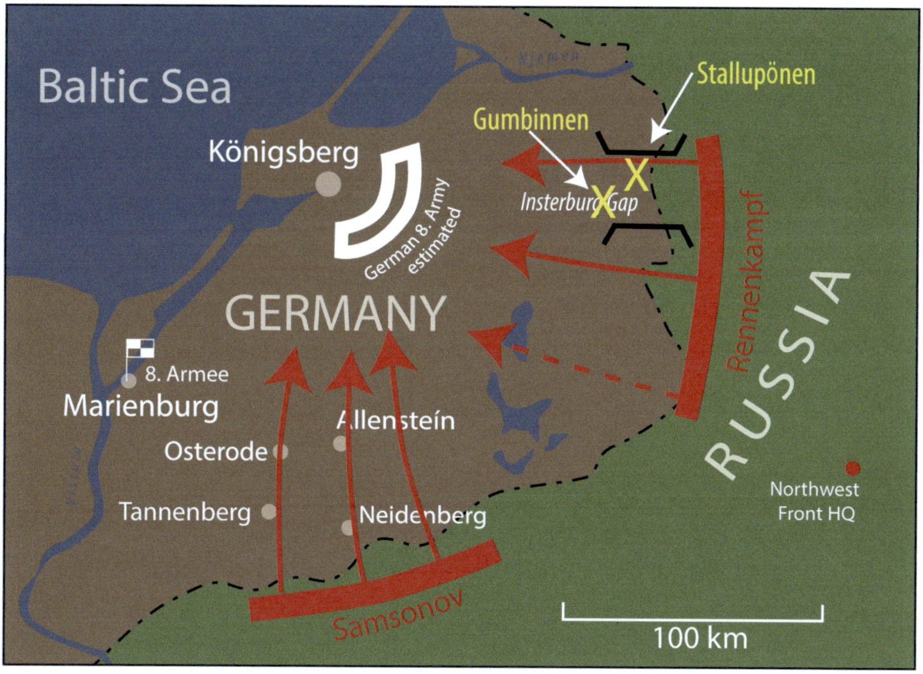

The Russians planned a two-prong offensive on Königsberg through the Insterburg Gap; striking 8. Armee from the south, where the Germans could not escape west to the Vistula River. Source map: Helmut Jäger

The first objective of Rennekampf's First Army was to lure the *8. Armee* forward into a Russian entrapment. Russian cavalry represented the first forces in Germany to conduct their reconnaissance mission. Rennenkampf's Instruction No. 1 guidance to his cavalry commanders did not address enemy forces or instructions for grouping the forces to insure successful maneouvers. He was under the mistaken impression he was able to send daily orders to the cavalry, possessing the most important mission to acquire accurate information on the enemy dispositions.[8] Once combat commenced, First Army intended to keep *8. Armee* in place and engage them in major combat. Russian forces were to envelop the weaker German army, while the Russian Second Army advanced from the south and attacked *8. Armee*'s vulnerable rear echelons. The Russian directive had the two armies forming a 120 kilometer front [Insterburg-Angerburg-Rastenburg-Seeburg]. The combined Russian armies' strategy was to advance on this shared front and push *8. Armee* back to the Vistula River.[9] *Gen.kav* Samsonov's Second Army planned to advance into *Ostpreussen* from the south and conduct operations on

THE RUSSIAN COMMANDERS

General ot kavalerii Paul Georg Karl Edler von Rennenkampf, First Army commander.

General ot kavalerii Aleksandr Vasilyevich Samsonov, Second Army commander.

Russian Cossacks on the move.

8. Armee's flank and rear areas.[10] *Gen.kav* Zhilinsky allowed Rennenkampf to divert forces from the southern flank, in an effort to strengthen the advance on Königsberg. Samsonov received authorization to protect lines of communication into Warsaw, by detaching troops from the main body heading into *Ostpreussen*. The gap created between the two Russian armies proved critical in the evolving campaign. Only one Russian corps covered the 80 kilometers of marshes and lakes that separated the two Russian armies. The reconfigured Russian First Army and Russian Second Army lacked the depth to effectively confront *8. Armee*. When Samsonov's Second Army crossed the *Ostpreussen* border five days after the Russian First Army,it lacked the necessary force to divert the Germany army. [11]

At this critical time, aviation represented a low priority for *Gen.kav* Zhilinsky. The catastrophic results of August 1914 were largely attributed to this fatal oversight. Zhilinsky rarely took initiative and instead, was constrained by his adherence to regulations, in an effort to avoid conflicts with his superiors.[12] He directed Rennenkampf and Samsonov "... to launch decisive attacks with the objective of annihilating the adversary, in order to cut off Königsberg and thereby block a German retreat to the Vistula River." Additionally, "... the Russian First Army was directed to advance north of the Masurian Lakes towards Insterburg - Angerburg, having already crossed the Angerapp River. The goal was to

The Hughes alphabetic transceiver was a printing telegraph, unlike the Morse code of dots and dashes, it was comprised of a printing wheel and a keyboard of 26 keys, with alphabet and punctuation characters at both the transmitter and receiving units.

outflank the Germans from the left and cut them off near Königsberg ... the Russian Second Army was to proceed towards Lötzen - Ortelsburg, and envelop the enemy from the rear and flanks of the Lake Area."[13] The invasion demonstrated that the two armies required closer coordination, using a network of available communications. This necessitated providing each other constant updates, as well as informing headquarters at the Northwest Front. The Germans had the advantage in employing existing homeland networks, while the Russians remained dependent on limited communication devices and messengers. *Gen.kav* Samsonov's staff codenamed their operation *Samsonovskaya*.[18] The operation was executed with a few telephones, a handful of telegraphic devices, and a teleprinter capable of discharging 1200 words per hour, known as a Hughes apparatus. *Stavka*'s Baranovichi headquarters included one Hughes apparatus. As the two armies entered *Ostpreussen*, secure communication proved difficult to maintain. Samsonov's Second Army lacked sufficient communication wire. Subsequently, all communication to and from subordinate corps headquarters necessitated employing a more vulnerable wireless connection—one that proved lethal in the coming battle.[19]

RIVVF Support to the Russian Invasion of Ostpreussen

RIVVF's four KAO's comprised a total of 14 aeroplanes to support Russian First Army's advance.[14]

- *2nd KAO* (commanded by *Shtabs-kapitan* A.V. Zalessky; 2 Nieuport IV)
- *3rd KAO* supported III Corps (commanded by *Shtabs-kapitan* Georgy Alexandrovich Myachkov; 5 Nieuport IV)
- *4th KAO* (commanded by *Shtabs-kapitan* V.N. Orlov; 3 Nieuport IV)
- *10th KAO* (commanded by *Shtabs-kapitan* Alexander Georgievich Topchev; 4 Farman HF.16)[15]

RIVVF had 24 aeroplanes supporting *Gen.kav* Samsonov's Second Army.[16]

- *1st KAO* supported *General ot infanterii* Leonid Konstantinovich Artamonov's I Corps (commanded by *Shtabs-kapitan* Christopher Fridrikhovich Prussis; 5 Farman HF.16)
- *23rd KAO* supported *General ot kavalerii* Sergey Mikhailovich Scheideman's II Corps (commanded by *Shtabs-kapitan* Peter Semenovich Prischepov; 5 Farman HF.16)
- *13th KAO* supported *Genleyt* Klyuev's XIII Corps (commanded by *Poruchik* Nikolai Vladimirovich Kokayev; 5 Nieuport IV)
- *15th KAO* supported *General ot infanterii* Nikolai Nikolaevich Martos' XV Corps (commanded by *Shtabs-kapitan* Anatoly Iosifovich Valnitsky; 6 Nieuport IV)
- *4th AR* and *21st KAO* supported *Gen.kav* Samsonov's Second Army staff (commanded by *Shtabs-kapitan* Viktor Rodionovich Poplavko; 3 Farman HF.16)[17]

As Russia quickly mobilized, the disorganized dissemination of vital materials resulted in new military ciphers and keys, that failed to be distributed to all Russian headquarters. *General-leytenant* Nikolay Alekseevich Klyuev's XIII Corps lacked the key required to read cryptograms from *General ot infanterii* Aleksandr Alexandrovich Blagoveshchensky's VI Corps. This shortfall caused *13th KAO* in support of *Gen.leyt* Klyuev's XIII Corps, to prioritize sorties, in an effort to deliver secure communications. To complicate matters further, Russian signalmen rarely enciphered messages and transmitted them in the clear.[20]

8. Armee Searches From the Skies for Russian Intentions

When war was declared, *Grosse Generalstab* became *GO* v. Moltke's *1.Oberste Heeresleitung (OHL),* the Kaiser's senior military headquarters, operating from Coblenz. On 8 August 1914, *GO* v. Prittwitz officially took command of *8. Armee* and set up headquarters at Marienburg. The *8. Armee War Diary* was initiated, citing aviation for the first time: "Army order No. 1 dated August 8, 1914, was issued at 2200 hours....The order committed the cavalry and air reconnaissance against the line Novo Georgievsk – Jurborg." Two German *Armee-Korps* became the main line of defense against the initial threat to *Ostpreussen* from Russia. *GdI* v. François' commanded I. *Armee-Korps* and *General der Artillerie* Friedrich von Scholtz commanded *XX. Armee-Korps*. Standing orders obliged both generals to keep all enemies off German soil.[21] *GO* v. Prittwitz cautioned his commanders not to exhaust the fleet of *Fliegertruppe* aviators and aeroplanes—a directive

Generaloberst Maxmilian von Prittwitz und Gaffron, 8. Armee commander in the opening weeks of World War I.

designed to sustain operations for the upcoming campaign. Prittwitz sent the following guidance:

> The general situation with respect to the first deployment and intent of the enemy, remains completely unclear. Therefore, *Armee-Korps* etc. will reach assembly areas as planned... Retreats will only be permitted by the command. Clearly, local advantages can and should be fully exploited and are encouraged... During this phase of war, we will:
>
> 1. Gain time, until the decision in the West is made, and our forces are freed up for important operations in the East.
>
> 2. Support the Austrian Offensive by keeping enemy forces down.
>
> 3. Secure the Vistula River Line as a basis for our operations. Those items are paramount to achieve our objectives. All units have to act in a coordinated manner even if separated and distant from each other. Any single advance of an *Armee-Korps* into enemy territory, will influence the situation of our Field Army, in an undesired manner. The mission is essentially a defensive one, for the time being. Clearly, success cannot be achieved by remaining in a fixed position. We cannot foresee at the present time, where the thrust of our Army will be directed. Depending on the results of our reconnaissance, our Army may use rail assets in order to increase direct cooperation among the units...[22]

Prior to Russia's invasion of *Ostpreussen, Generalmajor* Graf von Waldersee, *8. Armee* chief of staff, reported command intentions. These were derived from aerial reconnaissance sightings reported to *1. OHL*: "From the direction of the Narev River, no indications of an imminent attack were detected. Yet from the north, from behind the line Wirballen-Suwalki, a concentration of forces makes an attack from the east probable. *8. Armee*'s intent is to assemble behind the *Seen-Linie* [lake line], and further to the north along and behind the Angerapp River. From there, a blow at the advancing "Wilna" Army can hopefully be delivered." *Genmaj* v. Waldersee in accordance with *8. Armee* intentions reported to *GO* v. Moltke, *OHL*: "Reports (from agents) reconnaissance, wireless intercepts etc., have revealed that in the north *Gen.kav* v. Rennenkampf's Russian First Army is slowly approaching. A thrust against the Masurian Lakes appears to be imminent. We have not seen the Russians in the Narev River area. Aerial reconnaissance has confirmed limited sightings of Russians at the fortresses of Novo Georgievsk and other fortified locations on the Narev River. They are therefore manned to a limited extent. *Gen.kav* Samsonov's Russian Second Army has not been sighted near the south of *Ostpreussen* (although *8. Armee* anticipated Samsonov's army was heading in that direction.) This situation offers us the opportunity to deliver a sudden blow against the flank of *Gen.kav* v. Rennenkampf's Russian First Army."[23]

Rare look at Zeppelin Z-V [LZ 20 (production number)], based at Posen. On 7 August, Zeppelins commenced strategical reconnaissance over the region, with Zeppelin Z-V flying over Kutno. Russian infantry and artillery fired at Z-V and killed one of the crew.

AVIATORS SUPPORTING GEN.KAV SAMSONOV'S SECOND ARMY

Voennyi Letchik, Shtabs-kapitan Christopher Fridrikhovich Prussis commenced command of 1st KAO in 1913. He led 1st KAO in support of Samsonov's Second Army.

Voennyi Letchik, Shtabs-kapitan Peter Semenovich Prischepov, 23rd KAO commander

Voennyi Letchik, Shtabs-kapitan Anatoly Iosifovich Valnitsky, 15th KAO commander

Shtabs-kapitan Viktor Rodionovich Poplavko, 21st KAO commander.

Voennyi Letchik, Poruchik Nikolai Vladimirovich Kokayev, 13th KAO commander

Further defining the *8. Armee* objectives, *GO* v. Prittwitz established four intelligence fundamentals, that reflected defensive priorities:

1. Determine Russian force dispositions beyond the border from Schirwindt to Memel.

2. Determine Russian intentions of *Gen.kav* v. Rennenkampf's Russian First Army. Intelligence was required to confirm on which side of the Niemen River, Rennenkampf's Russian First Army was assembling.

3. Intelligence was required to assess if *Gen.kav* Samsonov's Russian Second Army was assembling behind or north of the Narev River.

4. An overview was required of Russian forces and intentions north of Warsaw.[24]

The First Fliegertruppe Aerial Reconnaissance

Fliegertruppe sorties during the first days of war crossed the border and reported no sightings of Russians' advancing formations or cavalry. Russian air activity was rarely mentioned. Several initial *Fliegertruppe* aerial sorties over Russian territory, in part, served *OHL* propaganda objectives against an unknown Russian enemy. Three Taube aeroplanes flew from *Ostpreussen* to Warsaw and dropped propaganda leaflets. In their first aerial sortie, *Leutnant* Alfred Mahncke and an observer were ecstatic as they crossed the border en route to Warsaw.[25] *8. Armee War Diary* recorded, "An aviator reports (unit not mentioned) the presence of a Russian cavalry regiment (apparently of 1st Cavalry Division), at Filipow, and on the hills east of Bakalarshevo, one or two batteries. The roads south, east, and north of Suwalki, for a distance of 20 kilometers, were found to be free of the enemy at 1800 hours."[26]

Zeppelins commenced strategical reconnaissance over the region. Zeppelin Z-V flew 7 August over Kutno and was subsequently fired upon by Russian infantry and artillery. One crew member was killed, but Z-V suffered minimal damage. Z-V reported to *8. Armee* (on the following day) that no Russians were reported west of the Vistula River.[27] On 9 August, Zeppelin Z-IV flew a southern mission to Lodz and reported on Russian positions. Z-V was repaired from the damage and resumed flight on 10 August. This time, Z-V dropped bombs on Mława south of the *Ostpreussen* border. Z-IV attacked Lodz the following day. *Fliegertruppe* aeroplanes and airships flew as far east as the Russian fortress at Kovno [Kaunas].[28] Initial *Fliegertruppe* aerial observations were plagued with errors, in both detecting and interpreting Russian maneuvers. On 8 August, *FFA 14 Fliegertruppe,* in flight from Deutsch-Eylau over the southern sector, reported at 11 AM, "Napierken-Pikielko-Unitzki, strongly occupied by cavalry. Three cavalry units advanced from Karpierken toward Kuklin-Mława. The sortie also covered Mława, Novogeorgiesk, Ziechanow, and Warsaw. No Russian forces of any significance were observed.[29] That night, *Gdl v.* François' *I. Armee-Korps'* staff telephoned *8. Armee*'s headquarters *Armee-Oberkommando 8 (AOK 8)* with a status report. *AOK 8* cited an 1800 hours report received from *FFA 14 Fliegertruppe,* "Close to and north of Filipow, Russian cavalry but not more than one regiment. One to two batteries on the hills east of Bekalargewo. The roads between Filipow-Suwalki and leading out of Suwalki are free of the enemy, for 20km. Nothing seen of the enemy in Suwalki."[30] That afternoon, the *FFA 14* aviator reported that the Kibarty-Maryampol-Prny-Pilwiscki-Schirwindt road was free of Russians. Suwalki was reported to be fortified for field service with fire trenches, strong points, and battery positions. *FFA 14*'s aircrews Flying Albatros-Taube and Rumpler-Taube monoplanes, reported brigade-strength camps, long columns of wagons, and massive troop movements, between Suwalki and the frontier. As a result of *FFA 14*'s mixed reporting, *8. Armee* believed the main body of the Russian First Army was south of the Rominte Heide.[31] Adding to their confusion, *Gouvernement Festung* Königsberg's *FestFA 5* mistakenly reported a column of 1. *Kavallerie Division,* as Russian reinforcements moving toward Pillkallen.[32]

On 9 August, *GdI v.* François affirmed his growing belief in the importance of aerial reconnaissance. *I. Armee-Korps* notified *AOK 8,* "About one Brigade Infantry with Artillery at Tauroggen. Dragoon Rgt.

19 and Hussar Rgt. 16 on both sides of the highway. As German patrols have met with fairly heavy losses, the commander of the *Landwehr-Brigade* considers that reconnaissance is possible only through aviators."[33]

AOK 8 received a report from *70 Landwehr-Brigade*, summarizing intelligence from cavalry in the Rypin area. Russian intelligence on *Festung* Graudenz was discovered by cavalry going through a cossack barracks in Rypin. "All the forts and defensive positions around Graudenz were marked on this map."[34] This was a cause for concern with *AOK 8*. How much did the Russians know about German defense in *Ostpreussen*?

Genmaj von Waldersee, *8. Armee* chief of staff, reported to *OHL* on 9 August, "As far as we know, the enemy's measures lead us to conclude that he is concentrating forces about two miles behind the line Wirballen-Lomsha." *AOK 8* expected the chief Russian threat existed behind the Wirballen-Suwalki front line. Both *GO* v. Prittwitz and *Genmaj* v. Waldersee shared the opinion that *8. Armee* was best served by striking the Russians from the Angerapp sector, north of Angerburg.[35]

Russia's Armies Advance to the German Border

Russian advances to the southern border between 10 and 14 August were unconfirmed by *Fliegertruppe* aerial reconnaissance. *FFA*'s were now operating from locations that directly supported their commanders. *8. Armee*'s own *FFA 16* now flew from Allenstein.[36] Zeppelins Z-IV and Z-V flew night sorties (10 and 11 August) conducting strategical reconnaissance south of *Ostpreussen* as well as dropped bombs on Mława and Lodz. Both airships confirmed the absence of major Russian forces along the Narev River.[37] On the evening of 10 August, *General der Kavallerie* August von Mackensen's *XVII. Armee-Korps* (supported by *FFA 17* flying from Deutsch-Eylau) reported aerial sorties covering southern *Ostpreussen*. His staff reported, "no troop movements, no railway traffic, no infantry seen in the area of Mława-Ziechanow-Nasielsk-Makow-Przasnysz. Apparently, there is still one cavalry division at Mława; a cavalry brigade with artillery has been recognized at Mława." *Fliegertruppe* observed "three new and large tents (seemingly for aeroplanes) have been sighted at the railway station of Ziechanow."[38] On 10 August, *8. Armee War Diary* recorded, "Aviators have observed one Division

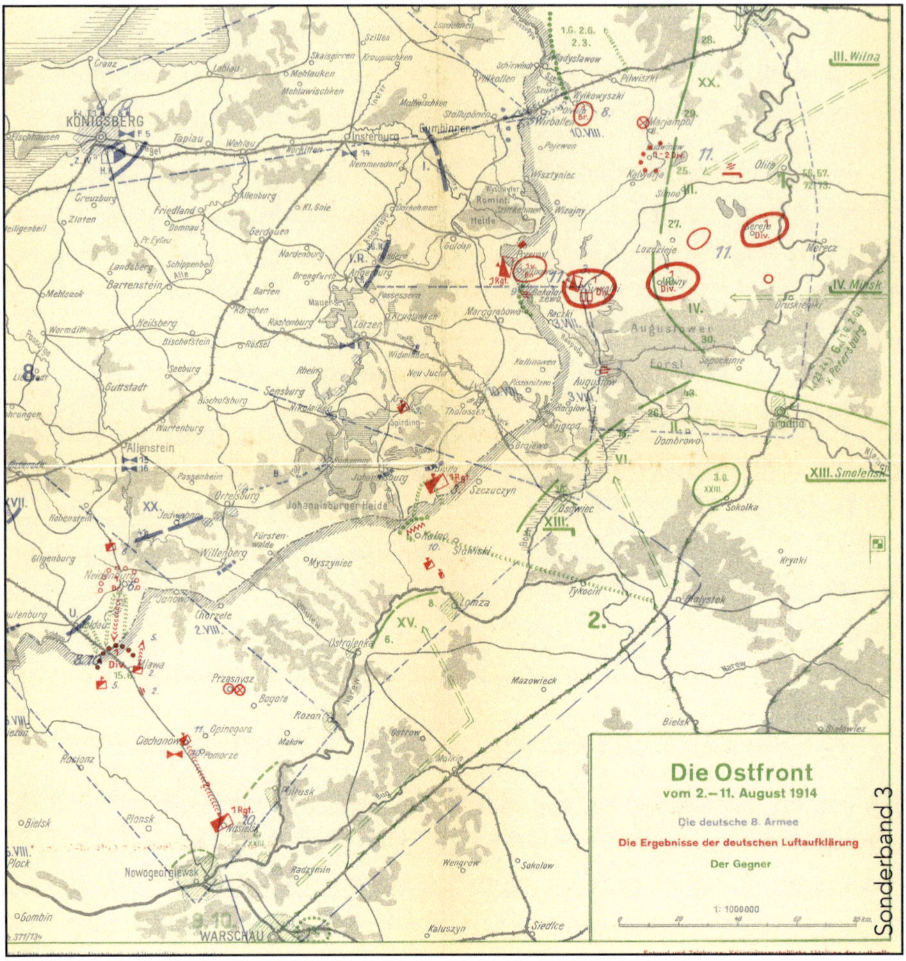

On 11 August 1914, Russian forces advanced to the Ostpreussen border. Larger map available at shootingthefront.com.

each at Filipow – Suwalki – Sejny, Shlavanty and weaker enemy forces near Lejpuny and west of Sereje, as well as on the Shiplinshka – Lodzieje road. 1 Cavalry Division at Suwalki."[39] As a result of a successful 11 August wireless intercept, *8. Armee* learned that *Gen.kav* v. Rennenkampf's headquarters (opposite northern *Ostpreussen*) was located at Vilna.[40] The following day, a wireless intercept discovered *Gen.kav* Samsonov's headquarters facing southern *Ostpreussen* "either at Volkovishk or Belostock."[41] On 11 August, *OHL*'s intelligence sent a detailed summary to *AOK 8*, concerning Russian force positions. The report showed one-half of *Gen.inf* Jepantschin's Russian XX Corps positioned north of the Niemen River, and the 1st Independent

Cavalry Brigade noted near Tauroggen. The remaining half of Russian XX Corps was observed between Kovno and Grodno. Additionally, *Gen.kav* Scheideman's II Corps, Russian III Corps, *Gen.art* Aliyev's IV Corps, and three cavalry divisions, were also sighted between Kovno and Grodno. Intelligence covering Scheideman was incorrect—the actual location was at the extreme southern end of the Kovno-Grodno line. II Corps was not a part of *Gen.kav* v. Rennenkampf's Russian First Army until 22 August. *OHL* alerted *AOK 8* that despite significant aerial reconnaissance reporting, the Russians were actually further to the north.[42]

AOK 8 sent out a message to all *Armee-Korps* and independent brigade commanders on 11 August, "Telegraph at once, brief summary of everything known about the enemy giving the number of Russian regiments involved. From now on, *Armee-Korps* will report all information as to the enemy and events, to *AOK 8* only, and not to Berlin. Signed *GO* v. Prittwitz." *AOK 8* was determined to be the singular source for intelligence reporting and were not going to be upstaged by *OHL*.[43] That day, *Ltn* Mertens (from *I. Armee-Korps' FFA 14*) flew a sortie over Filipow and Suwalki and reported two Russian divisions were sighted. His detailed account of the fortifications in Suwalki, proved to prescient, that trenches were soon to become the standard for the future of combat. "The fortifications consist of fire trenches with traverses, strong points, and concealed batteries."[44] On 12 August, *I. Armee-Korps* staff telephoned *AOK 8* with updated information (from *Ltn* Mertens) regarding the previous day's sortie. The aircrew flew at a height of 1400 meters and estimated Russian organizations observed according to "peace strength." Peace strength reflects the prewar estimates of Russian units. Mertens was also questioned, as to the validity of his earlier 9 August sortie. *I. Armee-Korps* remarked over the telephone, "He is certain that the works were non-existent at Suwalki on 9 August."[45] A second *FFA 14* aerial observer, *Obltn* Körner, filed a report that day, covering Merecz. *GdI* v. François' staff vouched for Körner's aptitude and ability, which was subsequently backed by *Ltn* Hesse.[46]

On 12 August, *Gen.kav* v. Rennenkampf's First Army assembled at the border. *OHL*'s view of the evolving situation on the Eastern Front was shared with *GO* v. Prittwitz and *AOK 8*. They had misjudged Russia's military strength and failed to take into account the effect of

General der Infanterie Hermann Karl Bruno von François, led I. Armee-Korps at Stallupönen, Gumbinnen, Tannenberg campaign (Usdau, Neidenburg), and Masurian Lakes. His use of aviation – Feldfliegerabteilung 14 – was superb. Information gleaned from those sorties shaped the battles that resulted in Germany's victory.

the Russian First and Second Armies' strategically planned envelopment maneuver. *OHL* estimated that a Russian offensive into Germany comprised two groups of forces north of the Bobr River, which consisted of five Russian army corps plus three available cavalry divisions. Actual Russian forces committed at the Northwest Front were twice *OHL*'s estimate. The initial *Ostpreussen* offensive entailed 19 Russian infantry and cavalry divisions in the First Army, and 18 infantry and cavalry divisions in the Second Army. *OHL* mistakenly calculated that only Rennenkampf's Russian First Army, represented a serious threat to *Ostpreussen*.[47]

GO v. Prittwitz issued a command to establish a defensive perimeter, 20 km west of the border, along the Angerapp River. This typified Prittwitz's predisposition towards exercising caution. Francois defied these orders, moving *I. Armee-Corps* forward to Stallupönen. *FFA14* deployed forward to Insterburg. Francois' decision, inadvertently placed *8. Armee's* forces in a weakened position. Francois was naturally pugnacious. His insubordination towards his superiors, reflected his ardent need to take action and defend Germany against the Russians. He displayed an arrogance in stark contrast to standard Prussian military tradition. Only his brilliance on the battlefield saved his career.[48]

On 13 August, *Fliegertruppe* aerial reconnaissance confirmed Russians massing on the border. Russian trains and rolling stock were sighted arriving at Ostrolenka on the Narev River. *AOK 8* accurately interpreted the formation as the initial assembly of the Russian Second Army. That evening, aerial reconnaissance in the north, confirmed a major

Russian force was assembling on the Niemen River.[49] *GO* v. Prittwitz's senior intelligence officer on the *8. Armee* staff, *Oberstleutnant* Max Hoffmann, received an overview of Russian wireless messages intercepted by Königsberg. "The consolidation of all the wireless messages received by the radio station at this location covers the following: At the present time it seems as if the Russians have three field radio stations…three field stations carry on a lively intercourse with each other." Hoffmann also learned that all troop movements were reported in cipher and *Gen.kav* v. Rennenkampf was now at Kalwarja with his staff.[50] On 14 August, *Leutnant* Ernst Canter from *FFA 14*, flew a morning sortie over northern *Ostpreussen* towards Kovno and reported "six long and empty trains without locomotives at the railway depot of Wylkowyski. Troops not seen…No traffic on the railway stretch." Canter further reported no Russian troops were observed on the Kovno-Marjampol-Kalwarja, Wladislawow-Railway depot Wilkowischki-town of Wilkowischki-Ludwinow. Russians fired at Canter's Taube when they flew over Marjampol. *I Armee-Korps'* headquarters staff telephoned *AOK 8* with the results.[51]

8. Armee War Diary for 14 August stated, "The assembly of strong enemy forces at and east of Filipovo, seems to indicate the possibility of a thrust from the enemy south of the Rominte Heide [Rominter Heath] against the lake line. To meet such a thrust offensively, the concentration of the German army toward the left flank has been ordered." Russian First Army units were observed at Wersholowo – Kibarty – Wylkowysky – Marjampol. *GO* v. Prittwitz executed his plan of operations based on OHL's estimate of Russian forces and intentions. *Gdl v.* François' *I. Armee-Korps* was to remain for the present by Gumbinnen-Insterburg. *I. Armee-Korps* was directed to have *FFA 14* cover "to the Niemen and via Tilsit against Szawle." *Genltn* Otto von Below's *I. Reserve-Korps* moved to the Angerapp line. *I Reserve-Korps* had one *Parseval und Sigsfeld Drachen-Ballon* [kite balloon]. *GdK* v. Mackensen's *XVII. Armee-Korps* was ordered to move by rail via Allenstein to Insterburg-Bokellen-Norkitten.[52]

In the south *GdA* v. Scholtz's *XX. Armee-Korps* advanced to the Ortelsburg region. *XX. Armee-Korps* was the single *Armee-Korps* defending the south with their *FFA 15* assigned to fly reconnaissance sorties "toward and across the Narev." Both *8. Armee's FFA* 16 L.V.G. aeroplanes and *XX. Armee-Korps FFA 15* aeroplanes flew along the

Shtabs-kapitan Prussis (left), 1st KAO commander, supported Gen.inf Artamonov's I Corps as it moved north towards Ostpreussen. Flying Farman HF.16 aeroplanes on 17 August, 1st KAO reported extensive German infantry and artillery activity in Mława as well as major rail movements heading north from Soldau.

southern border from Mława to the Narev River. *FFA 16* reconnaissance covered Ciechanow-Nadielsk-Luyencloe east of Warsaw-Pultusk and reported no activity. On the Warsaw-Nowominsk road, *Fliegertruppe* spotted a 5 kilometer long column with guns and cavalry. The head of the column was 17 kilometers west of Nowominsk.[53]

RIVVF Supports the Advance from the South

On 14 August, *Stavka* tasked Grand Duke Alexander and *Gen. kav* Kaul'bars to "make aviation and aeronautic affairs in the combat armies more orderly..." *Stavka* reinforced the doctrine that aviation detachments were subordinate to army, corps, and fortress headquarters.[54] At the eleventh hour prior to the actual invasion of Germany, *Stavka*'s directives to both *RIVVF* aviation leaders were ambiguous and confused. The limited awareness by military leaders on how to employ *RIVVF* in battle, was disturbing. Logistic support to *RIVVF*, quickly overwhelmed existing networks, when Russian armies entered *Ostpreussen*. The absence of aerial photographic resources at this stage, reflected poor planning and vulnerabilities to come.

At the commencement of hostilities, *Sh.kap.* Valnitskiy's *15th KAO*, received initial guidance from *General-leytenant* Vladimir Oranovskii, senior *Stavka* staff officer. *Stavka* needed rumors confirmed, that a sizable German force occupied the Kutno - Gostynin areas (100km west of Warsaw). On 2 August 1914, *Poruchik* Alexander Iosifovich Kulvinsky flew to Kutno and Gostynin and reported a few German patrols in the area.[55] Three days later, *15th KAO* deployed to Zambrowo to support *Gen.inf* Martos' XV Corps. They shipped five Nieuport IV monoplanes via rail. Their sixth Nieuport IV was damaged when loaded on the train in Warsaw. When *Sh.kap.*Valnitskiy arrived in Zambrowo ahead of his unit, he was surprised to find there were no XV Corps forces at the location. On 7 August, Valnitskiy reported to Martos' XV Army Corps chief of staff. His five Nieuport IV aeroplanes required maintenance to be effectively operational, resulting in repairs and test flights conducted over the next five days. *Gen.inf* Martos conducted a review of the *15th KAO* on the evening of 10 August. On 13 August, they were ordered to relocate to the border in support of XV Corps' advance into Germany. Their five Nieuport IV's flew ahead to provide reconnaissance.[56]

On 17 August, *Gen.inf* Artamonov's I Corps proceeded north towards Mława and Soldau. His *1st KAO*'s Farman HF.16 aeroplanes flew over southern *Ostpreussen* and reported:

1. Mława was full of activity. At the outskirts of the town, 1-2 regiments of infantry and artillery;
2. Along Soldau – Mława, movement of carts and separate groups of people;
3. In Soldau, no large concentrations of troops noticed; at the station there were several trains with covered carriages (up to 200 — 250 carriages), three trains were sighted with steamed locomotives heading in the direction of Deutsch-Eylau;
4. At 8 a.m., 1 — 2 cavalry squadrons came out heading to Mława from Soldau;
5. No movement of trains on Soldau – Lautenburg line;
6. On Lautenburg – Zalun highway - movement of individual carts;
7. Zalun -Turzha highway was free, in the town Turzha, 1 cavalry squadron was detected.[57]

Fliegertruppe Sorties Report More Russian Forces Advancing

Flying over northern *Ostpreussen* ceased on 15 and 16 August due to inclement weather. Ground observation reported Russians were not sighted north of Wirballen.[58] In the south, GdA v. Sholtz's *XX. Armee-Korps* reported to *AOK 8* (on the evening of 15 August) that *FFA 15* aeroplanes flew a sortie south of Lomja and observed an infantry brigade and logistic supply column on the road. Further coverage confirmed the Lomja-Kolno-Myschyniec area was free of Russian troops.[59] *AOK 8* departed Marienburg on 15 August for Bartenstein. *GdK* v. Mackensen's *XVII. Armee-Korps* commenced rail movement to Dankehmen.[60] *8. Armee War Diary* issued the statement, "Otherwise, generally no news of the enemy. An aviator reports the presence of one infantry division (XV Army Corps) south and southeast of Lomja."[61]

Despite the inability to fly aeroplanes due to inclement weather, other intelligence sources helped decipher Russian intentions at this critical moment. Perfect intelligence on Russian operations was acquired on the morning of 15 August. A *Leutnant* from *Radfahr Kompanien* [Bicycle Company] rode 15 kilometers behind the front lines and stopped a Russian staff car in Marggrabowa south of Suwalki. The German officer killed the three Russian officers and seized numerous documents that included orders for the Russian 1st Cavalry Division to advance. The *Leutnant* then drove the automobile into the lake. He successfully returned to the German lines with the Russian documents and delivered them to the German *3. Reserve Division*.[62] The next day, a senior officer from the division, personally delivered the translated papers to *AOK 8*. The first paper, *Order for the Detachment Suwalki*, (signed off by *Genltn* Kurko) directed reconnaissance "to ascertain the nature and extent of the enemy's forces, located at Marggrabowa." Subsequent text in the orders outlined Russia's objectives for initial skirmishes against the Germans in the sector. "To ascertain, through fighting the forces of the enemy assembled at and around Marggrabowa, and to gather all possible facts as to the general stability of those forces. Upon completion of this task, the detachment will retire over the same route over which it advanced, and move east of the line of the

Bakalarezewo Lakes."[63] A subsequent document, *Compilation of Information about the enemy on August 10*, "Airplanes [German] are flying daily in the direction of Kovno.[64] The captured documents revealed to German staff that Russian troop commanders were concerned that *Fliegertruppe* was engaged in an aggressive campaign of aerial reconnaissance over their forces.

8. Armee chief of staff *Genmaj* v. Waldersee wrote *GO* v. Moltke, summarizing the intelligence acquired to date on the Russian advance:

> Intelligence of all kinds, reconnaissance and intercepted enemy radio messages, confirm strong enemy forces behind the line Dubeningken-Filipow-Bakalarshewo-west of Raczki and in the neighborhood of Rajgrod. According to captured commands and reports, there is barely any doubt that Rennenkampf's army is slowly approaching our borders. At present, stronger forces have not been reported at Wilkowischki and Kalwaria. Airmen have discovered that Novo Georgievsk and other fortified places along the Narev River, are still strongly held. Beyond this line, Russian forces are not observed to be ready. North of Niemen, no other forces have been discovered. We have the impression that the Russians will conduct an early offensive against our Masurian Lake area. There is a possibility that *8. Armee* could strike a sudden blow against the Russian northern wing's flank.[65]

Genmaj v. Waldersee assessed that *Gen.kav* v. Rennenkampf's First Army intended to invade *Ostpreussen* from the central Masurian Lakes region. The failure of *Fliegertruppe* aerial reconnaissance to report Russian forces north of Kalwaria, factored into the assessment. Compounding the absence of information on Russian locations at this time, there was only one wireless intercept of *Gen.inf* Smirnov's Russian XX Corps transmissions, related to the Kovno-Königsberg rail line.[66] On the night of 15 August, *Gdl v.* François' *I. Armee-Korps* reported to *AOK 8* that Russian forces moving in the direction of Schidkehmen had crossed the border south of Wistita-Lake. At Bilderweitochen, a Russian infantry regiment of Russian *III. Corps*, attacked and was repelled by a battalion

Mobile German Radfahr-Kompanien [Bicycle Companies] were able to cross into Russian territory and acquire valuable information. A portrait of two Radfahr soldiers shows their essential kit to include an elaborate bicycle of the time. This very mobile and dynamic ground reconnaissance element proved very lucrative in acquiring critical information before the Russians invaded Ostpreussen.

of 41. *Infanterie-Division.*[67] On 16 August, subsequent aerial reconnaissance and secret agents reported that the situation "points to the massing of stronger forces of the enemy, apparently II. Army Corps, IV. Army Corps, III. Army Corps and parts of the XX. Army Corps (28 Infantry Division) between Suwalki and Kibarty."[68] The invasion of Germany was about to commence.

CHAPTER IV

Generaloberst von Prittwitz Loses Command

Russia Invades Germany

Gen.kav Zhilinsky launched a major offensive at 0800 hours on 17 August, in ordering *Gen.kav* v. Rennenkampf's First Army to cross the *Ostpreussen* border west of Suwalki. This was the Russians' first offensive operation on German soil. *Gen.kav* Samsonov's Russian Second Army was south of the southern *Ostpreussen* border. Favorable weather on 17 August, spurred *Fliegertruppe* aerial reconnaissance to engage in numerous sorties in search of Russians. *Gdl v.* François' *I. Armee-Korps' FFA 14* aviators first flew from Insterburg; their objective was to monitor the Russian First Army's advance from Suwalki to the *Ostpreussen* border between Filipowo and Kibarty. The immediate intelligence collection priority, was to determine the speed of the Russian advance. *FFA 14* aviators now observed from their Rumpler-Taube aeroplanes, that there were Russian forces massing at the border, moving slowly west into *Ostpreussen*.[1] That morning, *FFA 14*'s additional sortie (flown by *Oberleutnant* Sommayer) dropped a *Flieger-Meldung* on *I. Armee-Korps* headquarters at Frempen. The report read:

During the first battle between Russian First Army and 8 Armee, the armies could see, for the first time, an aeroplane flying during warfare. The picture is mostly symbolic. A Rumpler Taube is seen from the ground— suggesting to the Russians, that the enemy held an advantage in shaping the battle to come.

> Right Russian flank at Wyschtyniez—1st Cavalry Brigade, at 0845 am, 1 km southeast of Wyschtyniez on road

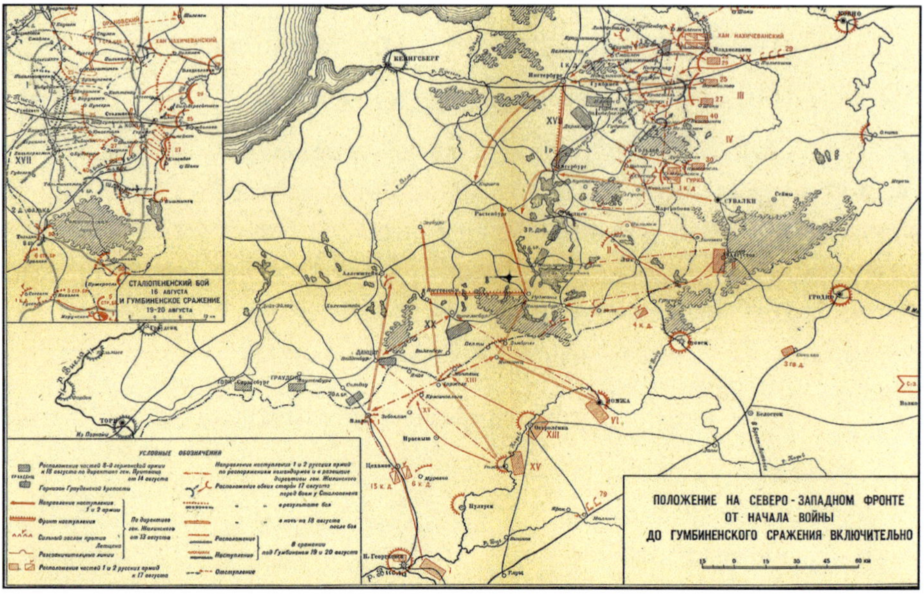

Russian map showing forces entering Germany around 17 August 1914. The separation between First Army and Second Army was 70 kilometers. Source: Carl J. Bobrow Larger maps available at shootingthefront.com

> Wyschtyniez-Wizajny. One cavalry division resting on the road W. to Powisztajcie, a second cavalry division resting at Czyzyczki, on the road W. to Wizajny. Columns marching or resting besides the road from Boksgyozki via Wizajny to Slobodkal. I am of the opinion that the total of the Russian forces advancing here, consist of at least two cavalry divisions and one infantry division. Signed by Schulz, Observation officer.[2]

At 0915 hours, another *FFA 14* aviator reported to *AOK 8*, "One Infantry Division resting on the road Suwalki-Filipow, the head of the column at Filipow. A Rifle Brigade with advance guard at Mierunsken is also resting. 1 Cavalry Brigade at Brosdowen, 9 cavalry troops with artillery or machine guns were counted. The trenches west of Suwalki have been abandoned…. Except for the cavalry troops mentioned, the vicinity and region of Sejny is free from Russian troops."[3]

Fliegertruppe used ongoing aerial reconnaissance to confirm a 70-kilometer gap between the two Russian armies. *AOK 8* determined the Russian armies required at the minimum, a three-day march to unite troops. This proved to be one of the most critical moments in the first few weeks on the Eastern Front. *GO* v. Prittwitz now held an advantage due to observations that were recorded by aerial reconnaissance. This led to his concluding that *8. Armee* could strike Rennenkampf's First Army, then maneuver south to attack and destroy Samsonov's Second Army.[4] Prittwitz's immediate challenge was to determine the optimal moment to initiate the most effective maneuvers against advancing Russian armies. First, he needed to assess and analyze, any and all information he could obtain with respect to Rennenkampf's First Army, as they posed the greatest threat to Königsberg.[5]

On 17 August, aviation's role in influencing command decision at that pivotal moment, was further reinforced by *I. Armee-Korps* headquarters. Staff telephoned *AOK 8,* reporting an urgent issue with *Feldfliegerabteilungen* aeroplanes. "*Armee-Korps* has at its disposal only three serviceable aeroplanes of which two are in the air today. The promised Rumpler monoplanes (3) have not yet arrived; they are urgently needed." The message was signed off by *GdI* v. François.[6] François personally informed *8. Armee* that aviation was critical to analyzing their maneuvers, down to the most minute detail.

In southern *Ostpreussen, GdA* v. Scholtz's *XX. Armee-Korps'* headquarters staff telephoned *AOK 8* (the morning of 17 August) with an important update from *FFA 15* aviators. "Road Mława-Ziechanow-Grzanysz free from Russian troops. Observed only a few cavalry troops at Ziechanow."[7] *8. Armee's FFA 16* reported at 1100 hours several trains sighted that morning. At Tauroggen, the German aeroplane drew fire from Russian infantry.[8] That evening at 1830 hours, *Genltn* v. Below's *I. Reserve-Korps* notified *AOK 8,* "According to an aviator report, one Russian Army Corps advancing against Filipow; a mixed Russian Brigade with much artillery marching from Prsherosl to Tubeninken."[9] The fusion of intelligence sources confirmed the Russian Second Army had initiated its movement towards *Ostpreussen.*

Battle of Stallupönen – Aviation Supports the "Holy Duty"

G*dl v.* François had always placed his trust in Kaiser Wilhelm II as the kaiser's "holy duty" was to protect his Prussian subjects. Three of the five *Armee-Korps* comprising *8. Armee, I. Armee-Korps, XVII Armee-Korps* and *XX. Armee-Korps*, were *Ostpreussen* and *Westpreussen* natives. All three were bound by the standing border defense order to protect Prussia's core provinces from any Russian incursion. François refused to condone any Russian violation of Prussia's "sacred soil." On 17 August, it spurred him to lie to Prittwitz about drawing minor forces back, and leaving the bulk of his divisions far forward from the Angerapp River. *I. Armee-Korps* then moved eastward to the border on *8. Armee*'s northernmost left flank. They assumed positions to enable a 'reconnaissance in force' along the Königsberg– Kowno railroad, south of the Stallupönen village.[10] François' actions posed a real threat to *AOK 8* planning. He had a strong tendency for insubordination, which would have been perceived as a court-martial offense at any other point in time. Back in Coblenz, *OHL* favored his actions—they felt that he was fighting the battle they wanted fought.[11] *8. Armee*'s standing border defense order, clearly entitled François to attack.[12] After reviewing the status of forces facing the invading Russian armies, the *War Diary* stated, "*I. Armee-Korps* stands in opposition to the Army Order of 14 August… (1st Brigade as far as north of Stallupönen). The entire Russian III Corps and apparently, parts of the 28th Division, advanced to the attack against *I. Armee-Korps*."[13] *AOK 8* was concerned about their entire front line. "Without consideration of all relevant factors operating on the front of *8. Armee* (which was still closing up to its left flank), *I. Armee-Korps* has permitted itself to be drawn into a battle where it failed to be supported by neighboring troops. The commanding general of *8. Armee* has ordered *I. Armee-Korps* to withdraw during the night in the direction of Gumbinnen-Niebudzen."[14]

At 1520 hours, *GO* v. Prittwitz received the following telephone message from *I. Armee-Korps*, "To His Excellency von Prittwitz. The order transmitted by Count v. Waldersee to break off the engagement at Stallupönen has been sent by me to the corps commander. General von François reports to your Excellency that at present, he

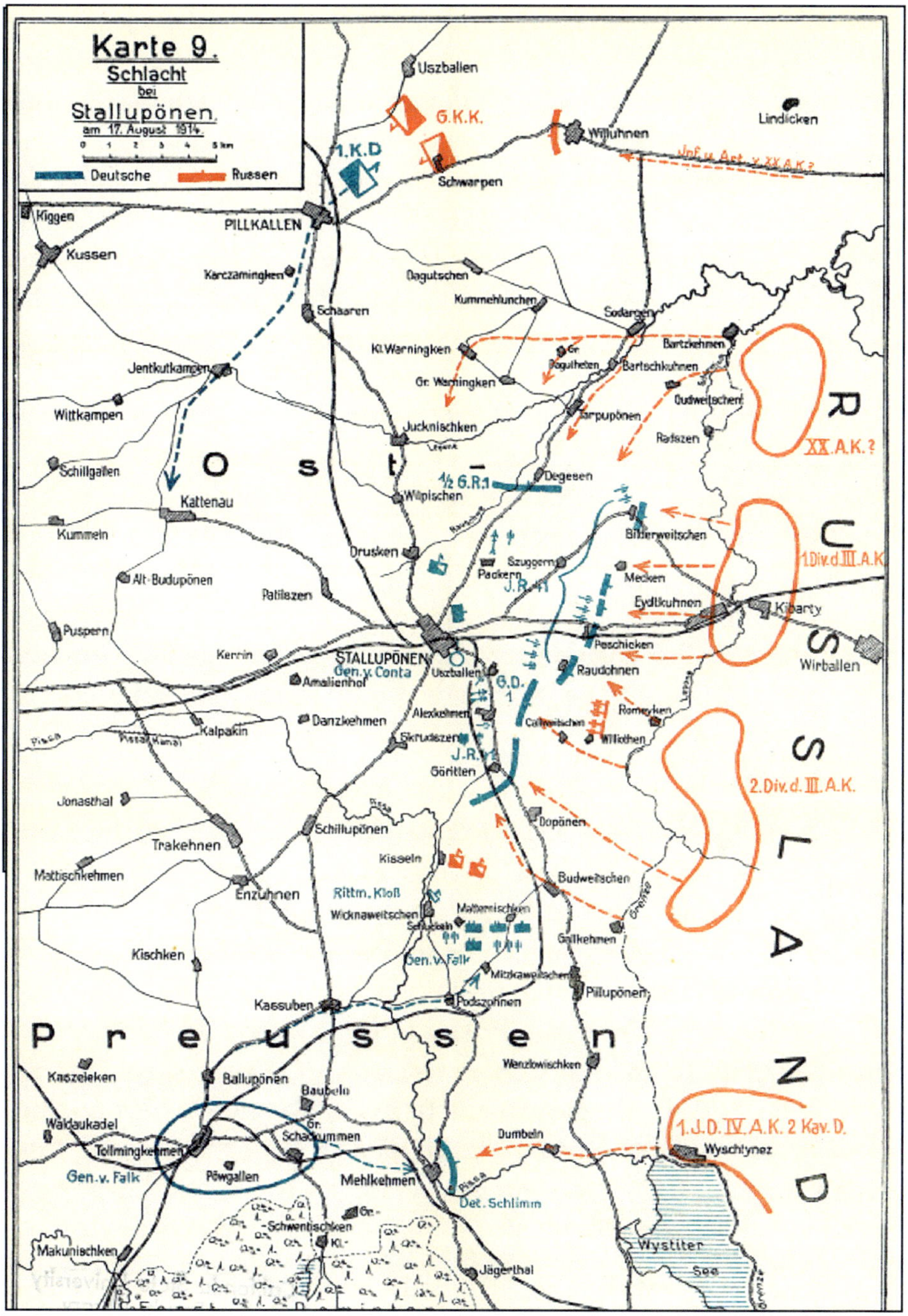

Gdl v. François' post-war memoirs include a map of the battle of Stallupönen (17 August 1914) – the first major battle between Russia and Germany in WW1. Source: François, Hermann von. Marneschlacht und Tannenberg
Larger maps available at shootingthefront.com

was unable to break off the engagement east of Stallupönen, as the entire *1. Division* was now engaged. I remain here in Insterburg." With the bold pronouncement to *AOK 8*, "*Gdl* von François will withdraw when he has defeated the Russians!" François (on his own initiative) maneuvered *I. Armee-Korps* far to the east of Gumbinnen. This action gave the impression that he was focused, primarily, on the defense of the *Ostpreussen* border, "in order to save Prussian people and soil against the Russian invasion."[15] This was but another act of insubordination. At 1915 hours, *GO* v. Prittwitz responded with a telephonic order to François. "The cipher telegram addressed personally to General von François, will be taken by an officer competent to decipher, by auto, to *General* v. François. Genl F. will certify its receipt and report what action he has taken. Signed *GO* v. Prittwitz.[16] At 2000 hours, François responded with "The Russians have been beaten on our whole front and have retired beyond the border. Two thousand prisoners taken and will be transported to Gumbinnen. *Gdl v.* François intends to remain as of today, in line Goritten-Stallupönen and withdraw tomorrow toward Gumbinnen, in compliance with orders received from Army Headquarters."[17]

Gdl v. François reported to *GO* v. Prittwitz at 2315 hours, summarizing what transpired in the battle. *1. Infanterie-Division* and half of *2. Infanterie-Division*, fought Russian III Corps as it advanced on both sides of the Kibarty-Stallupönen road. The left flank of *1. Infanterie-Division* held ground against a vicious Russian attack and suffered heavy losses of personnel and seven artillery pieces. Both divisions retired to the west at Brakupöhnen – Gumbinnen. François' *FFA 14* aviators reported the Russians retiring to Filipow.[18] *AOK 8* sent a telegram to *OHL* in Coblenz stating, "*I. Armee-Korps*, against orders from *AOK 8*, advanced alone from Gumbinnen to beyond Stallupönen, and was attacked by the Russian III Corps… A strong Russian advance is taking place, estimated from 4 to 5 Army Corps from the front Augustowo-Wirballen."[19]

An aviator report filed by *AOK 8*, described the battle fought by *I. Armee-Korps*. "General events: Enemy apparently not advanced beyond Filipow, camped at Przerossl. The aviator observed the fight of *I. Armee-Korps*, apparently two groups engaged at Stallupönen and west of Wystiniecz. West of Wystiniecz, he observed from 3 to 4 batteries firing against each other. Apparently, engagement favorable to us. Many farms are burning on both sides of the railway Kibarty-Stallupönen, to the left and on the frontier."[20]

François' maneuver resulted in a minor German success on the battlefield, but did little to upset the Russian timetable. Rennenkampf's massive First Army continued to advance cautiously to the west. *FFA 14*'s Rumpler-Taube aeroplanes continued to monitor their movement. François had a confidence in his aviation arm that proved facilitative in the ensuing weeks, and was amplified by German commanders growing trust in the utility of aerial reconnaissance. Both the Russians and the Germans declared Stallupönen a partial victory. Rennenkampf and his commanders were assured that the German's will to hold ground, was waning.[21]

Demands for increased aerial reconnaissance support from *8. Armee* also originated from François' southern flank. *Genltn* v. Below's *I. Reserve-Korps* was in the process of defending the Masurian Lakes region. His staff telephoned *AOK 8* with an update on Russian forces. "Marggrabowa is strongly garrisoned by Infantry and Artillery. Cavalry reported to be marching from Marggrabowa to Lyck...An aviator reported that a Russian Army Corps is advancing toward Filipowo, and a division with one to two cavalry divisions was at forenoon on the march from South-East toward Wistiniez. It is urgently requested that air [reconnaissance] be initiated east of the line Marggrabowa-Goldap."[22]

Fliegertruppe Monitors the Advance in the South

On 18 August, *AOK 8* conducted a review of known Russian positions based on updated intelligence. Five Russian army corps were presumed in position between Augustow and Schirwindt. Between Augustow and Lomza, no Russian forces were observed. Opposite the central *Ostpreussen*'s Masurian Lakes region, the Russian II Corps was noted south of Marggrabowa; *Gen.art* Aliyev's IV Corps had forces near Rominte Heide. To the north, Russian III Corps was in place on both sides of the Gumbinnen-Kibarty road. *Gen.inf* Smirnov's XX Corps occupied Schirwindt.[23] *AOK 8* also learned from *Gdl v.* François (following the engagement at Stallupönen), "The Russian Infantry makes good use of the terrain but shoots badly and surrenders very quickly."[24]

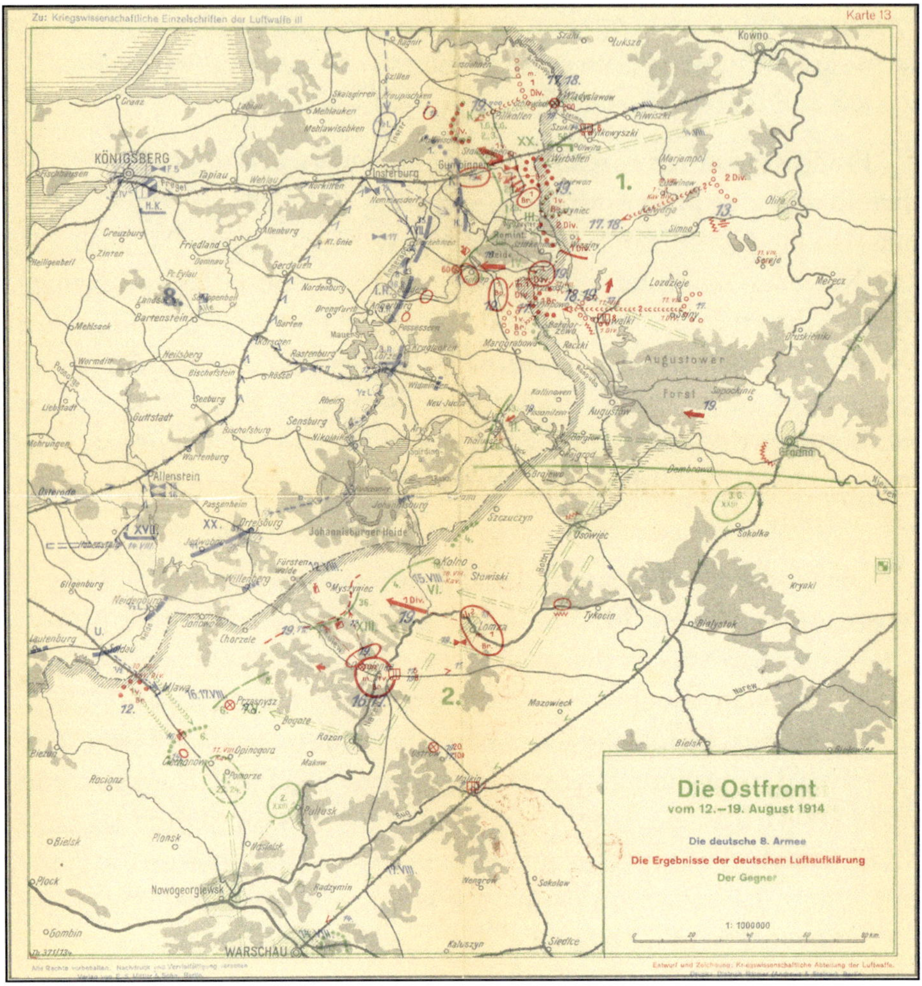

8. Armee on 18 August, reviewed the evolving battlefield and increased Fliegertruppe sorties to southern Ostpreussen, in an attempt to carefully monitor the advance of Gen.kav Samsonov's Second Army. Source: 3. Sonderband. Larger maps available at shootingthefront.com

8. Armee tasked their *FFA* fleet with surveying and evaluating the area to confirm Russian positions. To the north, François' *FFA 14* was tasked in covering northeast *Ostpreussen*. They reported brigade-strength columns, massive troop movements, and logistic tail moving, between Suwalki and the frontier. *Festung* Königsberg's *FestFA 5* was directed to fly north of the Niemen River. Opposite central *Ostpreussen*, *GdK* v. Mackensen's *XVII.* Armee-Korps with *FFA 17* was assigned Grodno-Olita. Moving further south, *GdA* v. Scholtz's

XX. Armee-Korps (with *FFA 15* in support) was to cover "beyond the Pultusk-Bialystock-Augustow line." *8. Armee's FFA 16* was committed to a more strategically important area south of *Ostpreussen*, charged with reporting Russian activity in the Warsaw-Makow region.[25]

That morning, in the Narew-Bohr sector to the southeast of *Ostpreussen, Fliegertruppe* sorties discovered two Russian corps. XIII Corps was moving west from Ostrolenka and VI Corps was observed heading northeast out of Lomsha. Both corps were heading towards the *Ostpreussen* border. The report forced *GO* v. Prittwitz to carefully consider which Russian army to fight first.[26]

The surge in aerial reconnaissance activity confirmed the 70-kilometer gap between Rennenkampf and Samsonov's armies, and that it was expanding in the vicinity of Osowiec and Filipow. This discovery gave *8. Armee* greater confidence to proceed with the strategy of fighting one Russian army at a time.[27] On the evening of 18 August, *8. Armee* received an aerial reconnaissance report covering Russian territory southeast of *Ostpreussen. Gen.inf* Blagoveshchensky's VI Corps was observed heading to Ortelsburg. The next day, *Fliegertruppe* aviators observed Russian forces still maneuvering and not in position to fight *8. Armee* forces. This allowed the Germans more time to position their forces before the full strength of Russia's armies moved into *Ostpreussen*.[28]

On 19 August at 0845 hours, François led *I. Armee-Korps* in a follow up attack. His staff alerted *AOK 8* that subordinate divisions, (*Division Brodrück* and *1. Infanterie-Division*) were fighting Russians on the Kullig-Kehmen-Niebudozen-Mampupönen line. They estimated Russian forces were at one-division strength south of the Gumbinnen-Stallupönen highway. *AOK 8* accumulated intelligence from contacts in the area, that indicated François' *1. Infanterie-Division* was vulnerable to Russian maneuvers already in progress. *I. Armee-Korps* staff ignored the threat and informed *AOK 8* that they were ready to launch an enveloping attack.[29] *AOK 8* received information that afternoon from *Fliegertruppe* (flying over *Ostpreussen*'s eastern sector) that columns of infantry and cavalry were now in Germany. In the Masurian Lakes region, a three-kilometer column was advancing towards Lyck. To the north, a seven-kilometer column was noted on the Grodno road marching to Augustow.[30]

At 1600 hours on 19 August, François contacted *GO* v. Prittwitz via telephone and outlined his plan to attack the Russian's northern flank early the next day. *GdI* v. François estimated his *I. Armee-Korps* would be attacked that evening by at least three Russian divisions. *AOK 8* assessed that Russian forces facing *I. Reserve-Korps* (in the village of Gumbinnen) were separated and isolated from the main body of *Gen. kav* v. Rennenkampf's Russian First Army.[31]

François felt strongly that both *GdK* v. Mackensen's *XVII. Armee-Korps* and *Genltn* v. Below's *I. Reserve-Korps* needed to move to new strategic positions opposite the Russians, and extend *8. Armee*'s battle line of forces, further to the south. At 1640 hours, *AOK 8* informed *GdK* v. Mackensen's *XVII. Armee-Korps* that the Russians had advanced on Mackensen's left wing, heading to the villages of Augustupönen and Grünweitschen. *XVII. Armee-Korps* was ordered to move from the Angerapp River to a point south of *I. Armee-Korps*.[32] Later that afternoon, *FFA 17*'s Albatros B.I aircrews sighted substantial Russian troops approximately 16 kilometers northeast of Gumbinnen. The aerial observer estimated the strength to be one division.[33] This was relayed via telephone to *XVII. Armee-Korps* at 2000 hours. Subsequent analysis proved the aviator report, incorrect. At 2220 hours, *AOK 8* sent *XVII. Armee-Korps* a message indicating Russian III Corps and 40th Division were sighted at the Walterkehmen-Grunweitschen-Augustupönen line. At 2300 hours, a subsequent message confirmed the intelligence.[34]

AOK 8 informed *GdK* v. Mackensen's *XVII. Armee-Korps* that the Russian First Army's left wing was positioned at Gruenweitschen (8 kilometers southeast of Gumbinnen). When Mackensen's *XVII. Armee-Korps* advanced the following morning, the intended flank attack confronted an unexpected Russian frontal assault. Rennenkampf's orders had preempted Mackensen's maneuver by extending the line of Russian forces an additional 5 kilometers southeast of Gruenweitschen at Sodehnen.[35]

A Prelude to Gumbinnen

Prior *8. Armee* intelligence had assessed the composition of the Russian First Army and determined it consisted of *Generalleytenant* Nikolai Nikolaevich Kaznakov's Guards Cavalry, *Gen.*

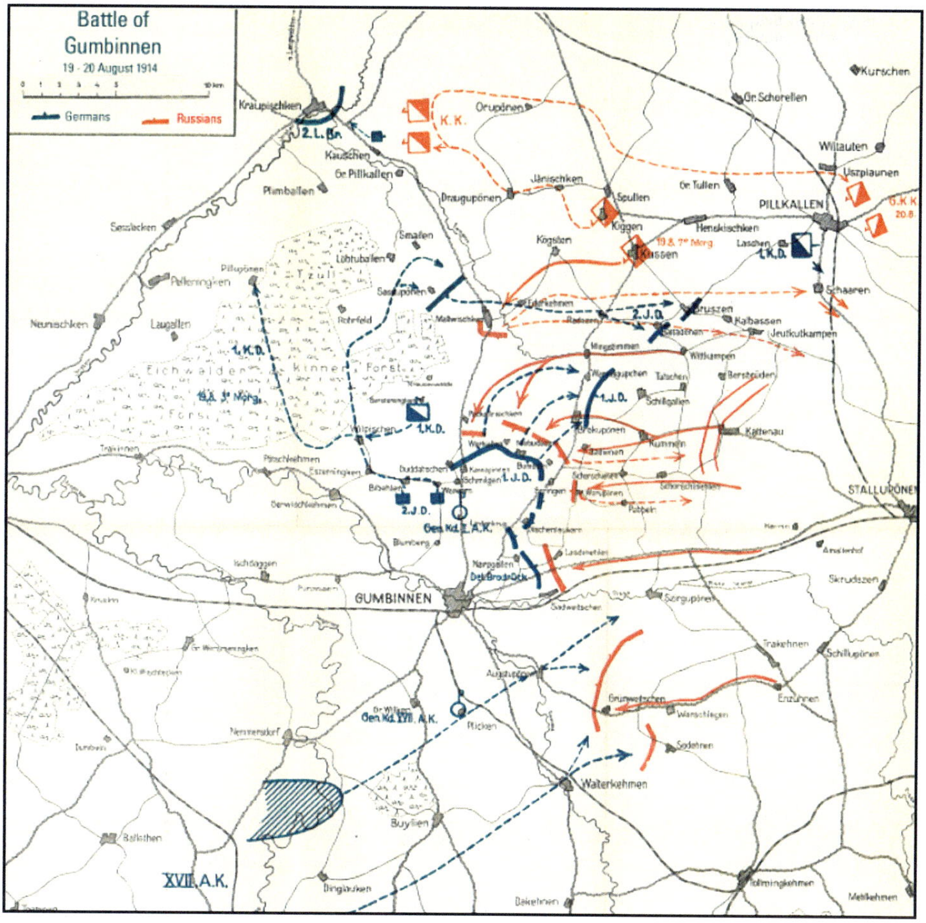

On 20 August 1914, the battle of Gumbinnen took place. Russia's victory, resulted in GO v. Prittwitz ordering 8. Armee to retreat behind the Vistula. His request for more forces to fight the Russians, prompted OHL to pull an Armee-Korps from the Western Front, when they needed all their forces to successfully advance on Paris. Source: François, Hermann von. Marneschlacht und Tannenberg Larger maps available at shootingthefront.com

inf Artamonov's I Corps, *General ot infanterii* Nikolai Alekseevich Epanchin's III Corps, and *Gen.art* Aliyev's IV Corps. Initial fighting uncovered the presence of *Gen.inf* Smirnov's Russian XX Corps. As *Fliegertruppe* flew over Lyck (between the Russian First and Second Armies) they reported a new corps in the area. Prittwitz now expected the Russian First Army possessed considerable reserves not yet engaged in battle—at least one corps on the northern flank and at least one corps south of the Rominte Heide.[36]

GO v. Prittwitz had three *Armee-Korps* primed for the attack on Gumbinnen: On the left northern wing, François' *I. Armee-Korps* lay east of the town of Gumbinnen. The right flank of the attacking *I. Armee-Korps* was to be protected by *GdK* v. Mackensen's *XVII. Armee-Korps*, and ordered to march to the north. South of Rominte Heide, *Genltn* v. Below's *I. Reserve-Korps* and *Generalmajor* Kurt von Morgen's *3. Reserve Division* now defended Goldap.[37]

The traditional role of cavalry reconnaissance was nearly absent at this critical juncture. As *Gen.kav* v. Rennenkampf's army advanced west, vital intelligence was limited. Khan Nakhichevan's cavalry corps comprised no less than five cavalry divisions —20,000 horsemen—but to date none had taken on a serious role in the battles, fought. A minor skirmish to the north (against an inexperienced *Landwehr* force) sent German militia fleeing. Khan's cavalry failed to follow up. They neither advanced west nor turned south to outflank *8. Armee* forces in place near Gumbinnen. Instead, Kahn's cavalry proceeded to rest and resupply. The absence of rigorous and cogent reconnaissance by cavalry, proved to be fatal.[38]

Gen.kav Zhilinsky was disappointed in Samsonov's slow advance and sent a telegram on 19 August stating: "First Army has been fighting at Stallupönen for two days. Any delay by Second Army invading from the south puts First Army at great risk." Samsonov replied to Zhilinsky: "The Army has been advancing from the time of your orders, non-stop covering over 32 kilometers over sandy terrain; conditions that make rapid movement impossible."[39]

Gen.kav Samsonov's Russian Second Army's slow advance clarified for *8. Armee,* that they could strike first against Rennenkampf's First Army. Despite this apparent advantage, *GO* v. Prittwitz remained intensely focused on the possibility that Samsonov could turn west and cut possible retreat lines to the Vistula River. *Fliegertruppe* aerial reconnaissance reported Samsonov's forces extending further southwest from Mława, than was previously understood. The most critical aerial reconnaissance reports indicated both Rennenkampf's and Samsonov's armies had not filled the 70-kilometer gap between them. It was now time to commit *8. Armee* against Rennenkampf's First Army at Gumbinnen.[40]

15th KAO aircrew (Podporuchik Chekhutov studying map, unknown observer facing camera with map folder in hand) prepare for their Nieuport IV (Number 2) sortie over Ostpreussen as Gen.inf Martos' XV Corps crosses the border.

On 19 August, *Gen.inf* Blagoveshchensky's VI Corps staff observed numerous *Fliegertruppe* aeroplanes overhead, tracking its movements. While covering Russian ground forces, *Fliegertruppe* discovered three tents that housed aeroplanes, suggesting Russian aviation assets were co-located near Blagoveschensky's headquarters. Additional sightings included two divisions in parallel columns, southeast of Ortelsburg.[41]

RIVVF Support to Russian Second Army

During the first few weeks, *RIVVF* aerial reconnaissance failed to sight any events of great enough import to alert Russian commanders as to specific German force maneuvers. Aerial reconnaissance on 18 and 19 August, validated older intelligence. There was no new intelligence confirming German intentions, which signified that *RIVVF* lacked the clout (at the time) to influence Russian commander deliberations. Mishaps, added to *RIVVF*'s operational challenges. *15th KAO's Starshiy Unter-Ofitser* S.

SORTIE DATE: 21 AUGUST 1914

AEROPLANE: HF 16

PILOT: Shtabs-kapitan Victor Rodionovich Poplavko

GROUND TRACK: Ostrolenko - Myshinets - Ortelsburg Dombrovo - Pelty - Friedrichstal - Altchaiken - Lipowitz Ortelsburg

SOURCE: Eliseev, Appendix 9

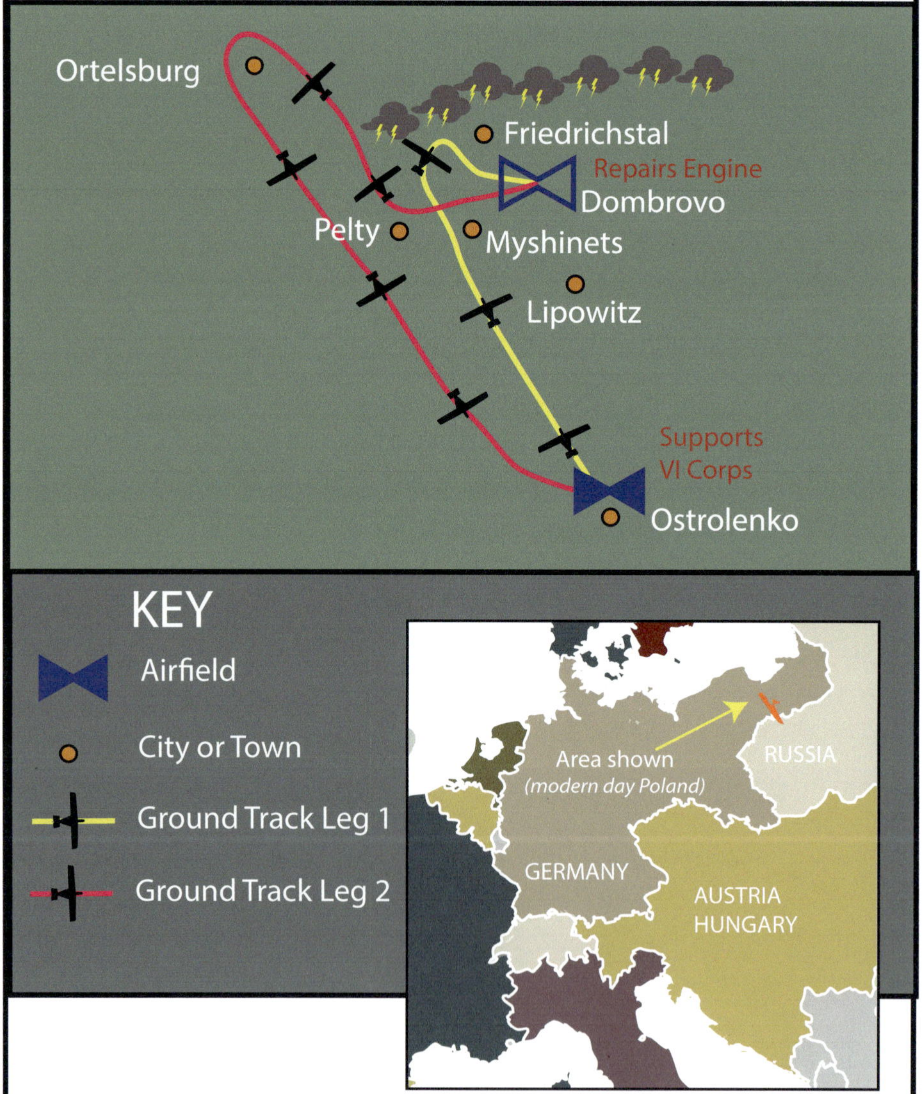

SORTIE DATE: 21 AUGUST 1914

To Quartermeister General

8 August 1914, 9 h30 min. [21 August – Gregorian]

5: 40 a.m. P. flew off with V. - a pilot, Sh. Kap. Poplavko to Ostrolenko -Myshinets - Ortelsburg. Approaching the border at an altitude of 1900 meters, we got into thunderstorm twice. Then, after buffeting we had to fly back on the half way to Ortelsburg, because the engine was not operating properly. Having landed at village of Dombrovo, we repaired the engine and made the second attempt to fly to Ortelsburg. At the border we saw the fires: customs building located northward of Dombrovo, the Northern edge of the villages of Pelty, Friedrichstal, Altchaiken, and 9 places... village of Lipowitz were burning. We approached Ortelsburg at the altitude of 1000 meters. The city was surrounded from the West and South with cross-tunnels. No blockhouses observed.

The opposite German troops were not observed.

There were fortifications with three barbettes for guns, communication trenches and fire trenches 2 versts eastwards the city, on the north of high road.

The trenches were absolutely not ...[illegible - S.Ye.] of the area. Troops and ammunition were not observed. Small trenches were observed near to the village of Gamerudau. Troops were not found in the city. There was a train set and a non-steaming locomotive at the railway station. There were a lot of new barracks near to the station.

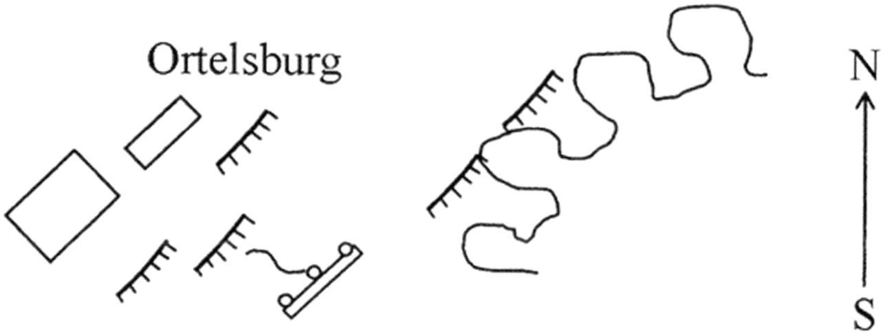

On the way back, flying over left column of our corps VI building, we were attacked by rapid salvo fire, though our maneuvers could not be observed. The fire was dangerous because we flew at low altitude for clear vision.

SORTIE DATE: 21 AUGUST 1914

Despite there were forests nearby, our troops were camped near villages.

Locals of Dombrovo did not observe the German troops. There were 25 dragoons yesterday. From hearsay, there were about four German companies in Ortelsburg yesterday.

Crew: Voennyi Letchik Sh.Kap. Poplavko

Letchik-Nablyudatel Sh.Kap. Captain Vitkovsky

Shtabs-kapitan Victor Rodionovich Poplavko

Menichenko (*Poruchik* Mamayev was the observer) crashed into bushes bordering the field and his Nieuport IV flipped over.[42] With each accident, less operational sorties were launched. Russia's army logistic trail lacked the ability to support and supply critical maintenance, as well as any and all necessary repairs. Russia's aviation debut on the Northwest Front proved to be inadequate.

Early in the morning of 20 August, *GdI* v. François' *I. Armee-Korps* surprised the Russians and drove *Gen.inf* Smirnov's Russian XX Corps back eight kilometers—in the direction north of Gumbinnen. *1. Infanterie-Division* and *2. Infanterie-Division* breached the Brakuponen-Wamagupohen line and reached Wittkampen, capturing 4000 Russian prisoners and many machine guns.[43] A German cavalry attack followed, creating panic among the Russian transport at Schwirgallen.[44] *Fliegertruppen* aerial reconnaissance took place that

General der Kavallerie August von Mackensen, XVII. Armee-Korps commander

morning, as the weather was suitable for flight, but initial sorties failed to discover Russians digging in their positions.[45]

GdK v. Mackensen's *XVII. Armee-Korps* marched throughout the night, arriving at Gumbinnen battlefield that morning. Early morning reports from Mackensen's *FFA 17* aviators (flying south of Gumbinnen) only reported sightings of wagon columns and deserted camping places.[46] Later sorties identified Russian forces at Sodehnen, that comprised the left wing of the Russian maneuver opposite *XVII. Armee-Korps*. As they flew further to the southeast to Rominte Heide, no other significant sightings were noted. *FFA 17* aerial observers inadvertently missed the area between the Gumbinnen railroad and Rominte Heide. It was at this specific location, that *Gen.inf* Epanchin's III Corps (25th and 27th Divisions from III Corps and three quarters of the 40th Division) had troops stand directly in the path of *GdK* v. Mackensen's *XVII. Armee-Korps'* advance. The Russians occupied strong defensive positions. *Fliegertruppen*'s failure to warn *GdK* v. Mackensen, proved to be a fatal oversight.[47]

At 0800 hours, *XVII. Armee-Korps* commenced a frontal attack of *Gen.inf* Epanchin's III Corps without the element of surprise and the protection of artillery. The Russians subsequently fired a devastating artillery barrage, driving *XVII. Armee-Korps* back and into a state of complete disorder. When German artillery commenced fire they added to *XVII. Armee-Korps*' confusion. This friendly fire hit German-held positions.[48] At 1500 hours, *XVII. Armee-Korps* reported approximately 8000 casualties. Units were in retreat as far back as the Angerapp River.[49] *8. Armee War Diary* recorded the confrontation: "involved in an exceedingly serious and heavy loss engagement."[50] François' prospects for and promise of success was eroded as a result of *XVII. Armee-Korps'* losses.[51]

At first, *GO* v. Prittwitz and *AOK 8* were optimistic about monitoring the evolving Gumbinnen battle. *8. Armee* received aerial reports indicating two Russian corps were south of Goldap, approximately 20 kilometers from *Genltn* v. Below's *I. Reserve-Korps. I. Reserve-Korps* also marched 30 kilometers through the night to the Gumbinnen area. The advanced guard arrived at Kleszowen at 0830 hours followed by *36. Infanterie Reserve Division*. The 5th Russian Rifle Brigade and 30th Russian Division awaited their arrival.[52] Instead of committing *I. Reserve-Korps* into the Gumbinnen battle, *8. Armee* ordered them to hold their position in order to defend against the Russian threat coming from Goldap.[53] *I. Reserve-Korps'* only aerial reconnaissance source was a single captive balloon belonging to *Feldluftschiffer-Abteilung 8*. The balloon generated reports throughout the day, despite being threatened by heavy shrapnel fire.[54]At 0910 hours, *AOK 8* ordered *Genltn* v. Below to halt at Kleszowen and have *36. Infanterie Reserve Division* cover *GdK* v. Mackensen's *XVII. Armee-Korps'* right flank. Unexpectedly, at 1000 hours, Below's *I. Reserve-Korps* was attacked from the south by *Gen.art* Aliyev's IV Corps. That afternoon, the battle witnessed *I. Reserve-Korps* repelling the Russian attack at Uszball, and *36. Reserve-Division* seizing terrain northwest of Kumehnen.[55]

Intelligence Alerts AOK 8

During the ensuing battle, *GdI* v. François' *I. Armee-Korps'* soldiers discovered a notebook on a dead Russian officer. It revealed that Rennenkampf's First Army planned to advance towards the Insterburg – Angerburg sector. The intelligence data from this source, implied First Army intended to attack the German positions that Rennenkampf's staff believed were behind the Angerapp River – not where the Germans were actually located. The captured notebook also uncovered Russia's intention to have Samsonov's Second Army move from the southwest to the northwest. This set up the Russian Second Army to encircle German positions in *Ostpreussen*.[56] The intelligence also revealed to *AOK 8*, that Rennenkampf was hesitant to make bold moves. This spurred *8. Armee* to take the advantage and maneuver more aggressively.[57]

In southern *Ostpreussen*, *Fliegertruppe* reports confirmed Samsonov's Second Army still moving north. On 20 August, aerial reconnaissance observed an infantry regiment with carts and artillery on the Mława to Soldau road. When Second Army reached Pelty and Mława, Samsonov telegrammed that terrain delayed the advance, but his army was making a great effort to support Rennenkampf's First Army.[58]

Prior *8. Armee* intelligence addressed the fact that the Russian First Army was comprised of *General-leytenant* Nikolai Nikolaevich Kaznakov's Guards Cavalry, *Gen.inf Artamonov*'s I Corps, *Gen.inf* Epanchin's III Corps, and *Gen.art* Aliyev's IV Corps. The presence of the *General-leytenant* Kasnakov's Guards' Cavalry on the Russian northern flank was known; therefore, the commander became more inclined to believe the army corps must also be present. *Fliegertruppe* flying over Lyck [between the Russian First and Second Armies] reported a new corps in the area. Prittwitz now expected that the Russian First Army had considerable reserves not yet engaged in battle—at least one corps on the northern flank and one south of the Rominte Heide.[59]

General der Artillerie von Scholtz, commander XX. Armee-Korps – the only German force facing Gen.kav Samsonov's advancing Second Army during the Gumbinnen battle.

As *AOK 8* closely monitored the decisive engagements at Gumbinnen, staff received a report from *Festung* Posen, through wireless intercept (at 1433 hours). It revealed that

Poruchik Alexander Borisovich Minchenok, Observer for 3rd KAO supporting First Army, recalled (in the map below) the missions he flew over Ostpreussen. Note the Farman HF.16 symbology.

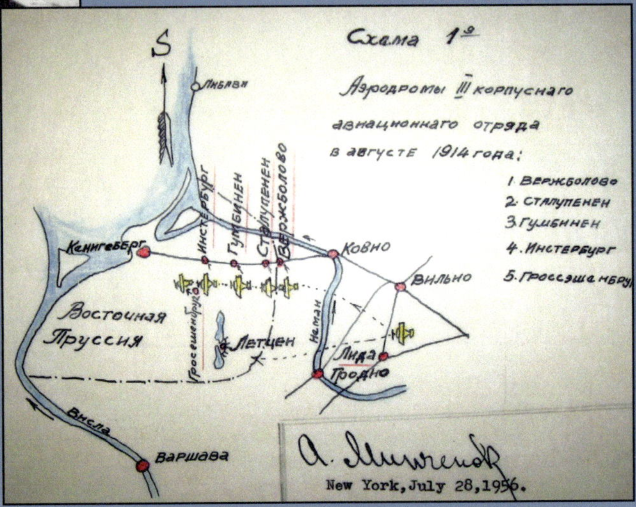

Gen.kav Samsonov's Russian Second Army in total, contained five corps (I Corps, VI Corps, XIII Corps, XV Corps, XXIII Corps) and a single cavalry division (15 Cavalry Division). In actuality, First Army only had III Corps, IV Corps, and XX Corps. That evening, *AOK 8* issued their *Armeebefehl* [Army Operations Order] for the following day. The report mistakenly confirmed Rennenkampf's First Army as comprising five corps and only assessed Samsonov's Second Army as comprising three corps. Effective *AOK 8* planning at this critical stage, required greater attention to the intelligence. Both aerial reconnaissance and wireless intercepts clarified the following argument: Samsonov's Second Army possessed more corps and was further west than initially determined.[60]

8. Armee only had *GdA v Scholtz's XX. Armee-Korps* defending southern Osterpreussen.[61] Scholtz reported to *AOK 8* that he held the advantage as a result of his *FFA 15 Fliegertruppe* effectively providing accurate aerial reconnaissance. The number of corps facing his forces was dramatically underestimated.

> Excellent air reports reveal the enemy's advance is estimated to be 2-2½ corps with some rifle brigades. Its right wing directed towards Ortelsburg. Near Ziechanow and to the west (the reconnaissance area of the aviation section attached to Army Headquarters) still remains unexplored. The aviator sent there from the XX Corps' section could not complete his mission.[62]

At 1430 hours, *GO* v. Prittwitz telephoned *XX. Armee-Korps* and spoke with Scholtz's chief of staff about Russian Second Army's advance. *FFA 15*'s report mentioning 2-2½ corps and some rifle brigades was repeated. A *FestFA 6 Graudenz* aviator landed near the command post serving the *Festung* commander. The report read:

> Marching column on road from south to Mława. Head at Slupsk at 1425, end at Pawlowo, behind that detachment of all weapons, one regiment after rest at Pawlowo from where it departed at 1410. One column on road Pultusk – Ziechanow, head 10 km east Ziechanow. On road Warsaw – Pultusk long dust cloud, head 10 km south of Pulusk, estimated one brigade strength. At Nadielsk strong enemy resting.[63]

The combined intelligence alerted *AOK 8* that Samsonov's Russian Second Army was advancing more rapidly than initially anticipated. It also indicated that his army was further west than previously estimated.[64] *8. Armee War Diary* recorded:

> Due to extraordinary losses by *XVII. Armee-Korps* [losses incurred that day in the Gumbinnen battle], it is uncertain as to whether or not a decisive success can be expected during the next days. The commanding general of *8. Armee* concludes to disengage the army (even during the night) from the enemy. And, to assemble it on both sides of the *XX. Armee-Korps* in Westpreussen. This will enable a decisive offensive against enemy forces advancing from Warsaw-Lomza.[65]

As *GO* v. Prittwitz came to terms with news of *GdK* v. Mackensen's defeat at Gumbinnen, he learned that *Gen.kav* Samsonov's Second Army had crossed the border. Aerial reconnaissance made clear to the Germans that the primary Russian thrust was south of Gumbinnen. This resulted in their underestimating the center mass of the Russian attack. The error prompted Germans to plan for an outflanking maneuver over the main railroad between Königsberg – Kowno. Additional *I. Armee-Korps* intelligence from ground observations indicated *Gen.kav* v. Rennenkampf's First Army had forces further north of the railroad. *GdI* v. François' *I. Armee-Korps* staff urged *GdK* v. Mackensen's *XVII. Armee-Korps*, to press forward in an effort designed to cut off an expected Russian retreat. *8. Armee* fully grasped that trying to outflank Rennenkampf's forces at this stage, was far too risky.[66] Besides Russians already positioned at Stallupönen, a column of infantry was marching to Pillkallen. Prittwitz interpreted the report as confirmation that Russian reinforcements were moving forward in a serious attempt to strengthen the Russian northern wing. An earlier agent report, alerted *AOK 8* that the Russian Guard Corps was entering the battle. In reality, the Russian Guard Corps was near Warsaw at the time. What aerial reconnaissance probably observed, was an infantry regiment sent to recapture Pillkallen from *1. Kavalerie Division*.[67] Earlier *Fliegertruppen* reports revealed a corps-strength unit at Lyck. *AOK 8* staff thought the unit was Russian I Corps, a fresh reserve entering the battle. In fact, it was Russian II Corps having relocated units from the Gumbinnen battleground. Prittwitz and staff now estimated Russian First Army in place and ready to deploy more troops—one corps to the north of the main body and one to the south at Rominte Heide.[68]

The prospects for defending *Ostpreussen* appeared slim, according to numerous intelligence updates on the Russian Second Army's advance.[69] Prittwitz feared that *8. Armee* was vulnerable to being cut off in attempting to escape to the Vistula River.[70] Prittwitz lost confidence at this critical moment, believing setbacks at Gumbinnen meant the *8. Armee* would fail to prevail against the Russian First Army. At 2000 hours, correspondence between Prittwitz and *GdK* v. Mackensen highlighted the situation:

> The situation has entirely changed. New Russian forc-

> es have arrived to the north. Moreover, I do not know where my cavalry division is. Russian forces of equal strength are fighting the *Reserve Korps*. A new Russian force—about a corps in size—is near Lyck. The Russian Second Army in strength, is advancing on Ortelsburg and Soldau. With this situation now coming together, I am retiring behind the Vistula.[71]

Prittwitz decided that he had only one option—to engage in a hasty retreat to the Vistula River, in an effort to save *8. Armee*. Without discussing the situation with his *AOK 8* staff, he called *GO* v. Moltke at Coblenz and informed him of his decision. He also indicated that holding the position was questionable, unless reinforced.[72] At 2020 hours, Prittwitz issued the order to break off the battle and retreat to *Westpreussen*. At this stage of the war, the stature of *GO* v. Prittwitz, was rapidly deteriorating. His subordinate commanders perceived his decision to break off fighting at Gumbinnen and retreat west, as inappropriate. *8. Armee*'s corps began to pull back. This critical maneuver was obscured from the Russians.[73]

At 2130 hours, *AOK 8* issued an *Armeebefehl* ordering a "retrograde movement" to *Westpreussen*. *Genltn* v. Below's *I. Reserve-Korps* was ordered to Darkehmen-Nordenburg. *GdK* v. Mackensen's *XVII. Armee-Korps* was ordered to proceed to Allenstein via Nemmersdorf-Tempen-Lugowen. *Gd I* v. François' *I. Armee-Korps' 1. Infanterie-Division* and *2. Infanterie-Division* were directed to proceed to Königsberg via Insterburg, with the *2. Infanterie-Division* covering the north bank of the Pregel River.[74] That evening, a Zeppelin Z-IV flew over the Gumbinnen battlefield to further ascertain Russian positions.[75]

Russian command developed a confidence at Gumbinnen, which had no logical rationale. This misplaced confidence, resulted in false perceptions of the character, conditions, and intentions of the German enemy. It also influenced *Gen.kav* Zhilinsky to direct Samsonov's Second Army to advance further to the West and not focus on the North—far-

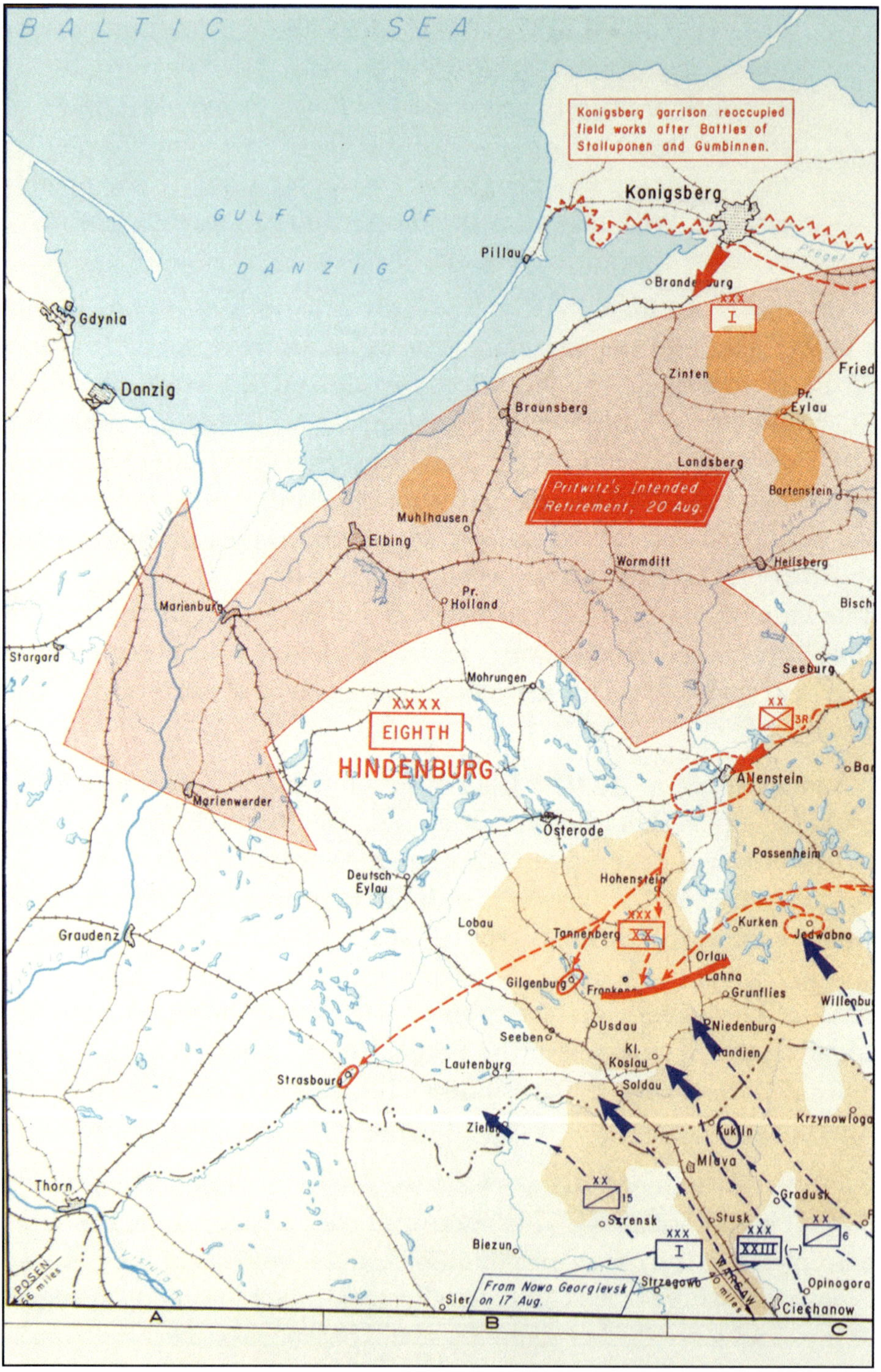

Map shows the intended "Retrograde Movement" ordered by GO v. Prittwitz—resulting in his dismissal as commander of 8. Armee. Source: West Point

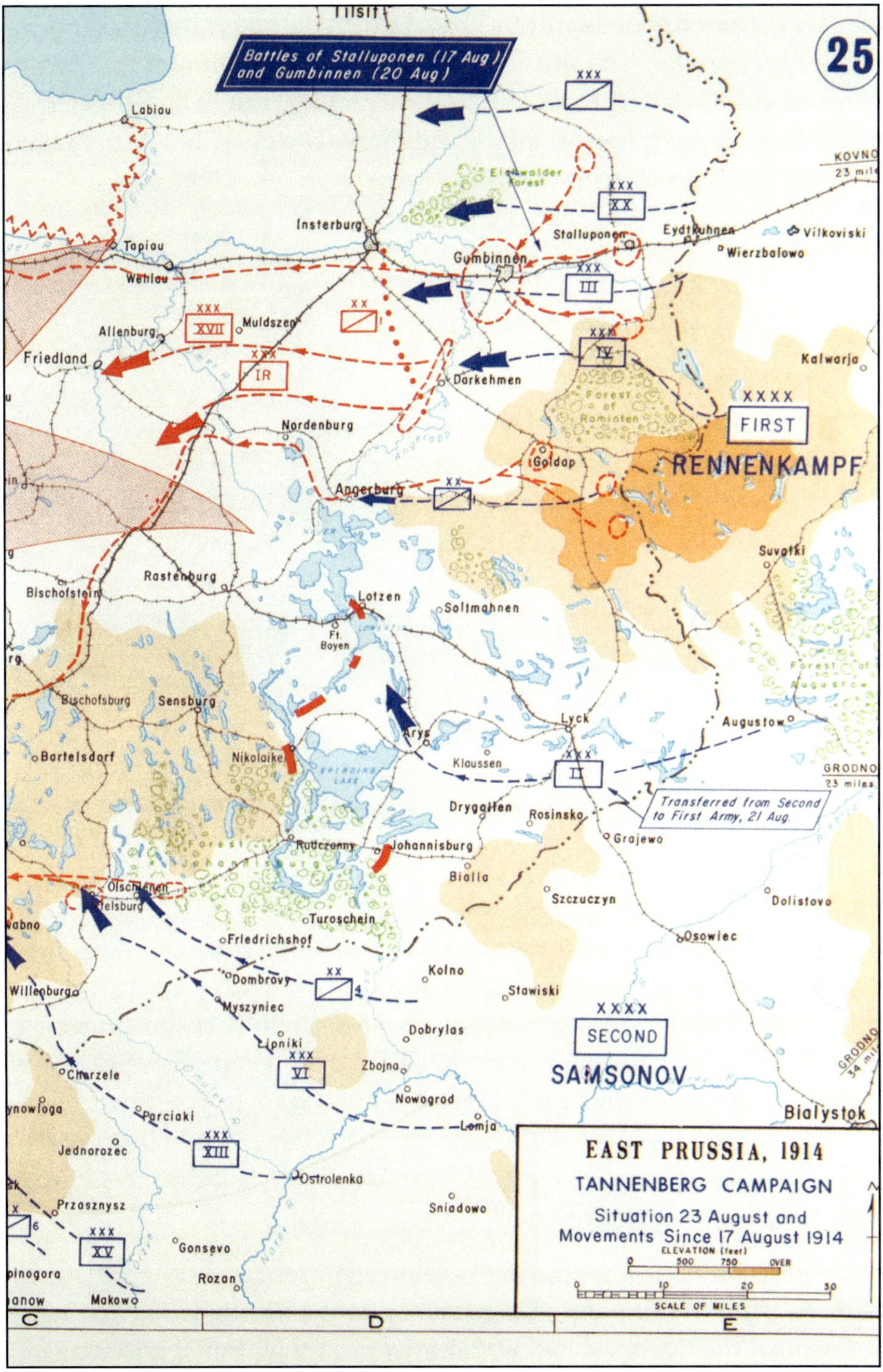

Larger map available at shootingthefront.com

ther away from Rennenkampf's First Army. This maneuver was reported by *Fliegertruppe*. The illusion of success at Gumbinnen, resulted in Rennenkampf's losing three critical days on the battlefield. Samsonov's forces lost a golden opportunity to advance, because First Army failed to move west. As Winston Churchill eloquently stated, "It led Zhilinsky to acquiesce in his strategic inertia."[76]

3rd KAO aviator Donat Aduevich Makijonek carried the important message but crashed his Nieuport IV on the return. Source: Marat Khairulin

8. Armee's Retrograde Movement - 21 August 1914

On 21 August, *GO* v. Prittwitz and *AOK 8* arrived in Mühlhausen by rail. *8. Armee War Diary* proclaimed, "the extrication of the army from the enemy (ordered by the commanding general of *8. Armee* last evening) is being accomplished on the entire front without any difficulty. This is the result of a victorious advance of both the German right and left flank." The Russians did not pursue *8. Armee*. They lost an opportunity at Gumbinnen from which they never recovered.[77]

The Russians continued west, assuming the Germans were centered on the Angerapp line. They held on to this belief in spite of evidence to the contrary—including detailed *RIVVF* aerial reconnaissance alerting them that the Angerapp line was void of German forces. *RIVVF* aerial reconnaissance noted forces withdrawing from the border while German forces were observed pulling back from Ortelsburg to Passenheim. Trains were noted heading in the direction of Deutsch–Eylau. Infantry regiments were discovered near Soldau.[78]

Armed with this information, Russian commanders now erroneously believed that the Germans were retreating to the Vistula River.[79] Aviators determined the Germans had withdrawn from their initial positions but failed to ascertain which direction they were heading. Königsberg or the lower Vistula River area, appeared to be a logical point for the German "retreat." The Russian command never accounted for the fact that *8. Armee* had shifted forces to the south in an effort to fight Second Army. Compounding their erroneous assessment, Rennenkampf's First Army's cavalry failed in its mission to locate the retreating forces. The absence of aggressive reconnaissance proved fatal. Rennenkampf's obsessive focus on the Vistula River, resulted in a myopic and pivotal strategic blunder.

RIVVF also functioned as message carriers, at this time. *Gen.inf* Epanchin's III Corps tasked *3rd KAO* to deliver secret instructions to *Genleyt* Khan Nakhchivanski's cavalry corps. Pilot *Podporuchik* Leonid Georgievich Ustyantsev and observer *Poruchik* Shanin, flew the first sortie in an attempt to deliver the message to the cavalry, but were unsuccessful. That evening, *St.Uo.* Donat Aduevich Makijonek proceeded to search for Khan Nakhchivanski's cavalry corps. Makijonek succeeded in delivering the message but nearly failed to return, as German ground fire forced him to land behind the Russian front lines. His plane was destroyed by Russian soldiers. *3rd KAO* now had three Nieuport IV aeroplanes to support III Corps.[80]

Generaloberst von Prittwitz is Relieved of Command

Kaiser Wilhelm and *GO* v. Moltke were alarmed in learning of Prittwitz's intent to leave *Ostpreussen* to the invading Russians, who were reportedly destroying everything in their path. Moltke shared his opinion with *AOK 8* and *Armee-Korps Generalstab* officers—that Gumbinnen's first day of battle was far from decisive. An opportunity for victory still remained. What was required, was a change in thinking and attitude. Prittwitz was reluctant to continue the attack on advancing Russian armies. He wanted forces pulled from the Western front to reinforce his single *8. Armee* and establish parity. *GO* v. Moltke mistakenly assumed that the ongoing thrust towards Paris was succeeding. A change in command was required to stabilize the Eastern Front. Upon hearing the news the Kaiser relieved both Prittwitz and *Genmaj* v. Waldersee of their command. He appointed *General der Infanterie* Paul von Hindenburg as *8. Armee* commander.[81]

Hindenburg and his new chief of staff, *Generalmajor* Erich Ludendorff were ordered to proceed to the Eastern Front. Their presence in the coming days, shaped Germany's military leadership for the remainder of the war. *OHL* fully expected Hindenburg to make a major difference in salvaging the deteriorating situation. The rapidly evolving Western Front in France and Belgium dominated Moltke's thinking. He needed strong leadership to hold the course on the Eastern Front. Gumbinnen threatened to turn a tactical reverse into a strategic catastrophe. The shifting forces, from the Western to Eastern front, set in motion several chains of causation that violently and decisively affected the whole course of the Great War. From its inception and up to the present day, few know that Gumbinnen played a pivotal role in the Great War.[82]

PART III

AVIATION OVER TANNENBERG

CHAPTER V

Are the Germans Retreating?

A Momentous Occasion and a Quickly Evolving Threat

In the annals of German military history, 23 August 1914 was a momentous turning point. *GdI* v. Hindenburg and *Genmaj* Ludendorff arrived at *AOK 8*'s Marienburg headquarters at 1400 hours to take command of *8. Armee* from *GO* v. Prittwitz. *8. Armee* staff was deeply engaged in maneuvering their forces to confront a more confident Russian force. Hindenburg and Ludendorff arrived and both generals immediately began work. Updates were provided from *Generalmajor* Grünert, *8. Armee Oberquartiermeister*, and *Oberstleutnant* Hoffmann, *8. Armee* Operations Officer (1a). Intelligence from *Fliegertruppen* aerial reconnaissance confirmed *Gen.kav* v. Rennenkampf's Russian First Army had not crossed the Angerapp River.[1] Initial intelligence discussions prompted *AOK 8* to modify the retrograde maneuver; forces were ordered to hold a line about 100 kilometers northeast of the Vistula River along the Passarge (Pasleka) River.[2] This decision was communicated to *I. Armee-Korps, I. Reserve-Korps,* and *XVII. Armee-Korps*.[3]

General der Infanterie Paul von Hindenburg assumed command of 8. Armee, on 23 August 1914.

GdI v. Hindenburg and *Genmaj* Ludendorff chose to concentrate nearly all forces south, in an attempt to strike the Russian Second Army with a quick, decisive blow. German battle plans now followed a basic concept: stop the Russian advance of Samsonov's Second Army south of Allenstein by conducting a pincer movement. *GdI* v. François' *I. Armee-Korps* was

Oberstleutnant Max Hoffmann served as AOK 8 Generalstab Ia (Operations) under the command of GO v. Prittwitz and GO v. Hindenburg. His astute understanding of the battle strategy, made him a key player in AOK 8's role throughout the Ostpreussen campaign.

ordered to move his entire *Armee-Korps* south by rail to merge with *GdA* v. Scholtz's *XX. Armee-Korps. XX. Armee-Korps* was to hold their position. The two *Armee-Korps* were to attack Samsonov's Russian Second Army from Gilgenburg. The remaining two *Armee-Korps*, (*Genltn* v. Below's *I. Reserve-Korps* and *GdK* v. Mackensen's *XVII. Armee-Korps*) were ordered to leave positions near the Angerapp River and attack from the northeast from the Allenstein-Passenheim-Ortelsburg line.[4] All that remained of *8. Armee*'s original three *Armee-Korps* (opposite Rennenkampf's First Army) was a cavalry division and 1½ infantry divisions. This token force was tasked with harassing and confusing Rennenkampf's forces. That evening, *AOK's Oberstleutnant* Hoffmann recalled, "it was the most nerve-wracking period of the entire battle."[5]

GdI v. Hindenburg's decision to move most of *8. Armee* south to fight Samsonov's Russian Second Army, was strategically astute despite obvious risks. The absence of effective Russian cavalry employment meant Rennenkampf had no knowledge German forces had withdrawn. Both Russian armies could have immediately launched fatal attacks, destroying the token force in front of Königsberg and overwhelming *XX. Armee-Korps* (the single corps in the south). Though Russian cavalry was negligent, both *I. Reserve-Korps* and *XVII. Armee-Korps* risked detection by *Gen.inf* Blagoveshchensky's VI Corps as they moved south. On his first day as *8. Armee* commander, *GdI* v. Hindenburg wired *OHL* in Coblenz clarifying his battle plan. He announced, "*8. Armee* will assemble all

Generalmajor Erich Ludendorff, assumed chief of staff role for 8. Armee on 23 August 1914.

forces at *XX. Armee-Korps* for an enveloping attack on 26 August." He also informed *OHL* that headquarters were moving from Marienburg to Riesenburg.[6] To all *8. Armee* soldiers, Hindenburg proclaimed, "Let us have trust in each other and do our duty!"[7] Confidence in his forces overshadowed any lingering doubts.[8]

Hindenburg arrived in *Ostpreussen* with only 34 serviceable aeroplanes in place, for aerial reconnaissance. Despite the limitation, *Fliegertruppe* operationalized a successful battle plan. The threat posed by Rennenkampf's First Army was monitored by *Festung* Königsberg's *FestFA 5* and *I. Armee-Korps' FFA 14*. Russian forces merged into a narrowly aligned formation that made it easy for *Fliegertruppen* to monitor.[9] Their constant surveillance of Russian positions allowed *AOK 8* to better prepare for the upcoming battle. *Festung* Königsberg's *FestFA 5* played the most important role by flying Taube sorties from Königsberg over *Gen.kav* v. Rennenkampf's Russian First Army. The remaining *Feldfliegerabteilungen* and *Festungfliegerabteilung* delivered on key sightings, and aided *AOK 8* in finalizing maneuvers against Samsonov's Second Army. Aerial reconnaissance confirmed Second Army had established an 80-kilometer front from Lautenburg to Ortelsburg. *Fliegertruppe* succeeded where *RIVVF* failed, in providing valuable updates to *Genltn* v. Below's *I. Reserve-Korps* and *GdK* v. Mackensen's *XVII. Armee-Korps*. They located the VI Corps of *Gen.inf* Blagoveshchensky, marching north by way of Ortelsburg.[10] Despite other successful sightings on 23 August, *8. Armee* failed to discover that an entire army corps (*Genleyt* Klyuev's XIII Corps) had moved northeast of German positions at Orlau, in the villages of Jedwabno and Omulefofen.[11]

Russian Reconnaissance is Challenged

G*en.kav* Samsonov's Second Army advanced an impressive distance in a short period of time. *Genleyt* Klyuev's XIII Corps arrived at the battlefield after marching 250 kilometers in eleven days. *Gen. inf* Martos' XV Corps reached 215 kilometers after ten days. Once arriving on *Ostpreussen* soil, Samsonov's Second Army contended with traversing challenging terrain and limited communication networks to support operations. Further complicating command planning and execution, was lack of leadership and guidance from *Gen.kav* Zhilinsky's Northwest Front headquarters. Zhilinsky possessed a flawed perspective on both German intentions and positions, as well as the Gumbinnen battle. He believed the German army was still highly demoralized and in retreat. He subsequently directed Samsonov, "You are to intercept the enemy as he retreats before the advancing army of *Gen.kav* v. Rennenkampf, with the primary objective of cutting off his escape across the Vistula."[12]

General ot infanterii Martos, commander of Russian XV Corps.15th KAO provided aerial reconnaissance support to his forces.

The absence of effective reconnaissance from cavalry and aerial resources proved fatal. *Gen. kav* v. Rennenkampf had lost the German army. Seven German divisions were nowhere to be seen and the location of *GdK* v. Mackensen's *XVII. Armee-Korps* and *Genltn* v. Below's *I. Reserve-Korps*, was unaccounted for. He both believed German forces were thinning out in front of his advancing armies but at the same time, that François and Mackensen were willfully choosing to withdraw their *Armee-Korps* to Königsberg. Rennenkampf assumed he was now facing a deliberate siege of Königsberg—an arduous task requiring more infantry and heavy artillery. Despite the ability to maneuver great distances in a short period of time, his First Army cavalry divisions had only covered 77 kilometers over the previous three days. A more vigorous effort would have confirmed the rearguards of *GdK* v. Mackensen's *XVII. Armee-Korps* and *Genltn* v. Below's *I. Reserve Armee-Korps* been confirmed facing First Army.

Rennenkampf's cavalry arm neither disrupted Germans (with deep attacks behind their lines) nor provided adequate intelligence on *8. Armee*'s withdrawal.[13] *Gen.kav* Zhilinsky outlined the importance of the aggressive use of cavalry "to screen and hide movements of our corps from the enemy, garrison the most important points, seize river crossings, and wreak havoc deep in the rear, in order to disrupt railroad transportation." In spite of his mastery in cavalry, Rennenkampf never utilized them to track or harass Germany's movements or retreats. This failure contributed to the evolving chaos. On 23 August, when he finally made full use of his cavalry resources, troops advancing into empty space found no Germans.[14] It was too late.

Compounding this failure, Russian aviators were unable to find their way around *Ostpreussen*. Confusing and contradictory reports of German concentrations and advances became the norm. *Gen.kav* Samsonov's Second Army employed *RIVVF* aerial reconnaissance, while on the ground, Samsonov's cavalry divisions were fully occupied, securing the flanks of the ongoing advance. *Shtabs-kapitan* Prussis' *1st KAO* flew Farman HF.16s, supporting *Gen.inf* Artamonov's I Corps, with three sorties over Mława and Soldau, that same day. Near Gilgenburg, they observed encamped German divisions. *RIVVF* aerial observers believed them to be the German *II. Armee-Korps*—but in fact, the unit was not in *Ostpreussen*. Both road and rail traffic in the area, was very active. Subsequent aerial sorties by *15th KAO*'s Nieuport IV aircrews, confirmed infantry and artillery massed at that location. At the Lautenburg rail station, extensive activity was underway with troops unloading and marching into the city. To the northwest of the border at Deutsch-Eylau, two trains with troops from *I. Armee-Korps* were observed heading toward Lautenburg. *Gen.inf* Martos' XV Corps' *15th KAO* was under the command of *Shtabs-kapitan* Valnitskiy, and their Nieuport IVs flew two missions covering Neidenburg and Gilgenburg. They also sighted extensive rail activity in the region, particularly around Hohenstein and Allenstein. *Gen.leyt* Klyuev's XIII Corps' *13th KAO* under the command of *Poruchik* Kokayev flew two sorties over Willenberg – Jedwabno – Allenstein – Passenheim, and reported an infantry camp at Jedwabno, active rail movement from Allenstein to Passenheim, and resupply units leaving Allenstein. *13th KAO* reported extensive rail activity in the region, particularly around Allenstein. Finally, *23rd KAO* flew one Farman HF. 16 sortie near Ortelsburg and Bischofsburg and took note of German logistic elements. No rail activity was observed between Ortelsburg and Bischofsburg. Samsonov's aviation arm gathered intelligence on a buildup of German forces (near Lautenburg) opposite the left flank of *Gen.inf* Artamonov's I Corps. Reports that German forces were vast in scope, were submitted to both army headquarters and met with skepticism. The prevailing view was that aerial reconnaissance failed to corroborate agent reports. "Whence the Germans could have brought such large forces? According to intelligence as obtained by agents in corps headquarters and Army headquarters, it is clear that the enemy has a maximum of 1½ or 2 corps." *Gen.kav* Samsonov was adamant about employing his forces, without the benefit of adequate intelligence on the rapidly building German posture.[15]

When Rennenkampf consulted with Zhilinsky, they determined Samsonovskaya had to take initiative to cut off Germans from retreating across the Vistula River. To meet that objective, Samsonov was ordered to swing his left wing even farther away from the First Army, expanding the gap and increasing vulnerability to *8. Armee* maneuvers.[16] *Gen.inf* Blagoveshchensky's VI Corps' role as gap-filler between the two armies failed. His divisions were ineffective in employing reconnaissance assets. *Polkovnik* Konstantin Ivanovich Serbinovich, chief of staff, Russian 4th Division, had a lack of imagination for determining the battleground his division was to face. He stated, "no orders for reconnaissance in this direction had been received." Compounding Russia's vulnerability, *RIVVF* suffered from serious maintenance and repair problems that resulted in very low commission rates.[17] To the north, First Army's III Corps *3rd KAO* pilot Donat

Aduevich Makijonek and observer Vasily Yakovlevich Makeichik, flew a sortie in the Masurian Lakes region and discovered a large concentration of German artillery. Despite cogent and timely intelligence, commanders and staff ignored the report. The next day, Makijonek crashed his aeroplane leaving *3rd KAO* with just two Nieuport IVs.[18]

First Encounter at Lahna-Orlau

On the afternoon of 23 August, the villages of Lahna and Orlau (north of Neidenburg) witnessed the first battle between *8. Armee* and Samsonov's Second Army. Two divisions belonging to *Gen.inf* Martos' XV Corps attacked entrenched positions that were occupied by *XX. Armee-Korps' 37. Infanterie-Division.*[19] On 24 August, *GdI* v. Hindenburg and *Genmaj* Ludendorff left Marienburg to meet up with Scholtz's *XX. Armee-Korps* headquarters in Riesenburg. Their goal was to figure out how to gain control in the upcoming battle.[20] *Gen. inf* Martos' XV Corps received messages from *Genleyt* Klyuev's XIII Corps that morning, indicating XIII Corps was moving west from Jedwabno. They strategized how best to assist Martos' forces in combat.

FFA 15 was the primary Fliegertruppe unit to monitor Gen.kav Samsonov's army as it crossed into Ostpreussen. A Jeanin "Stahltaube" takes off to reconnoiter the advancing Russian army.

Königsberg intercepted two of *Gen.inf* Klyuev's XIII Corps messages. *AOK 8* received the wireless intercept at 0845 hours.[21] *AOK 8* alerted *GdA* v. Scholtz that a sizeable Russian force was approaching his sector. Two Russian corps were within reach of enveloping *8. Armee*'s only *Armee-Korps,* defending southern *Ostpreussen*.[22] Scholtz was expecting *3. Reserve Division* to augment his *XX. Armee-Korps*. Fearing envelopment by the Russians, Scholtz decided to withdraw.[23] *3. Reserve Division* marched north from Orlau to Hohenstein and prepared to fight the Russians from defended positions.[24]

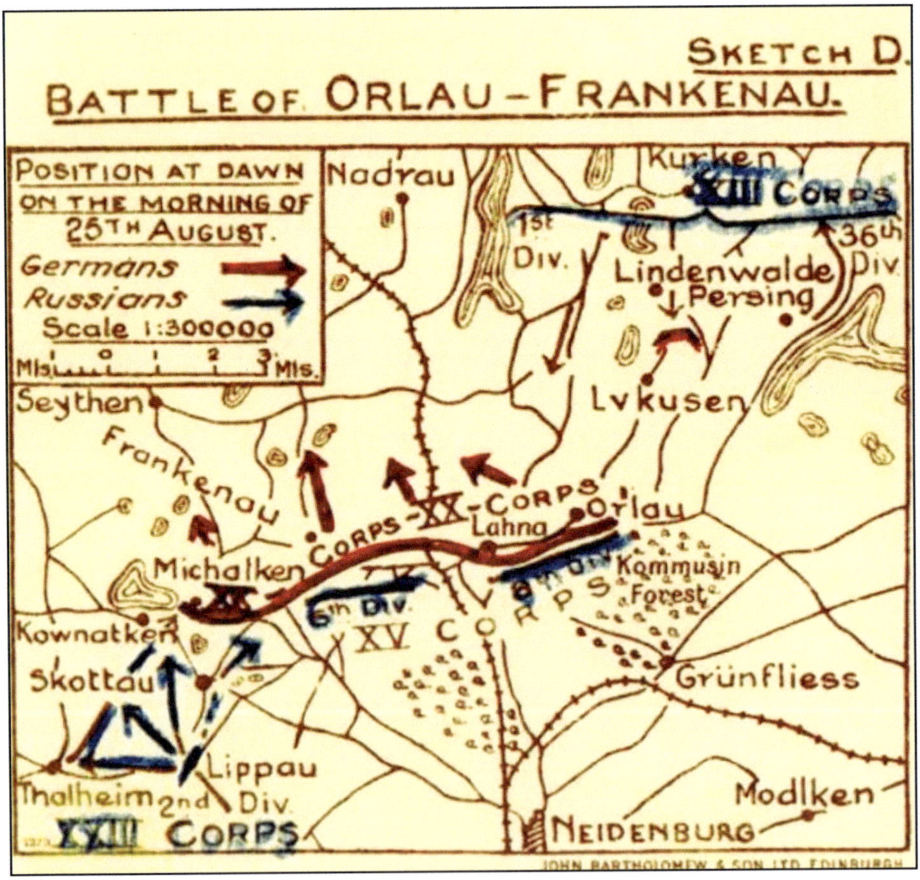

The first major battle of the Tannenberg campaign at Orlau is portrayed through the Ironside maps and examined by American soldiers learning the art of war at Leavenworth, KS. The map was used to teach the art of war at the US Army's Command and General Staff school in the 1930s to show forces in motion.

Two *Feldfliegerabteilungen* (*XX. Army-Korps' FFA 15, 8. Armee's FFA 16*) and two *Festungfliegerabteilungen* (*Festung* Graudenz's *FestFA 6* and *Festung Boyen / Lötzen* (*FestFA 7*) were active over southern *Ostpreussen*. *Fliegertruppe* patrolled Lautenburg, Soldau, Usdau, Ortelsburg, Bischofsburg and Mława. That morning, *Ltn* Franz Hagen and his aerial observer *Oberleutnant* Richter (*FFA 16*), reported an enemy column from Koschlau to Usdau (*General ot infanterii* Kiprian Antonovich Kondratovich's XXIII Corps).[25] It was the onset of the entire invading Second Russian army.

Königsberg Wireless Intercepts and the Evolving Battle

The initial battles in *Ostpreussen* highlighted the critical role Germany's collective intelligence played in constructing an effective defensive campaign against the Russian invasion. Military historians have recounted the confounding Russian lapse in security. Combatants routinely transmitted their positions, intent, and strength of forces, in the clear. Historians attribute this lapse to the lack of code books, inept procedures distributing ciphers and keys, and a sense of urgency.[26] Germany's strong communication defense contributed to Russia's challenge. German cavalry and civilians cut existing land lines in occupied Russian territory, forcing commands and units to employ wireless. Stunned German operators hurriedly recorded the intricacies of troop movements and battle plans, that originated from extensive wireless communications between Russian regimental and corps headquarters. German cryptologists broke new Russian codes (old and new cipher codes being used simultaneously by Russian army groups) and they employed a methodology that compared orders transmitted via the older (already broken) code, with new transmissions. This was usually effective because the encryptions were not that dissimilar.[27] The challenges of encrypting wireless also impacted *RIVVF* mission priorities. *Genleyt* Klyuev's XIII Corps lacked the keys necessary to read cryptograms from *Gen.inf* Blagoveshchensky's VI Corps. This shortfall caused *13th KAO* (in support of XIII Corps) to prioritize sorties, in an effort to deliver secure communications.[28]

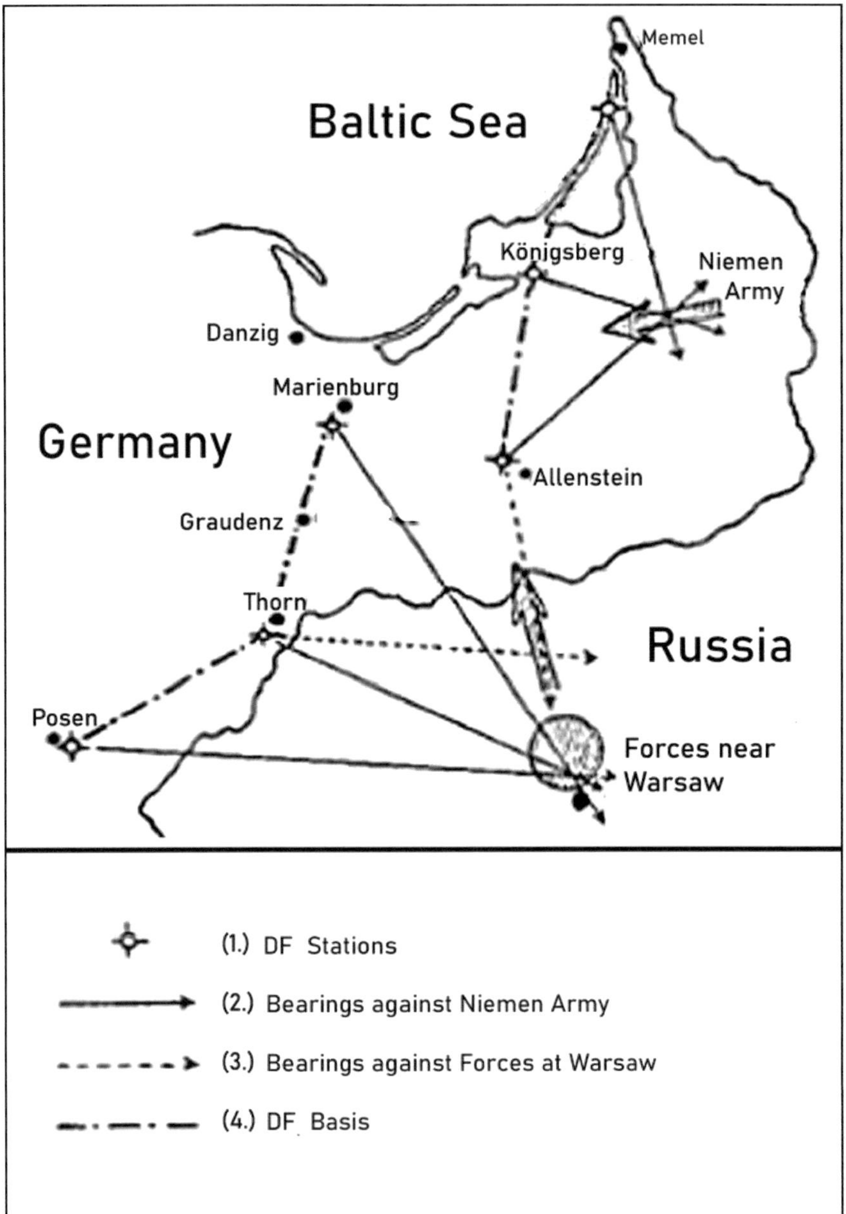

Example of Direction Finding against the Russian Forces of the Niemen Army and near Warsaw

Major Randevig illustrated the wireless intercept architecture of Ostpreussen that resulted in important acquisition of messages transmitted by both Russian First Army and Russian Second Army. Source: Die deutsche Funkaufklärung in der Schlacht bei Tannenberg

In the early morning of 25 August, Königsberg intercepted and copied two orders from two Russian army headquarters, that spelled out intentions for the coming battle. The first intercept revealed that Rennenkampf's plans for advancing on Königsberg had not changed. This confirmed to Hindenburg that Rennenkampf's Russian First Army would not take part in Samsonov's battles. Königsberg radio sent a telegram to *AOK 8* at Riesenburg at 0530 hours, outlining Rennenkampf's objectives for *Gen.art* Aliyev's IV Corps, operating at the southern wing of the Russian First Army.[29]

> The army will continue attack. On August 25 it will reach the Wirdeln-Saalau-Norkitten-Kl. Potauren-Nordenburg line; on August 26 Dammerau-Petersdorf-Wehlau-.Allenburg-Gerdauen. The Pregel river separates sectors of XX. and III. Corps. Road Schwirbeln-Kl. Potauren-Allenburg separates III. and IV. Corps. Road will be completely used by III. Corps. Khan Nachitchevansky will proceed in direction Allenburg ahead of Army front into area between Pregel river and line Darkehmen-Gerdauen-Bartenstein with own Division, south of Gerko. Crossing the Pregel is XX. Corps task. Rennenkampf.[30]

At 0600 hours, Samsonov transmitted a priority message directing *Genleyt* Klyuev's XIII Corps to continue advancing to the northwest. *Gen.kav* Samsonov's Second Army march route and objectives were spelled out.[31] The essence of the copy read:

- After the battle on the front of [Russian Second Army] XV Corps, the enemy corps [*XX. Armee-Korps*] has retreated to Osterode on 24 August.
- Near Gilgenburg, Russian [First] Army continues pursuing enemy retreating towards Königsberg Rastenburg.
- Russian Second Army will continue advance to the Allenstein - Osterode line on 25 August. Russian Second Army corps will reach the following objectives: XIII. Corps: Gimmendorf-Kurken line; Russian XV Corps: Nadrau – Paulsgut; Russian XXIII Corps: Michalken-Gross Gardienen.
- Borders between advancing Russian corps: Between Russian XIII Corps and Russian XV Corps: the line covering Muschaken-Schwedrich. Between Russian XV Corps and Russian XXIII Corps:

Professor Ludwig Deubner, a classical philologist who spoke fluent Russian mastered the Russian's cipher system employed by Northwest Front.
Source: Ein deutscher Professor war der erste Codeknacker - WELT

the line covering Neidenburg-Wittikwalde. Russian I Corps will remain in the designated 5. Sector, protecting the left flank of Russian Army and Russian VI Corps will move to Bischofsburg-Rothfliess area to secure right wing.

- Near Rastenburg, Russian 4. Cavalry Division (assigned to *Gen.inf* Blagoveshchensky's Russian VI Corps) will remain in Sensburg and reconnoiter the area up to the Rastenburg – Bartenstein and Seeburg – Heilsberg line.

Signed—Army Staff Ostrolenka[32]

At this pivotal point in time, these two wireless intercepts were considered a decisive turning point in military historians' analysis of the origins of the Tannenberg battle. These two intercepts also informed their ultimate assessment of *AOK 8*'s effectiveness in executing their mission. A contemporary expert, Theobald Schäfer, cautioned (in *Deutsches Offziersblatt*) that German leaders' knowledge about Russian force deployment, could not have emanated from randomly intercepted communications. Both messages of 25 August, reinforced Hindenburg and Ludendorff's vast body of prior knowledge. Sources, including aerial reconnaissance intelligence, clearly indicated what to expect from near-term Russian maneuvers.[33] *Genmaj* Ludendorff's recollection of the critical moment, illuminated the point: That the decision to commit to the Gilgenburg assembly area for battle, had already been made prior to receiving new intelligence.[34] The negligence of the network of Russian wireless operators was inexcusable, but this failure was never deemed the singular or primary factor, responsible for *8. Armee*'s victory at Tannenberg.

Germany's network of wireless intercept stations was extensive. "*Radiogross-Stationen*" [Main Radio Stations] operated from fortresses at *Festung* Königsberg, *Festung* Thorn, *Festung* Graudenz and *Festung* Posen.[35] A larger intercept operation maintained by *AOK 8's Radiogross-Stationen 8* [Heavy Radio Station 8], generated many significant reports.[36] *8. Armee* also operated five mobile stations (two mobile sta-

The infusion of important wireless intercepts of First Army and Second Army made validation by Fliegertruppe a priority. The location of Fliegertruppe units on 24 August in Ostpreussen is portrayed on the map. FFA supporting Armee-Korps relocated the following day to support the upcoming Tannenberg battles. Known landing ground locations for RIVVF aviation is also shown.

XX. Armee-Korps' FFA 15's Ltn Baron Elard von Löwenstern (fourth from left) flew as an Jeannin Stahletauben observer. His memoirs Der Frontflieger, provided a rare glimpse of Fliegertruppe operations during the Ostpreussen campaign.
Source: Der Frontflieger. Bernard & Graefe, Berlin 1937

tions at *AOK 8* and three with *1. Kavallerie. Division*).[37] Königsberg's ability to decipher Russian transmissions was greatly facilitated by Professor Ludwig Deubner, a classical philologist who spoke fluent Russian. Deubner mastered Russia's cipher system employed by Northwest Front. He arrived at Königsberg on 18 August as a *Landsturm* soldier and immediately began translating clear text intercepts. By the final days of September 1914, Deubner and his team of codebreakers were sending translations to *Genmaj* Ludendorff each night at 2300 hours. Ludendorff became habituated to anxiously posing the question to staff, "Any radiograms?" His protocol included stalking the cryptanalytic section regularly, to anticipate any potential intelligence delays. This process insured that orders were not impeded and could be generated, expeditiously.[38]

Active Aerial Reconnaissance on 25 August

At this critical juncture, Germany's highest intelligence priority was to keep track of Rennenkampf's First Army by employing aerial reconnaissance. In the early morning of 25 August, numerous sorties were curtailed due to fog and low hanging clouds. As the weather improved, several sorties were generated. *Fliegertruppe* confirmed scattered Russian cavalry detachments (one sighting was a division strength unit) south of Strasburg-Lautenburg, out to the Vistula River. *AOK 8*'s initial fears, dissipated—as Russian cavalry had failed to discover and destroy François' *I. Armee-Korps* at the rail head. Ongoing German reconnaissance confirmed only scattered Russian cavalry detachments (at most one division) south of Strasburg-Lautenburg.[39] Additionally, Zeppelin Z-V flew a daylight mission over the *Ostpreussen* southern border and reported Russian cavalry at Rypin, Zuromin, Biezun and Sierpe. Z-V dropped bombs at Sierpe while taking fire from Russian machine guns.[40]

Generál-kvartirméyster Danilov, at Stavka, provided insight into the RIVVF challenges during the Ostpreussen campaign in his post-war memoirs.

In northern *Ostpreussen*, *Fliegertruppe* and other German intelligence, confirmed the Russian First Army was slowly moving westward and retaining the bulk of forces at the Angerapp River. An *FFA 17* Albatros B.I supporting *GdK* v. Mackensen's *XVII. Armee-Korps*, flew west of Insterburg (16 kilometers west of Gumbinnen) and discovered a long column of Russian infantry (*Gen.inf* Smirnov's Russian XX Corps) heading west. The same sortie discovered a Russian column 32 kilometers south of Insterburg, heading southwest towards Nordenburg. Another *FFA 17* aviator reported at 1300 hours, that the Russian First Army had advanced its left flank from Nordenburg to Gerdauen. The logical inference from this intelligence, was that Rennenkampf had finally spurred his army to move out.[41]

Fliegertruppe discovered *Gen.kav* Scheideman's II Corps near Lyck (in the Masurian Lakes region) and confirmed to an anxious *AOK 8* that the corps was not heading southwest to reinforce *Samsonovskaya*.[42] A wireless intercept alerted *AOK 8*, that Scheideman's II Corps was moving to the southwest towards Drengfurth. This maneuver placed *XVII. Armee-Korps* and *I. Reserve-Korps'* at risk, in their advance to the south. That evening, *AOK 8* determined their primary attack would be launched by François' *I. Armee-Korps*, the following day.[43]

Fliegertruppen were especially active that day, covering southern *Ostpreussen*. *XX. Armee-Korps' FFA 15* aeroplanes reported Russian bivouacs, infantry, and artillery, between Gilgenburg and Neidenburg. Russian command posts were confirmed co-located to *RIVVF* landing grounds. *Festung* Graudenz's *FestFA 6* discovered seven Russian aeroplanes near Soldau, suggesting as many as three corps-level command posts were nearby. *8. Armee's FFA 16* aircrews reported that Russians were settling in place at their respective locations; large Russian bivouacs were sighted between Ziechanow north to Mława. Extensive railroad activity in the area showed the Second Army was reinforcing its western wing for the oncoming battle.[44]

Ltn Elard von Löwenstern was an aerial observer assigned to *FFA 15*. His observations from the Jeannin Stahltaube aeroplane he was flying included comments about the Russian troops:

> On 25 August we were transferred to Osterode. The enemy left wing is apparently at Heinrichsdorf-Usdau. The Russians sleep late, seldom break camp before 0700-0800, then march for 2-3 hours, then rest again until almost 1700. They then march until 2000. *XX. Armee-Korps* was assigned the duty of holding out for three days (against the 4-5 corps

> which we had reported advancing) until the *I. Armee-Korps* comes down through Deutsch-Eylau and *3. Reserve Division* through Allenstein.[45]

Löwenstern further reported:

> On the 25,[th] we had detected large bivouacs near Skottau, Lippau, Januschkau, Michalken, Neidenburg, Soldau, Seeben, Kroschken, and one cavalry corps in Lauenburg. The roads from the south were free. In the Kurken – Jedwabno – Wuttrienen, Maransen Heath, apparently one army corps was resting in bivouacs.[46]

RIVVF Support Prior to Battle

Samsonov's eastern flank comprised *Gen.inf* Blagoveshchensky's VI Corps' marching directly in the path of *I. Reserve-Korps* and *XVII. Armee-Korps*.[47] *Genltn* v. Below's *I. Reserve-Korps* was enroute to Allenstein, when tasked to march via Seeburg to attack the Russian VI Corps.[48] Russian aerial reconnaissance on 24 and 25 August, observed German forces massing nearing their own right flank. On the left flank of the Russian Second Army, a German division was observed in the center of Strasburg. At Lautenburg, a German infantry division with artillery, was discovered heading east to Usdau, in the path of *Gen.inf* Kondratovich's XXIII Corps. North of Lautenburg (near the town of Gilgenburg), another German division was moving east towards Usdau. In addition to heavy activity on the roads, *RIVVF* aviators reported extensive rail activity from Strasburg to Lautenburg, Deutsch-Eylau to Soldau, and west to Gilgenburg, as German forces formed their attack against XXIII Corps. *GdI* v. François' *I. Armee-Korps* was maneuvering to attack Samsonov's left flank. About 60 kilometers to the northeast near Bischofsburg, German infantry and artillery were observed—that were later confirmed as elements of *GdK* v. Mackensen's *XVII. Armee-Korps*. To the detriment of the rapidly evolving battle, Blagoveshchensky refused to trust his aviators and failed to respond to intelligence from aerial reconnaissance at Bischofsburg. Instead, he decided to move west to Allenstein. This placed *Gen. inf* Blagoveshchensky's VI Corps' northern flank at grave risk from

XVII. Armee-Korps, and subsequently resulted in the Russians suffering a lethal blow at Bischofsburg. [49]

RIVVF was the first to send a warning to *Gen.kav* Samsonov's Second Army and corps staff, that *8. Armee* had shifted position from northern to southern *Ostpreussen*. Aerial reconnaissance reports showed that the German force was far greater than initially presumed. The threat to *Gen.inf* Artamonov's I Corps became real, and caused Samsonov's staff to take notice. The rapidly changing situation propelled Samsonov to reevaluate the left flank of Second Army's center group. Samsonov requested permission from *Gen.kav* Zhilinsky to shift his center group to the northwest in an attempt to strike the Germans head on. *Gen.kav* Zhilinsky refused; at the time, he didn't have access to the same *RIVVF* reports and was subsequently unaware *8. Armee* posed a massive threat to Samsonov's left flank. Zhilinsky only permitted Samsonov to reposition Artamonov's I Corps, in an effort to shield against German forces on the Second Army's left flank.[50] Zhilinsky's teletype to Samsonov reflected his lingering bias towards Rennenkampf's recent victory to the north. It was a bias that proved fatal.

> After Rennenkampf's victorious battles, the German troops quickly retreated with heavy losses, blowing up the bridges behind them. In front of you, only weak forces remain. Therefore, you must employ only one corps at Soldau and with the bulk of your troops, push on towards the Allenstein-Sensburg line, to be reached day after tomorrow, at the latest.[51]

In his postwar narrative, Russian General Aleksandr Aleksandrovich Noskoff commented on the illusion clouding the thinking of Russian commanders. For example, Rennenkampf reported to Zhilinsky, that he "victoriously pursued the retreating enemy." Noskoff noted that Rennenkampf "pursued a spectre. Why did he do so? This remains one of the most confounding mysteries of the Tannenberg battle."[52]

The Russian commanders' limitations in exercising rational, strategic analysis, can in part be attributed to their under-utilization of military aviation support. This point of view was highlighted after the war, by *Generál-kvartirméyster* Danilov, *Stavka* chief of staff. "In the absence of aeroplanes, everything that took place with the Germans behind their advance guards remained unknown to us."[53] Danilov, further elaborated,

> As a consequence of our lack of sufficient aeroplanes, the enemy could maneuver as he wished by using a highly-developed railroad network, and be certain his plans could not be discovered by us, very quickly. At the same time, with considerably better aviation facilities, the enemy was able to observe each of our steps with complete impunity.[54]

Dissension Against AOK 8 on 25 August

8. *Armee* possessed a near-perfect opportunity to defeat both Russian armies separately. François' *I. Armee-Korps* remained in transition from the north to the south. Vital artillery and ammunition elements had not arrived. François's goal, was to shift farther south, in an effort to launch a preventive strike against the Russian advance. Despite mist and cloud cover hampering aerial observations, *Fliegertruppe* reported numerous Russian military trains originating from the direction of Warsaw, to the border stations of Illowo and Soldau. Auxiliary intelligence determined Russian 3. Guard Infantry Division and the 1st Rifle Brigade, arriving to reinforce *Gen.inf* Artamonov's I Corps. *AOK 8's Armeebefehl* directed *GdI* v. François' *I. Armee-Korps* (at 0400 hours) to take the heights around Seeben, advancing to Russian I Corps' positions on the Usdau high ground, that provided good observation and long fields of fire.[55]

On 25 August, *GdI* v. François, having achieved notoriety for his dissenting opinion versus *GO* v. Prittwitz, now challenged both *GdI* v. Hindenburg and *Genmaj* Ludendorff. He considered *AOK* 8's planned flank attack against Russia's Second Army left wing to be a replay of his Gumbinnen experience. An assault on Usdau meant putting his troops at great risk, without adequate artillery support. François argued for a delay, to enable artillery to arrive. In addition, he assumed he knew the best move for *I. Armee-Korps* against the advancing Russian army. *Fliegertruppe* aerial reconnaissance fostered a confidence in him, that led him to believe he understood the optimal direction to maneuver and defeat the Russians. Both Hindenburg and Ludendorff failed to accept François' argument. *AOK 8* believed an extension granted the Russians another day to prepare their offensive. *8. Armee* was going to attack the following day, irrespective of the argument. *AOK 8* made it clear that François better get back in line.

The next day, François disregarded *8. Armee*'s direction and decided to limit *I. Armee-Korps'* attack. The result was *8. Armee* did not defeat the Russian left wing as intended in their 26 August plan. François' forces did accomplish their assigned mission, albeit 24 hours behind schedule.[56]

CHAPTER VI

Russian Misconceptions

The Battle of Tannenberg Commences

At 2020 hours on 25 August, *GdI* v. Hindenburg notified *8. Armee* to fight at Tannenberg—one of the greatest battles of the 20th century. Orders were generated from his newly established *AOK 8* headquarters, at Löbau. His strategy focused on two separate battle groups: *Ostgruppe* [Eastern Group] and *Westgruppe* [Western Group]. *Ostgruppe* was comprised of *Genltn* v. Below's *I. Reserve-Korps* and *GdK* v. Mackensen's *XVII Armee-Korps*—marching south towards Allenstein. *Westgruppe* was comprised of *GdI* v. François' *I. Armee-Korps* and *GdA* v. Scholtz's reinforced *XX. Armee-Korps*.[1] The newly promoted *Generaloberst* v. Hindenburg calculated German forces had only three days of cover before Rennenkampf reacted to *8. Armee*'s new positions. Hindenburg wired *OHL* in Coblenz, "*8. Armee* in conjunction with *XX. Armee-Korps*, is planning to conduct an all-out attack on 26 August."[2] *Gdl* v. Francois' *I. Armee-Korps* was to occupy the Seeben heights no later than 1000 hours, and maneuver towards Usdau. *XX. Armee-Korps* was to support *I. Armee-Korps*' advance by moving in the direction of *Gr.* Grieben-Jankowitz and *3. Reserve-Division* was to advance to the Hohenstein region.[3]

Fliegertruppe was under great duress in the coming days, tasked with providing *8. Armee* vital information on Russian deployments and disposition. On 26 August, 120,000 combatants of both armies converged onto the Tannenberg battlefield. The majority of aeroplanes flying overhead and observed by combatants on the ground, were *Fliegertruppe*. The Russians were unaware that *Ostgruppe's GdK* v. Mackensen's *XVII. Armee-Korps* and *Genltn* v. Below's *I. Reserve-Korps* were rapidly advancing south in two parallel columns. *Ostgruppe*'s objective was to attack *Gen.inf* Blagoveshchensky's VI Corps immediately upon contact.[4]

Westgruppe's GdA v. Scholtz's *XX. Armee-Korps* stood alone against four Russian army corps (*Gen. inf* Artamonov's I Corps, *Gen.inf* Kondratovich's XXIII Corps, *Gen.inf* Martos' XV Corps, and *Genleyt* Klyuev's XIII Corps). Scholtz was advised to hold to the last man until *GdI* v. François' *I. Armee-Korps* arrived to fend off and battle Artamonov's I Corps.[5]

Lieutenant-Colonel Alfred Knox was appointed British Military Attaché in Russia in 1911 and personally witnessed the Northwest Front debacle in the destruction of Second Army.

On 26 August 1914, Lieutenant-Colonel Alfred Knox (a British attaché to Russian forces) recounted: "The Russians seem to muddle through in a happy-go-lucky-way."[6] *Gen.kav* v. Rennenkampf told *Gen.kav* Zhilinsky, "We still have no information as to where the defeated formations of the German I and XVII Army Corps have fled... our cavalry have not at this time uncovered the direction of retreat of the main German forces." Rennenkampf failed to answer the most important strategic question: Where had the *8. Armee* gone?[7] He held onto the false assumption, that the bulk of German forces were retreating to Königsberg and the remaining forces were retreating towards the Vistula River.[8]

Gen.kav Samsonov was far less prepared to lead a major battle, than *Gen.kav* v. Rennenkampf had been, five days earlier. His army had been deprived of rest for six days and were advancing over difficult roads at an average of 24 kilometers per day. Rennenkamp's campaign had the benefit of full support from rail resources that delivered troops and weapons, while Samsonov was challenged by complicated infrastructure issues. As one of Samsonov's adjutants remarked after inspecting the forces prior to battle, "I don't know how the troops can manage any longer... dome system of requisitioning must be organized."[9] Samsonov was left alone to face *8. Armee* gathering in southern *Ostpreussen*—an enemy superior in organization, communication, quality, and morale.[10]

Samsonov generated Army orders No. 6353 the morning of 26 August stating:

> The situation on the front of the XV [*Gen.inf* Martos] Corps and 2nd Division no change. There is one enemy division in Allenstein. The Second Army will advance on the line: Allenstein—Osterode. XIII Corps [*Genleyt* Klyuev] on Kellaren. XV Corps, Schönfelde—Gusnofen. 2d Division of the XXIII Corps to Hohenstein—Reichenau. I Corps, Soldau—Gilgenberg.VI Corps [*Gen.inf* Blagoveshchensky], Bischofsburg.[11]

Samsonov's Second Army's decision to move forward was predicated on Zhilinsky's false beliefs—that Rennenkampf had broken the German spirit at Gumbinnen. Samsonov failed to anticipate much resistance. His growing predicament unfolded, as the stark reality proved to be that the Russian Second Army had spread itself too thin (112-kilometer arc) in enemy territory. Samsonov was now entering a trap he couldn't escape.[12]

Two FFA 15 aviators supporting GdA v. Scholtz's XX Armee-Korps – prepare to launch on a sortie to track Gen.kav Samonov's approaching Second Army.

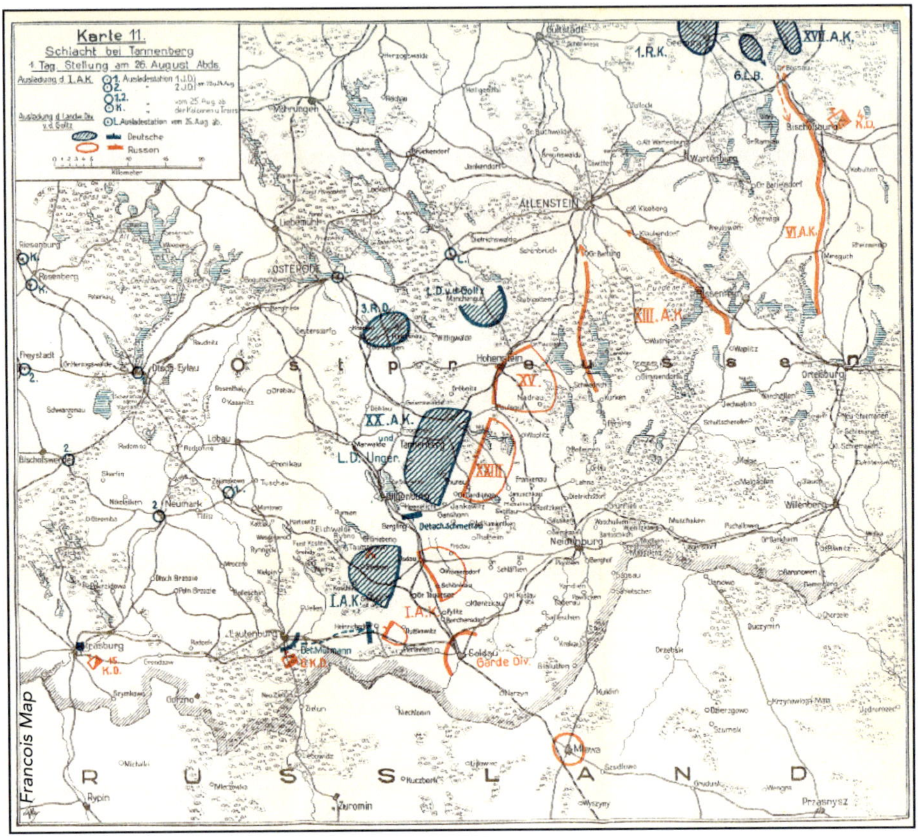

Forces in motion on 26 August 1914. Gdl v. Francois recalls the battle of Usdau in the southern Ostpreussen Sector. Note the route of the Russian withdrawal from Bischofsburg following the battle against XVII. Armee-Korps and I. Reserve Corps. This map available for download at shootingthefront.com

The Battle of Usdau

On 26 August at 0400 hours, *GdI* v. François' *I. Armee-Korps* began firing on *Gen.inf* Artamonov's I Corps lines at Seeben Heights, while *I. Infanterie-Division* launched a frontal attack at Usdau.[13] François reported to *AOK 8* that his attack had commenced, but subsequently predicted his initial objectives were unobtainable, due to lack of artillery. His insubordination towards *AOK 8* at Stallupönen, resurfaced during this battle. *AOK 8* expected François to take Seeben Heights, followed by Usdau, where the high ground permitted optimal observations of the battle, and a clear vantage point for engaging in long-range artillery fire. *I. Armee-Korps* took Seeben Heights six hours after the artillery commenced fire, but as *AOK 8* had no reliable confirmation, they considered the ground held by the Russians.[14] *FFA 14* aviators observed one infantry regiment and two long columns "without infantry" marching southward. Large bivouacs northwest of Mława (for at least two regiments) were spotted near Klentzkau, a division-level encampment noted near Kammersdorf, and several camps were sighted near Narym. A column about three kilometers in length was observed near Stupsk. These sightings were acknowledged by similar reports from *Festung* Graudenz's *FestFA 6*. All reports confirmed the Russian Second Army had more reinforcements, and were enroute to the battle.[15]

1st KAO flew Farman HF.16's supporting General ot infanterii Leonid Konstantinovich Artamonov's I Corps. He led I Corps against I. Armee-Korps at Usdau.

Westgruppe's XX. Armee-Korps fought on *I. Armee-Korps'* right flank. Scholtz's forces met fierce resistance at the onset, but advanced forward by afternoon. *XX. Armee-Korps'* staff continued to receive reports from *FFA 15*. At 1050 hours, two Russian infantry companies were detected marching south on the Mühlen-Michalken road near Januschkau. No Russian bivouacs were observed. Near Neidenburg, a brigade-strength unit was sighted along with a column of vehicles. At Kurken (between Neidenburg and Allenstein) some Russian camps and limited traffic was visible. At Gilgenburg (between Lautenburg and Soldau) columns of infantry, cavalry and ammunition columns, were noted. North of Lautenburg, up to three cavalry brigades were observed moving southeast. As *FFA 15* aviators flew east, they failed to spot any suspicious activity from Russians gearing up to attack *XX. Armee-Korps*.[15]

FFA 15's *Ltn* v. Löwenstern recalled his morning sortie (flying a Taube):

> Adverse weather impeded our air reconnaissance severely with fog, haze, and low clouds. Our aeroplanes returned with many bullet hits. In the evening, we reported columns of vehicles advancing on Hohenstein … and, most interesting for us, one Russian airfield at Neidenburg, with three biplanes [Farman] and one monoplane [Nieuport]; another airfield at Soldau with six aeroplanes. In the air we never saw one. The Russians preferred to use used the aeroplanes for messaging services.[16]

Generalleutnant v. Morgen, *3. Reserve-Division* commander, left Allenstein and was advancing on Hohenstein in compliance with *AOK 8*'s orders. Morgen learned from *Fliegertruppe* that Russian forces were observed to the east, near Bischofburg and Ortelsburg. The report suggested that a Russian maneuver to outflank his division was underway from the north. Concerned about a frontal attack and being outflanked from the north, Morgen delayed his division's advance, hoping to attack the outflanking Russian formations on the next morn-

Generalleutnant v. Morgen's 3. Reserve-Division advanced on Hohenstein in compliance with AOK 8's orders. Morgen learned from Fliegertruppe that Russian forces were observed to the east, near Bischofburg and Ortelsburg.

ing.[17] That evening, *FFA 15* reported the feared Russian outflank maneuver from the east was not in progress.[18]

At 1000 hours, *FFA 16* aviators observed a brigade-sized element [XXIII Corps] near Soldau. They also noted an infantry regiment at Borchersdorf and six transport trains at the Mława railroad station (later updated to 14 transport trains). An hour later, they reported a cavalry regiment south of the Strasburg railroad to the southeast of Gilgenburg. As they continued the sortie, they detected a cavalry detachment to the south in Russian Poland near Olszewo. Upon landing at noon, *AOK 8* learned from aircrews that Samsonov was receiving reinforcements from Mława.[19] A Russian surprise attack from the south was possible, since Mława served as the main staging base for Second Army, prior to their advancing into *Ostpreussen*. *8. Armee* took a decisive step with aviation assets. Three aviation units: *FFA 14*, *FFA 16*, and *Festung* Graudenz's *FestFA 6* now covered the southeast sector.[20]

Russian reinforcements were observed detraining in Soldau and Mława. *8. Armee* access to intercepted wireless transmissions, identified them as the Russian Guard (3rd Guard Division) sent to reinforce *Gen.inf* Artamonov's I Corps. *AOK 8* became confused over what had transpired. They believed the Russian Guard elements from *Gen.kav* v. Rennenkampf's First Army had arrived south. German agents reported that the Russian Guard had relocated from Wilna to Warsaw. In reality, it was elements of *Gen.inf* Kondratovich's XXIII Corps.[21]

8. Armee's communications suffered from intermittent signals despite functioning telephone networks. Late arriving situation reports, meant only *GO* v. Hindenburg had an accurate assessment of the battle's progress on the first day. Much to his chagrin, news was not positive. Objectives for *8. Armee* commanders were only partially achieved by *Westgruppe's* François and Scholtz. Although Seeben village had been taken, the most important target (Usdau heights) remained in Russian hands. *XX. Armee-Korps'* right wing was to split into two forces at Lake Mühlen.[22] *8. Armee* failed to successfully defeat the Russian left wing. By the end of the day, *I. Armee-Korps* and *XX. Armee-Korps* ceased forward operations towards Usdau.[23]

Ltn v. Löwenstern summed up the challenge all aviators faced at this time:

> During the Tannenberg Battle, when it was important to detect the whereabouts of the Russians, for some time an overcast cloud covered the battlefield. We were forced to cruise the area, flying above the clouds, and find the Russians by occasional penetrations. We learned how difficult it was to locate what we searched for and then produce a correct report, in spite of the distinctive lake features...Had the Russians not fired so nervously at us, we would often not have detected them. Thus, we knew by the bullet impacts into our "Taube" -wings, that they were there and we immediately hummed back into the clouds![24]

That day, as the battle evolved, *Ltn* v. Löwenstern reported frustration due to the lack of significant sightings:

> For the first time in my life, I saw a battle from above, while flying at low altitude...I was disappointed. I had imagined that quite differently. I had seen the masses of troops of the army corps approaching the days before and I figured them to be densely congested and thus, easy to recognize on the battlefield from my low flying altitude. But how little was really visible! There was empty ground below. Even the troop masses were dispersed and almost lost in the area.[25]

RIVVF Support to the Evolving Battle

At Northwest Front headquarters, *Gen.kav* Zhilinksy continued to insist that the Germans were retreating from their positions, as the Russian First Army advanced towards Königsberg. The intelligence Zhilinsky received from the front, failed to reflect reports *RIVVF* had generated.[26] At this critical stage, the Russians lacked a clear view of the location of Germany's main formation. *Gen.kav* Samsonov ordered *Gen.inf* Blagoveshchensky's VI Corps to leave a screen at Bischofsburg and proceed southwest to

15th KAO's Nieuport IV (number 3) transported by ground crew in Ostpreussen. Note Russian national markings (cockade with blue [outer], red [center] and white [inner].

15th KAO's Nieuport IV (number 3) taking off from an Ostpreussen landing ground flying a sortie in support of Gen.inf Martos' XV Corps.

Allenstein in support of two advancing Russian corps (XV Corps, XIII Corps). Samsonov soon reversed course and cancelled the order but the unfolding disaster was already underway.[27]

In support of *Gen.inf* Artamonov's I Corps, *RIVVF* reported François's *I. Armee-Korps'* arrival in Usdau. Throughout the regions of Lautenburg and Heinrichsdorf, *I. Armee-Korps* had struck the left flank and rear of I Corps. Gen.inf Artamonov made a critical error in withdrawing one of his brigades from his right wing, in an effort to counter an attack on his left flank at Lautenburg. When the brigade left its position north of Usdau, a seven-mile gap opened up between I Corps' right flank and *Gen.inf* Kondratovich's XXIII Corps and *Gen.inf* Martos' XV Corps.[28] *RIVVF* witnessed the Russian 6th Cavalry attacking *I. Armee-Korp*, over Heinrichsdorf and reported 1½ German infantry regiments moving from Gilgenburg in the direction of Gardinen, and two German infantry regiments moving from Seewalde towards Nadrau. They also observed one regimental infantry column with artillery [*3. Reserve Division*], heading from Hohenstein to Osterode.[29]

The Battle of Bischofsburg

As *Ostgruppe's Genltn* v. Below's *I. Reserve-Korps* and *GdK* v. Mackensen's *XVII. Armee-Korps* moved south, their primary Russian adversary, *Gen.inf* Blagoveshchensky's VI Corps, was ignorant of the impending threat. VI Corps had neither ground nor aerial reconnaissance to make appropriate tactical decisions. *23rd KAO* still remained under the control of II Corps. Mackensen's *XVII. Armee-Korps* advanced based on reports from *FFA 17*, indicating a Russian infantry division was sighted to the south at Ortelsburg. Despite having the correct aerial reconnaissance report, Mackensen's staff still held onto the false belief that the main Russian forces were east of the Masurian Lakes. *FFA 17* was misguided in reporting only one Russian infantry division at Ortelsburg. In fact, the entire VI Corps reinforced with two infantry divisions and one cavalry division, were at that location![30]

A *Fliegertruppe* aviator alerted *AOK 8* that *Gen.inf* Blagoveshchensky's VI Corps was heading towards Bischofsburg. As *AOK 8*

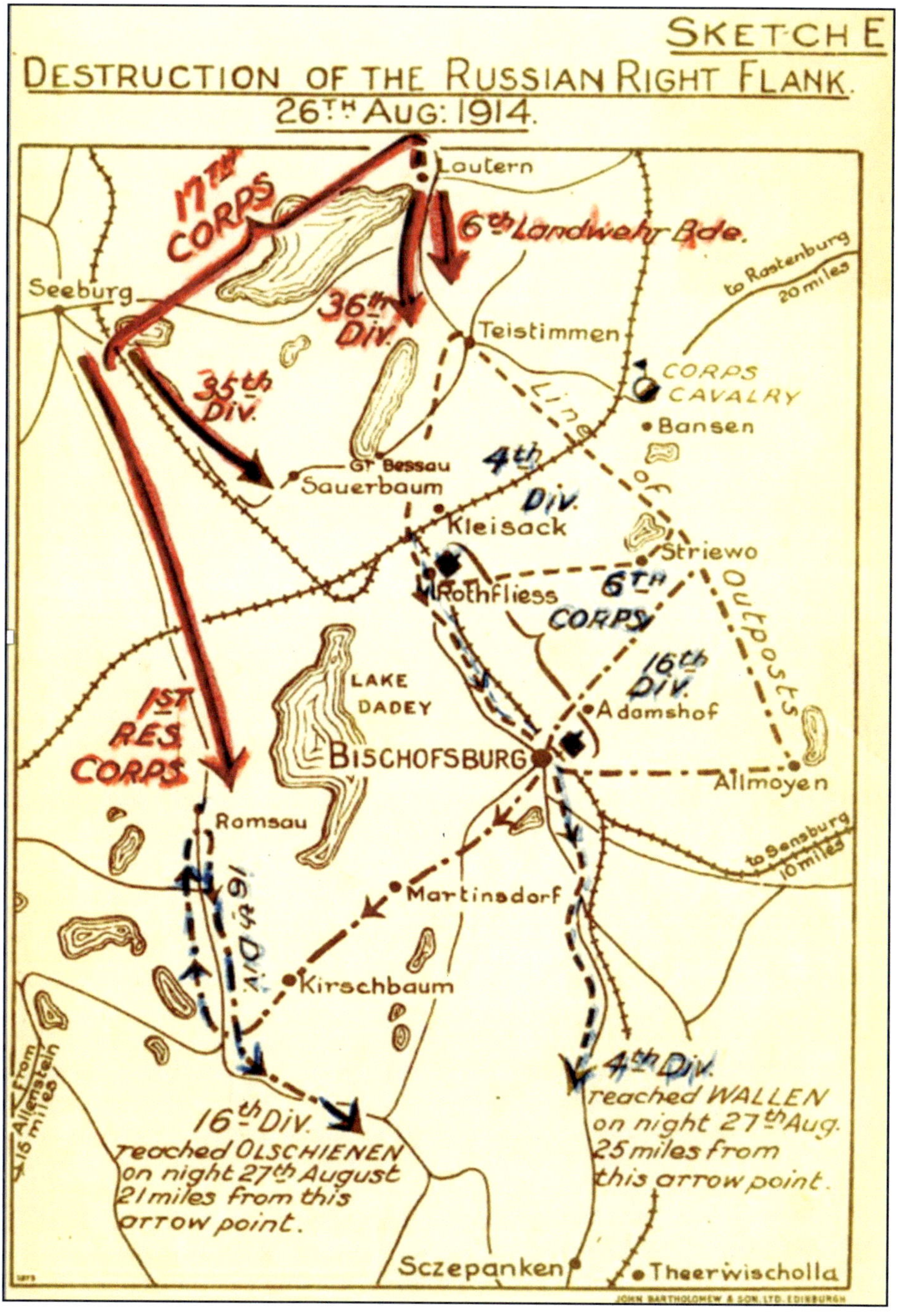

The battle of Bischofsburg between GdK v. Mackensen's XVII. Armee-Korps and Genltn v. Below's I. Reserve-Korps against the Gen.inf Blagoveshchensky's Russian VI Corps was a decisive victory for Germany. VI Corps commenced their retreat to the southeast, leaving Gen.kav Samsonov's Second Army to be enveloped by 8. Armee. Source: Command and General Staff school

Genltn v. Below's I. Reserve-Korps (which had no aeroplanes assigned) single Feldluftschiffer-Abteilung Parseval und Sigsfeld Drachen-Ballon was the first to confirm the Russian retreat from Bischofsburg.

tracked known Russian positions, they strategized maneuvers to strike the Second Army, where it was most vulnerable. The gap was widening between VI Corps and the center group of Second Army.[31] Blagoveschchensky was ignorant that *Ostgruppe* was quickly heading south. He took no precautions against the rapidly changing threat. Seizing the opportunity, *AOK 8* ordered *Ostgruppe* to expedite their move. It was time to annihilate VI Corps.[33]

On 26 August, *GdK* v. Mackensen tasked *FFA 17*'s Albatros B.I. aircrews with covering the enemy situation, particularly the rear and eastern sector on his *XVII Armee-Korps*' left flank. First and foremost, was the necessity of monitoring and reporting on the Russian First Army's direction and pace of advance. At 1010 hours, *FFA 17* noted a long Russian column marching west towards Wirballen (11 kilometers west of Insterburg) as well as Russian cavalry near the Pregel River. Russian artillery was sighted with two more columns of

Additional aerial reconnaissance on First Army was accomplished by Zeppelin Z-IV.

Aerial reconnaissance by Taube aeroplanes assigned to Festungfliegerabteilung (FestFA) 5 out of Königsberg kept watch over the Russian positions. The Jeannin "Stahltaube" (steel dove) was one of the Taube aeroplanes assigned to FestFA 5.

infantry moving towards Nordenburg, (16 kilometers southwest of Insterburg). West of Rastenburg, a cavalry detachment was sighted moving west. *FFA 17*'s aerial reconnaissance provided further confirmation to *AOK 8* and Mackensen that Rennenkampf's main force was not moving south. Mackensen tasked *FFA 17* with reconnoitering *GdA* v. Scholtz's *XX. Armee-Korps* area at Allenstein. The last phase of his initiative, instructed *FFA 17* to closely monitor Bischofsburg for any Russian activity. *Fliegertruppen* confirmed Russian traffic on the roads out of Ortelsburg heading towards Bischofsburg.[34]

Armed with aerial reconnaissance reports, Mackensen orchestrated the battle to his advantage. His *XVII. Armee-Korps* attacked an unsuspecting VI Corps at Bischofsburg. The Russians fought back so aggressively that Mackensen alerted *AOK 8* to send reinforcements from *Genltn* v. Below's *I. Reserve-Korps* (still marching south towards Allenstein). Below directed four battalions and three batteries of *36. Reserve-Division* and *6. Landwehr Brigade* to reinforce *XVII. Armee-Korps*. He ordered the remainder of his *I. Reserve-Korps* to continue marching south. Timely reinforcement by *I. Reserve-Korps* to Mackensen's forces, proved most effective. Blagoveshchensky's VI Corps was thrown back and forced into a state of disorganization.[35] As VI Corps fled Bischofsburg, Blagoveshchensky failed to inform either *Gen.inf* Martos' XV Corps or *Genleyt* Klyuev's XIII Corps. Now, three Russian corps found themselves at great risk.[36]

1. Kavallerie Division – the only 8. Armee force in place between Gen.kav v. Rennenkampf's First Army and Königsberg. The remainder of 8. Armee was fighting Gen.kav Samsonov's Second Army.

Generalleutnant Brecht. Kommander 1. Kavallerie Division [Commander 1. Calvary Division]

Genltn v. Below's *I. Reserve-Korps* command staff assigned their single aerial reconnaissance asset in support of *XVII Armee-Korps*. As he had no assigned aeroplanes, Below's staff worked with a solo captive balloon (*Feldluftschiffer-Abteilung Parseval und Sigsfeld Drachen-Ballon*). The stationary aerial platform provided constant updates on the battle's progress as well as information on Russian artillery batteries. At 1600 hours, the *Drachen-Ballon* observer was the first to report that *Gen.inf* Blagoveshchensky's battered VI Corps had ceased fighting and was retreating towards Ortelsburg.[37]

The following day, *GdK* v. Mackensen and *Genltn* v. Below both intended to annihilate what was left of Blagoveshchensky's VI Corps. They received indications from aerial reconnaissance that trenches were being dug by VI Corps rearguard elements. An *FFA 17* sortie reported no trenchwork or related activity. In reality, VI Corps was retreating as quickly as possible to the east. Mackensen's *XVII. Armee-Korps* was ordered to pursue VI Corps, while *Genltn* v. Below's *I. Reserve-Korps* was ordered to quickly head south to Allenstein.[38]

Northwest Front Blind to the Unfolding Drama

Neither *Gen.kav* Zhilinsky nor *Gen.kav* Samsonov were aware of Blagoveshchensky's VI Corps' disaster. Blagoveshchensky was unsuccessful in reaching them with updates on his forces. Without a full understanding of the turn of events, Zhilinsky ordered *Gen.kav* Rennenkampf to dispatch two corps towards Königsberg. The remainder of *Gen.inf* Artamonov's I Corps were to pursue "those enemy troops which do not take refuge in Königsberg." Zhilinksy was under the impression that Russian I Corps was going to cut off retreating German forces to the Vistula River. Therefore, Rennenkampf believed Königsberg remained his primary objective.[39]

The Russian advance towards Konigsberg, reminded AOK 8 that their greatest risk resided with Rennenkampf. Ludendorff

Radfahr-Kompanien [Bicycle Companies] work with 1. Kavallerie Division in distracting Russian cavalry and presenting the illusion that German forces are retreating west to the Vistula River.

wrote in his memoirs: "Rennenkampf's formidable host hung like a threatening thundercloud to the northeast. He need only have closed with us, and we should have been beaten."[40]

The evening of 26 August, *AOK 8* tasked *Festung* Königsberg's *FestFA 5* with aerial reconnaissance of Rennenkampf's First Army's known locations at Wehlau and Friedland. *FestFA 5* confirmed First Army still heading west towards Königsberg, as opposed to south in support of Samsonov's Second Army. Corroborating intelligence from both wireless intercepts and *FestFA 5* aerial reconnaissance, indicated the Russian First Army was proceeding to Allenburg, east of Königsberg. Armed with this intelligence, *8. Armee* now concentrated on fighting Samsonov's Second Army. The remaining concern was to determine Russian force disposition and where *Gen. kav* Scheideman's II Corps was heading. Wireless intercepts led the Germans to conclude that the Russians were heading south, though aerial reconnaissance did not confirm this.[41] The Airship Z-IV provided supplemental information indicating the Russians were still heading west to Königsberg. This propelled Z-IV to commence bombing Russian camps and assemblies along the Pregel River and Nordenburg area.[42] The second day of the Tannenberg battle, was to be a day of decision.

8. Armee assigned three mobile wireless intercept stations to work with 1. Kavallerie Division on responding to a timely manner to Gen.kav v. Rennenkampf's First Army.

CHAPTER VII

A Day of Decision

The Thinking of Commanders on the Eve

The evening of 26 August, *AOK 8's Armeebefehl* issued a summary of the enemy situation in stating: Russian Second Army held at their center group one or two corps at Gross-Gardienen—Waplitz with a reinforced group at Borchersdorf and one division at Usdau. Additional Russian forces were approaching Allenstein from Kurken. Hindenburg's order directed *Westgruppe*'s *GdI* v. François' *I. Armee-Korps* to close the Russian retreat to Neidenburg by taking Usdau. At the same time, *GdA* von Scholtz's *XX. Armee-Korps* was to conduct a vigorous flank attack on the Russian's center group.[1] Scholtz objected before the order was issued, due to his expectation that the Russians were preparing a major attack on his corps. He wanted to assume a defensive position. Reluctantly, Hindenburg accepted Scholtz's recommendation and looked to François to achieve the objective.[2]

Colonel Conger, U.S. Military Attache to Germany in 1927, offers a post-war analysis of this specific point in the battle while in discussion with *General* Hell, *GdA* v. Scholtz's chief of staff for *XX. Armee-Korps* as well as former *8. Armee* staff officers. Conger recorded

Prior to Usdau, Hindenburg, Ludendorff, Scholtz, and Hoffmann are together surveilling the battlefield. Source: Noskoff, Der Mann der Tannenberg verlor. Berlin 1938.

the prevailing opinion, "that after the commencement of the battle there was no specific plan from day to day; the only idea in the minds of the Higher Command was to make the best of a desperate situation. Let the future take care of itself."[3]

The concerns over *Gen.kav* v. Rennenkampf's intentions, clouded *GO* v. Hindenburg's assessment. "Danger threatens from Rennenkampf. One of his corps is reported to advance via Angerburg in the north."[4] *AOK 8* received an aerial reconnaissance report that evening, stating *Gen.kav* Scheideman's II Corps was marching around the northern shores of Grosser Mauersee. The Russians could arrive at the Tannenberg battle within two to three days. *AOK 8* was unaware *Stavka* had reassigned Scheideman's II Corps to the Russian Ninth Army, with instructions to march to Warsaw.[5]

Rumpler-Taube 4C flown by Vizefeldwebel Alfred Friedrich (above right) assigned to FFA14 during the Tannenberg battle. Promoted to Leutnant, Friedrich led Döberitz Geschwaderausbildung (squadron training) in August 1915.

That evening over dinner, *Gen.kav* Samsonov received word from *Gen.inf* Artamonov that the Russian I Corps at Usdau expected the Germans to attack with two or three divisions advancing from the northwest. Additional information from *RIVVF* revealed another German division was advancing against I Corps. Samsonov told Artamonov to hold on to the last man.[6]

Westgruppe Attacks at Usdau

AOK 8 altered their 27 August reconnaissance process by allowing *Armee-Korps* commanders to task their organic *Fliegertruppe* units. The sole exception was in the Usdau sector, where *AOK 8* held covering Samsonov's Second Army, as their highest priority. *8. Armee*'s *FFA 16* was now responsible for supporting both *Westgruppe* and *Ostgruppe*. *AOK 8* needed aerial confirmation of any Russian movements that might place *8. Armee* at risk from the flanks. Central areas of emphasis focused on Mława to the west, and Jedwabno to the east. Both *I. Armee-Korps' FFA 14* and *XX. Armee-Korps*'s *FFA 15* covered Kurken to track *Genleyt* Klyuev's XIII Corps. *Festung* Graudenz's *FestFA 6* covered the approach lines from the far point of the southwest border along the Rypin – Zielun – Herwilno – Gorzno road. In the north, *Festung* Königsberg's *FestFA 5* continued to monitor Russian First Army advances from the east, along the Pregel River.[7]

At 0400 hours on 27 August, *Westgruppe* attacked Usdau. *GdI* v. François' *I. Armee-Korps* launched a shattering bombardment with more than 100 artillery pieces firing on the entrenched Russian positions. *Gen.inf* Artamonov's I Corps suffered heavy casualties and pulled forces back. *I. Armee-Korps* entered Usdau at 1100 hours. *8. Armee* now commenced cutting off the Russian retreat line at Neidenburg-Willenberg. Military historians often consider this to be the inflection point of the Tannenberg battle.[8] *Westgruppe's* taking Usdau, proved the most important *8. Armee* victory. It fulfilled the goal of enveloping the Second Army's center group. Usdau heights were the springboard for closing *Gen.kav* Samsonov's Second Army's core center—exit route to the south—the key to *GO* v. Hindenburg's and *Genmaj* Ludendorff's battle plan.[9]

SORTIE DATE: 27 AUGUST 1914

AEROPLANE: A [Taube] 152

PILOT: Lt v. Winckler

Ground track: Dtsch.Eylau-Heinrichsdorf-Soldau-Mława-Soldau-Thalheim-Dtsch.Eylau.

SUBMITTED: Dtsch.Eylau 27.8. 6:30 pm. to Gen.Kdo.I (General Command I Armee-Korps)

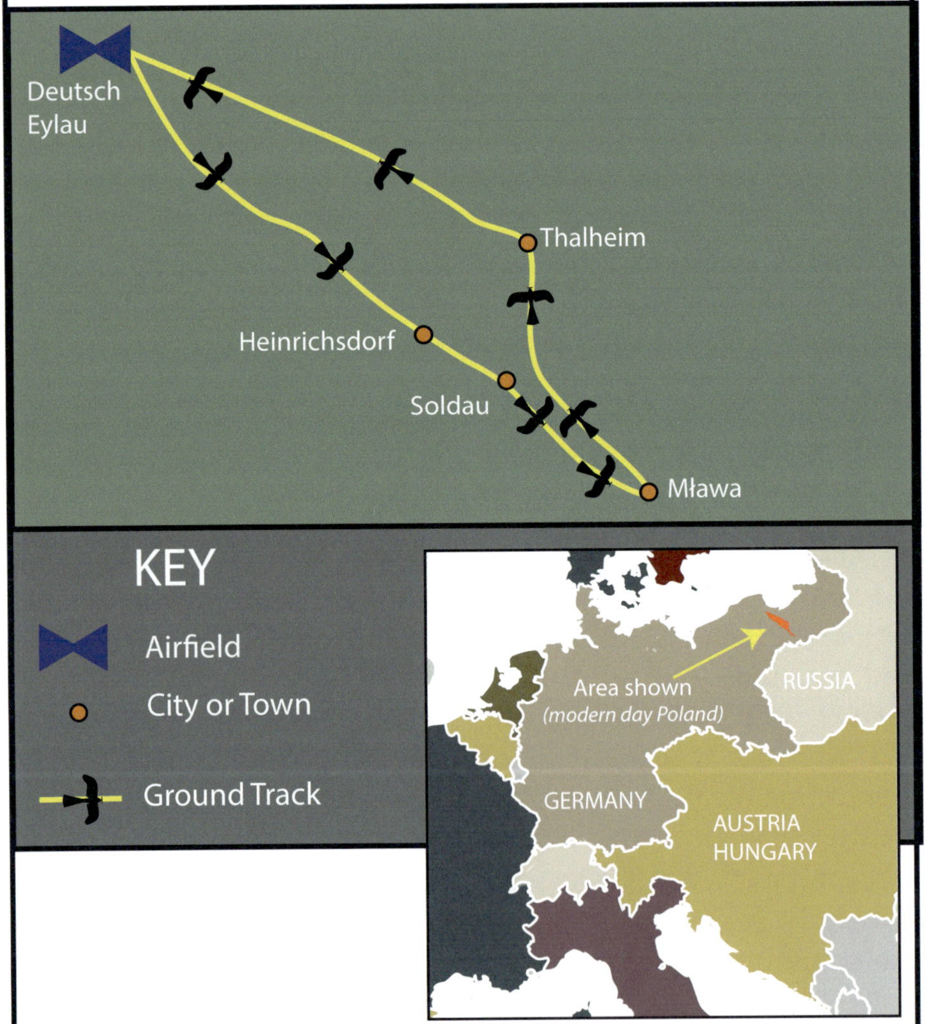

SORTIE DATE: 27 AUGUST 1914

FFA 14 flew a sortie the afternoon of 27 August to surveil the rear echelons behind Usdau. As GdI v. François was about to press on to Neidenburg, he was alerted to an FFA 14 aviator's report and wireless intercepts, indicating Russian forces at Soldau were being reinforced.

Source: RG 165

SORTIE AVIATOR REPORTS: 27 AUGUST, 1914

1. Strong enemy troop concentration at RR Station Mława and RR Station Illowo, estimated strength about 1 division. Numerous railroad cars at both RR Stations, though no railroad traffic.
2. Strong enemy columns short distance between each on road from Soldau to RR Station Mława, most of them seem to be vehicles. Apparently marching on Mława.
3 Infantry trenches between Soldau and Kurkan, occupied by Infantry and M.G. (machine guns)
4. Near Hohendorf and Priom west of Soldau about 2 Infantry Companies each, at and around Wierslau about 1 Btl.; roads Soldau-Gr.Kuslau-Usdau free of enemy.
5. Several aeroplanes at Illowo.

—v. Heyden, Observer

In the late afternoon of 27 August, *Festung* Graudenz's *FestFA 6* aviators (flying over southwest *Ostpreussen*) observed large columns of Russians, retreating. One of the highlights of their report, was in their noting columns south of Soldau to Illowo, as well as six Russian aeroplanes on railroad cars. This suggested the Russians were removing high-value aeroplanes from the battlefield.[10] *AOK 8* received *FestFA 6*'s report at 0135 hours on 28 August.[11] There was no confirmation that the document was forwarded to *GdI* v. François' *I. Armee-Korps* until after the events of 28 August. This impacted the significance of *FestFA 6*'s report. Subsequently, François was forced to make a decision, absent timely information. This resulted in a lost opportunity to cut off the Russian retreat and change direction (away from Neidenburg), which would have made victory more certain.[12]

13th KAO's Poruchik Alexander Viktorovich Nikolaevsky was shot down and captured by I. Reserve-Korps soldiers on 27 August while trying to deliver a message to Gen.inf Blagoveshchensky's VI Corps.

As *GdI* v. François was about to press on to Neidenburg, he was alerted to an *FFA 14* aviator's report and wireless intercepts, indicating Russian forces at Soldau were being reinforced. Logistical stocks were sighted at Illowo and Mława rail depots.[13] Despite delays due to inclement weather, *8. Armee War Diary* stated, "Aviators have observed about one Russian Army Corps on the road Mława-Soldau. Head of forces at Soldau and at the railway stations of Mława and Illowo. From 10 to 12 railway trains under steam at the railway depot Illowo."[14]

Subsequent *FFA 14* reports created confusion due to conflicting data. The first report estimated a Russian corps encamped near Soldau and Borchersdorf. The report that followed, covered the same sector, only observing a few battalions of infantry and vehicles in columns. Immediate clarification was sought, and *FFA 14* was ordered back to Soldau.[15] François considered it imperative to shift his priorities by turning *I. Armee-Korps* southwards to Soldau, vice Neidenburg.[16] *I. Armee-Korps'* maneuver propelled *Genmaj* Ludendorff to fear that François' actions were going to expose *8. Armee* to a fatal Russian attack to the flank.[17]

At 0700 hours, *8. Armee*'s *FFA 16*'s aircrews reported Russian troops moving westward out of Hohenstein. German postwar analysis indicated this force was *Gen.inf* Martos' XV Corps. Russian wireless was intercepted, revealing the XV Corps' mission was to encircle *8. Armee*'s left wing. Additional wireless intercepts revealed *Genleyt* Klyuev's XIII Corps was to support the attack on Martos' right. Klyuev's XIII Corps was detected 10 kilometers south heading in the direction of Allenstein.[18] Klyuev had intelligence that German troops were 15 kilometers east of Allenstein, but he did not trust his sources. One of his divisions had just entered Allenstein. *Gen.inf* Blagoveshchensky's VI Corps was supposedly in the Wartenburg area. In reality, *Genltn* v. Below's, *I. Reserve-Korps* occupied Wartenburg. In order to establish contact with Blagoveshchensky, Klyuev made an attempt to communicate with him using *13th KAO*.[19] *Poruchik* Alexander Viktorovich Nikolaevsky was tasked. He took off in a Nieuport IV, (No. 130) and did not return.[20] German sources recalled:

> About 1700 hours, a Russian aeroplane appeared over Wartenburg, where General Below stayed. It was the first one seen since the corps had departed to the west. Fierce firing began from every road and square and out of the windows and the marching troops. The flyer could have revealed the approach [of the Germans] and was shot down. The gas tank was hit and the pilot forced to land. He refused to give any statement.[21]

An Ally Flies for RIVVF

The French pilot, *Sous lieutenant* Alphonse Flavien Louis Poiré, was a unique figure in the unfolding drama of August 1914. His military career included an assignment with the *67eme regiment de infanterie*. When war was declared Poiré was in Russia. His services as an aviator were accepted by *RIVVF* and he commenced aerial observations in a Farman HF.16 for the Second Army with *21st KAO*. For the next two years, he flew with the Russians, accumulating over 100 combat sorties and 200 flying hours. Outside of Russia, *Sous lieutenan*t Poiré was one of the few well-known aviators. The other, was *Shtabs-kapitan* Nesterov, who flew over the Southeastern Front in 1914. As a result of his contacting allied liaisons (such as Lieutenant-Colonel Alfred Knox), his place in history was preserved. Knox mentioned Poiré in his account of the unfolding Tannenberg battle. On one sortie (the morning of 27 August) Poiré flew northwest of Neidenburg and calculated three Germans corps in the area. Knox and Poiré concluded Germans aimed to contain the Russian center and right flank while pushing back to the left, to cut lines of communication with Mława and Warsaw. Poiré related to Knox "…the German guns are in pits. It looks as if they are holding the Russians' center and right and perhaps pushing round their left to cut the line of communication from Prasnish-Mlava-Neidenburg."[22]

French Sous Lieutenant Alphonse Flavien Louis Poiré flew for RIVVF for most of the war. His initial assignment was with 21st KAO supporting Second Army. He is pictured in front of a Farman HF.16.

Aviation Loses an Icon

At this climactic moment, the German *Luftschiffertruppe* experienced their first major loss. Around midnight of 27/28 August, Zeppelin Z-V departed Posen for a reconnaissance and night bombing mission on Mława and Illowo.[23] It did not return. Lieutenant-Colonel Knox, British attaché, reported from Mława early that morning:

> Rifle-firing started all round the station. We ran out to see an enormous Zeppelin hovering at a height of about 900 to 1,000 metres in the sun. It looked so extraordinarily peaceful! Suddenly it threw four bombs, one after the other, in quick succession. The loss was six killed and fourteen wounded but it might have been far greater, for the station was crowded. I picked up a piece of one of the bombs. The Zeppelin hovered round and finally sailed away. Infantry firing proving useless, soon stopped, and a battery came into action, doing good work. One was filled with impotent rage against the machine, and it was with genuine delight that I heard it had been brought down and its crew captured.[24]

As Z-V approached the Mława railway junction early that morning, a barrage from an ensemble of Russian infantry, machine gunners, and artillery, struck the airship. Russian cavalry chased the slowly descending Zeppelin. Upon forced landing, Z-V was set on fire by the air crew so the Russians couldn't add it to their own fleet. On 29 August, *FestFA 6* Graudenz reported the skeleton of the wrecked airship near Liepowiec (Lipovitz).[25] *8. Armee War Diary* recorded, on the evening of 28 August, "Z-V after an emergency landing at Mława, has been burned by the crew. The crew has been captured."[26] The Russians salvaged substantial remains of Z-V and sent them to Petrograd for examination. *Hauptmann* Ernst Grüner, Z-V's commander, and his nine crewmen, were captured and sent to a Siberian prisoner of war camp. All but one of the crew died in captivity.[27] As fate would have it, *Major* von der Lieth-Thomsen, the first *Feldflugchef* [Chief of Air Service], was serving in the opening weeks of war as an observer and tactical commander for Z-V. The day before the ill-fated mission, he was transferred to serve as *Erster Generalstabsoffizier – I.a.* [Operations] of the newly established *XXIV. Reserve-Korps*. Z-V's loss prompted German commanders to rethink the utility of airships over the battleground. If aviation's goal was to enhance maneuvering army operations, it required more than a degree of stealth. Airships were visually vulnerable and less reliable than aeroplanes in supporting an ongoing battle. Their involvement inevitably resulted in the enemy gaining too great an advantage by being privy to the intelligence collection from the sortie's flight path.[28]

Absence of Valuable Information

Gen.kav Samsonov lacked a clear understanding of what his army faced. Both military plans as well as their implementation, failed to meet Samsonov's main objective –to destroy *8. Armee*. *RIVVF* aviators were in the air, reporting on German forces. They observed Scholtz's *XX. Armee-Korps* in motion, revealing regimental columns of infantry with artillery moving from Osterode to Hohenstein. Another regimental infantry column was observed moving from Seewalde to Waplitz. Extensive enemy traffic was sighted between Usdau and Gilgenburg. Despite these critical observations, Russian command failed to take advantage of *RIVVF* in their maneuvers.[29]

Samsonov's lack of knowledge of his own corps overshadowed his ignorance of enemy force locations. *Gen.inf* Artamonov's I Corps was in a hasty withdrawal following the defeat at Usdau, leaving

Samsonov's Second Army's left flank vulnerable. When Samsonov learned of Artamonov's failure, he ordered *Genleyt* Alexander Alexandrovich Dushkevich to take command of I Corps.[30] *Gen.inf* Blagoveshchensky's VI Corps was being defeated at Bischofsburg by *Ostgruppe*. Samsonov's staff assumed his VI Corps was marching on Allenstein.[31] Postwar analysis revealed *Gen.kav* Zhilinsky (on the evening of 27 August) was aware that *8. Armee* had achieved success at Bischofsburg, Gilgenburg, and Soldau. Zhilinsky then ordered Rennenkampf to maneuver his left flank to the south to aid Samsonov. Zhilinsky's command came too late. The Second Army's fate was sealed.[32]

Dramatic – but far from the truth. A contemporary British publication recalls the shooting down of Zeppelin Z-V near Mława. A barrage from an ensemble of Russian infantry, machine gunners, and artillery, struck the airship with cavalry chasing the slowly descending Zeppelin.

CHAPTER VIII

Envelop Through Attack!

Intelligence Updates AOK 8's Plan

8. *Armee*'s strategy was informed by new intelligence and subsequently evolved on 27 August. *FFA 15* aviators subordinate to *GdA* v. Scholtz's *XX. Armee-Korps,* reported that evening, that Russian columns were approaching from the east. Scholtz interpreted *FFA 15*'s observations as indications that the Russians were preparing for a northern outflanking move. He wanted to remain in a defensive posture but faced objections from *GO* v. Hindenburg and *Genmaj* Ludendorff. They wanted *XX. Armee-Korps* to launch an all out attack against Russian positions. Hindenburg and Ludendorff met with Scholtz and discussed the proposed plan. Scholtz recommended to Hindenburg that he shift one of his first line divisions north to counter the anticipated Russian envelopment. Hindenburg reluctantly agreed. *37. Infanterie-Division* was subsequently tasked to advance to the northern wing of *XX. Armee-Korps*. Hindenburg's preference was to launch a strong thrust against Russian troop concentrations around the southern and eastern banks of Lake Mühlen and ten kilometers south of Hohenstein. *GdA* v. Scholtz's *XX. Armee-Korps* was reinforced to attack with *Genmaj* Sontag's *41. Infanterie-Division, Genltn* v. Morgen' *3. Reserve-Divsion, Genmaj* v. Unger's *Landwehr-Division,* and *Genmaj* von der Goltz's *Landwehr-Division v.d. Goltz*.

An FFA 15 aviator's flight uniform reflects Fliegertruppe unit identification with the 15 sewn on the left shoulder. This unknown aviator also was a recipient of the iron cross.

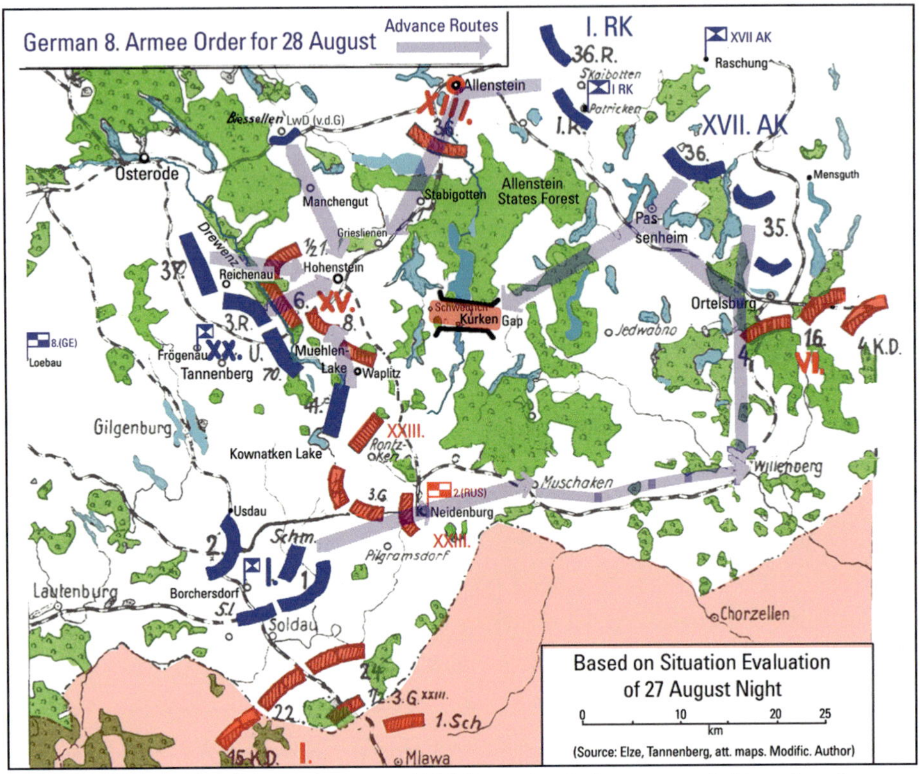

The envelopment of the XIII Corps and XV Corps by 8. Armee was initiated at the battle of Hohenstein. Note VI Corps was still retreating south. Source: Helmut Jäger Larger map available at shootingthfront.com.

The village of Waplitz, close to the eastern edge of Lake Mühlen, was to be *41. Infanterie-Division*'s first objective. Waplitz was an intermediate target on the way to Hohenstein. *Genltn* v. Morgen's *3. Reserve-Division* was to advance north of Lake Mühlen towards Hohenstein.[1] In the south, *GdI* v. François' *I. Armee-Korps* was heading towards Soldau - still not in position to close the exit from Neidenburg to Willenberg. He was still unaware of *Festung* Graudenz's *FestFA 6*'s report that had been generated the previous evening.[2]

That evening, *AOK 8*'s *Armeebefehl* stated: Reinforced *XX. Armee-Korps*, *Landwehr-Division* Biesellen and *I. Reserve.Korps* and *XVII. Armee-Korps* will envelop the Russian XIII. and XV. Army Corps through attack. *XX. Armee-Korps* and *3. Reserve-Division* will attack the enemy from the right flank near Hohenstein, at 0400 hours.[3]

In his memoirs, Hindenburg recalled the decision-making process that night. "The situation is clear for us; in the evening we issue the order to envelop the center core of the enemy, the Russian XIII Corps and Russian XV Corps."[4] Ludendorff's post-war memoirs addressed the plan in more detail and his statement exuded his confidence to win:

> Nothing remained to be done on the 28th, but to issue orders for the *I. Armee-Korps* to occupy Neidenburg. In the meantime, the Korps had already made a turning movement in that direction. *XX. Armee-Korps* was to carry out the attack, which had been fixed for the 27th and more specifically to push forward the *41. Infanterie-Division. Landwehr-Division v.d. Goltz* was to attack Hohenstein, due to the Russians occupying Allenstein; the division had to detrain at Biessellen and Osterode. [Goltz's forces were assigned to *8. Armee* and had left Schleswig-Holstein on 25 August.] *I. Reserve-Korps* and *XVII. Armee-Korps* were moved up west, covered against attack from Ortelsburg to positions on a line running from Allenstein to Passenheim."[5]

AOK 8 was focused on attacking Samsonov's Second Army center core. *Ostgruppe's* pursuit of *Gen.inf* Blagoveshchensky's Russian VI Corps required new *AOK 8* guidance. Hindenburg envisioned *GdK* v. Mackensen's *XVII. Armee-Korps* as the optimal *8. Armee* force, to effectively block the Russian escape routes from the impending 28 August battleground. At 2100 hours on 27 August, *AOK 8* staff at Löbau, was updated by a *I. Reserve-Korps* report. It detailed a reconnaissance patrol's discovery that a Russian force (the strength of one division) occupied Allenstein. More Russians were observed entering the town. The staff surmised that the force was part of *Genleyt* Klyuev's XIII Corps, waiting to join *Gen.kav* v. Scheideman's II Corps. *Genltn* v. Below's *I. Reserve-Korps* staff recognized *Ostgruppe* was at great risk from a decisive Russian maneuver, originating from Allenstein. The information was quickly forwarded to *AOK 8* and a last-minute update to the *Armee-Befehl* was generated for *Ostgruppe*. Allenstein needed to be purged from the Russians.[6]

In the early morning of 28 August, *Genmaj* Ludendorff and selected *AOK 8* staff officers, relocated to a forward operating location at the village of Frögenau, near Scholtz's *XX. Armee-Korps* command post. Ludendorff recalled, "A very ineffective field telephone connected us with *I. Armee-Korps* but no communications at all with the other forces existed." The absence of sufficient communications from Frögenau meant Hindenburg lacked up-to-date situation intelligence and was unable to control the evolving battle by direct orders. Ludendorff's communications with *Ostgruppe's XVII. Armee-Korps* and *I. Reserve-Korps* remained broken until later that afternoon.[7]

Ludendorff's first priority upon arrival at Frögenau, was to confirm through intelligence, whether or not there was any indication that *Gen. kav* v. Rennenkampf's First Army was diverting south to aide Samsonov's Second Army. *Festung* Königsberg's *FestFA 5* reported at 1000 and 1100 hours that First Army was moving west and that no forces were observed heading south to reinforce Second Army.[8] *8. Armee War Diary* reported, "Wireless messages confirm that the Russians assure that *I. Armee-Korps* [François] retiring to Königsberg and *XVII. Armee-Korps* [Mackensen] to Friedland."[9] First Army advanced infantry units were able to proceed to a line about 25 kilometers southeast of Königsberg. It was and remains, historically, Rennenkampf's deepest penetration into Germany. A single engagement occurred when Russian cavalry clashed with *8. Armee*'s *1. Kavallerie Division* about 15 kilometers southwest of Rastenburg.[10] Hindenburg could now proceed with confidence to achieve victory on the Tannenberg battlefield.

Russians Defeated at Hohenstein

In accordance with the direction of the 27 August *Armeebefehl, GdA* v. Scholtz reinforced *XX. Armee-Korps* with *Genltn* v. Morgen's *3. Reserve-Division,* attacked *Gen.inf* Martos' XV Corps near Hohenstein and Lake Mühlen at 0400 hours on 28 August.[11] *Genmaj* Sontag's *41. Infanterie-Division* advanced in dense fog into Waplitz and met intense fire from Russian infantry occupying the buildings. The battle for Waplitz became a furious hand-to-hand combat. As fog lifted, Russian artillery proceeded to decimate *Genmaj* Sontag's forces. *41. Infanterie-Division* broke and fled ten kilometers to the west. *AOK 8*'s initial strategic plan faced a catastrophic dilemma. Scholtz was hesitant to further commit his forces, fearing a strong Russian counterattack. He learned that *Genltn* v.

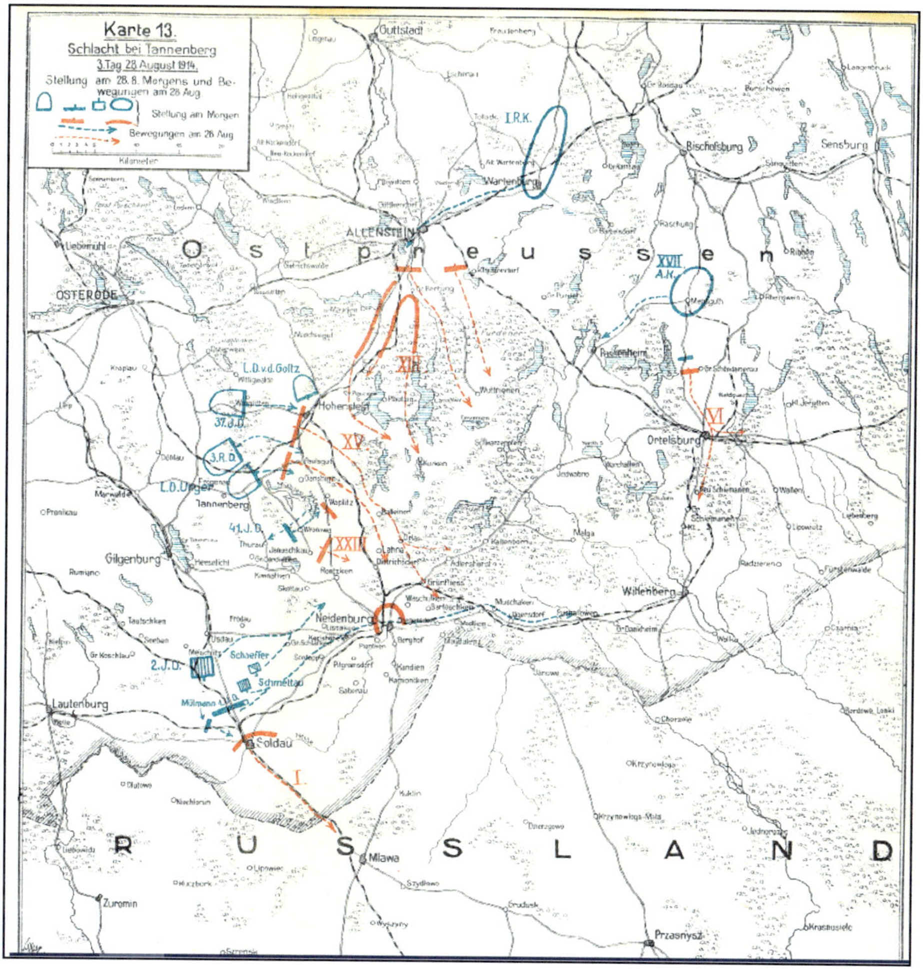

Gdl v. Francois portrays the battlefield for 28 August 1914—the Russian defeat was underway with 8. Armee's envelopment of XIII Corps and XV Corps. Source: Francois, Herrman: Mavneschlacht und Tannenberg. Berlin 1920. Larger map available at shootingthfront.com.

Morgen's *3. Reserve-Division* was successfully advancing at the northern shore of Lake Mühlen. Morgen's actions spurred Scholtz to order the remainder of *XX. Armee-Korps* to advance.[12] The largest engagement of the Tannenberg battle commenced with the fight to occupy and control Hohenstein.

Shortly after the *AOK 8* command staff arrival on the Frögenau heights, they received a first situation update through the only existing telephone line. *GdI* v. François reported that he had set his *I Armee-Korps* cavalry forces on Neidenburg, but his *1. Infanterie-Division*

SORTIE DATE: 29 AUGUST 1914

AEROPLANE: LVG B.1

PILOT: Lt. Kammetz

GROUND TRACK: Osterode-Hohenstein-Dorotkowo-Kurken-Adlershorst-Neidenburg-Hohenstein-Osterode

SUBMITTED: 9 am K.

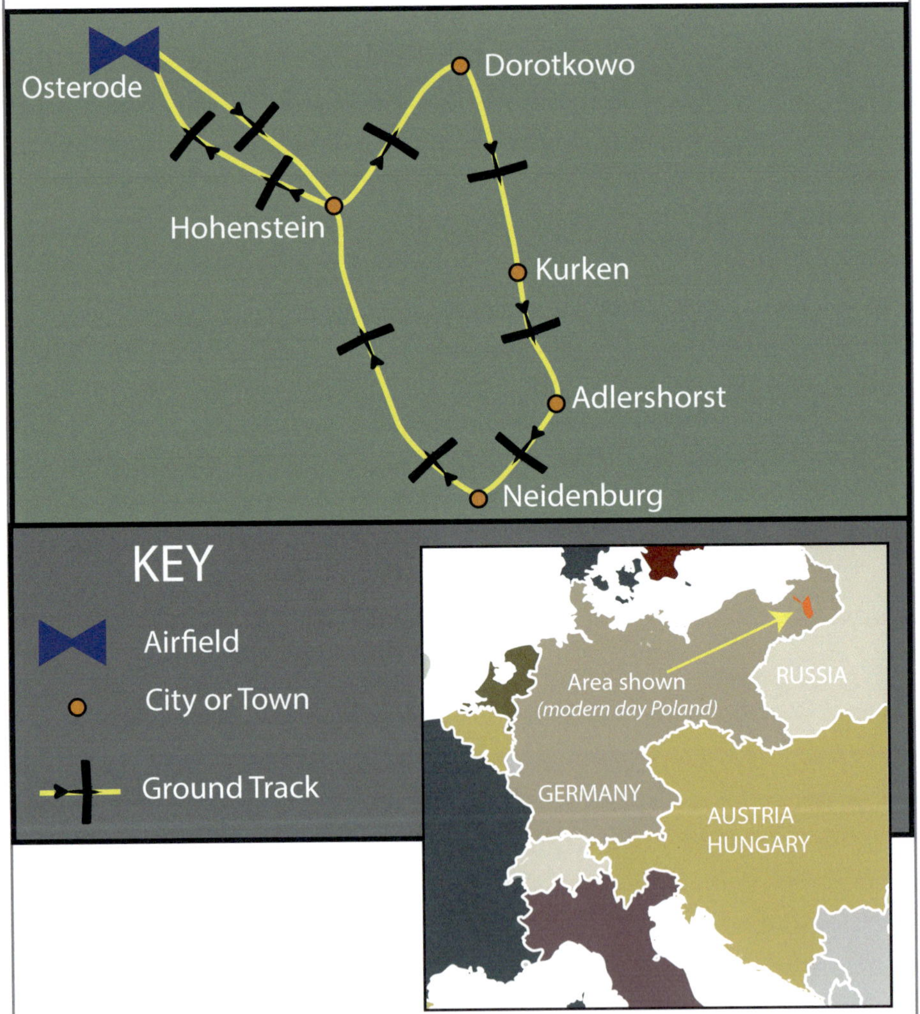

SORTIE DATE: 29 AUGUST 1914

AOK 8 now focused on cutting off any and all Russian attempt to escape. 8. Armee's FFA 16 flies a critical early morning sortie determining locations of XV Corps, XIII Corps and XXIII Corps prior to final envelopment. Source: NARA RG 165

SORTIE AVIATOR REPORTS: 29 AUGUST, 1914

1. 6:15 am. Situation at Hohenstein unable to view. Ground fog and Smoke.
2. 6:30 am. column of all weapons x) on road Allenstein-Hohenstein marching on Hohenstein with begin halfways Stabigotten-Grieslienen, ending at Dorotkowo. Shortly southwest RR-Stat. Stabigotten troop assembly strength 1 Inf Regt. seems to also begin march on Hohenstein.
3. In Forests Parden, Ramuker & Lanskerofen no troops visible (6:30 - 6:50 am)
4. On paths between Dietrichsdorf and Schwedrich leading in general direction to east, from 6:50 and 7:20 am troops maneuvering - columns of all weapons in marching-, assembling-, or combat-order - in action resp. either advancing from or retreating to east, main parts of our forces already east of railroad Neidenburg -Jakobsthal.
5. 8 am. Situation at Hohenstein (map 1 : 100 000) seams to be as follows:

   I. Artillery positions

   a) between Mörken and Amerika (at road Hohenstein-Grieslienen), facing to east and to northeast. [Amerika was the name of a mansion at that location.]

   b) west of Grünau and west of Grün-Mühle, facing to west and southwest.

   II. Infantry and baggage cars moving between small forests halfways Mörken-Grünen heading to Grünen, likely Russian. Own Infantry, seem to advance in lines to east, southeast of Amerika.

   III. Apparently 1 own Division developing for combat advancing to eastnortheast from line east corner Lake Mispel - Hohenstein.

   IV. Terrain northeast Hohenstein covered by smoke, impossible to look through.

—Leutnant v. Wagner, Observer

was still fighting at Soldau against weak Russian opposition. Hours later, Hindenburg and staff learned that *Genmaj* Sontag's *41. Infanterie-Division* was retreating from Waplitz. *AOK 8* now felt a heightened concern for *8 Armee*'s positions. The withdrawal of *Genmaj* Sontag's *41. Infanterie-Division,* created a gap from Lake Mühlen between *XX. Armee-Korps* and *I. Armee-Korps*. *Gen.inf* Samsonov's Second Army was now perfectly positioned to exploit the maneuver and upend *8. Armee*'s battleplan.[13] In order to remedy and counter the situation, *GO* v. Hindenburg immediately directed *GdI* v. François to move the *2. Infanterie*-Division and all unengaged available forces to the village of Rontzken, to close the gap.

Limited and ineffective communication between *AOK 8* and most of the subordinate corps' commanders, resulted in incomplete situation updates throughout the morning. *8. Armee*'s fate appeared to be

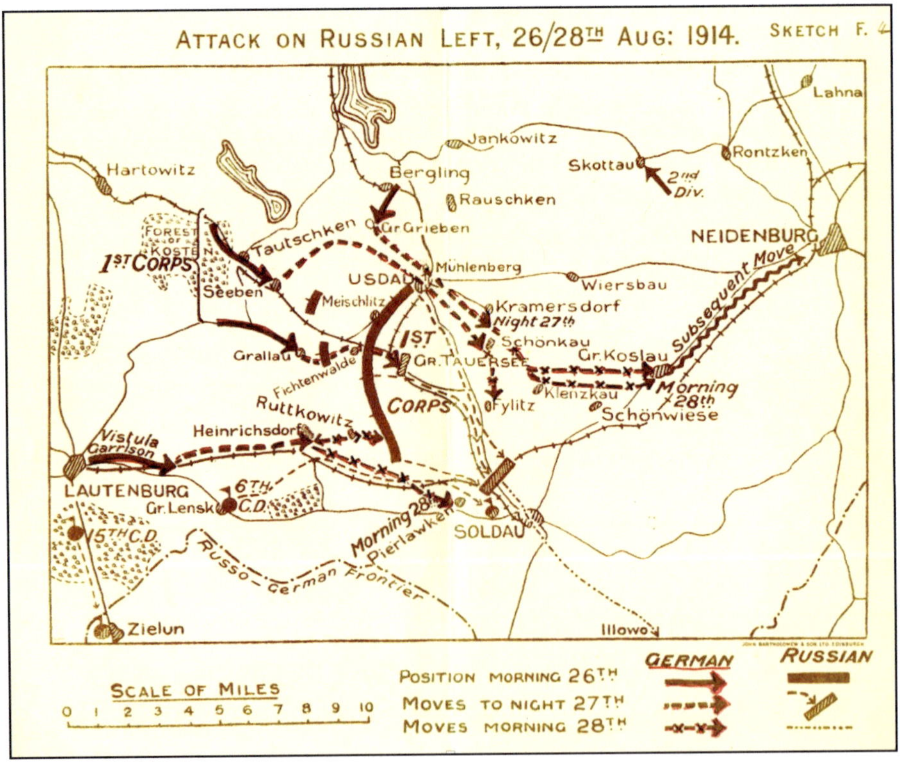

Westgruppe's I. Armee-Korps formed a thin line of pickets—25 battalions stretched over 60 kilometers cutting off the Russian escape route to the south. Larger map available at shootingthfront.com.

at risk. Hindenburg was concerned with *GdI* v. François' independent thinking and how it might impede, victory. As a result, he penned a personal handwritten request to François admonishing him that he: "...could render outstanding services to *8. Armee*, if he acted in accordance with the commander-in-chief's intent."[14] Hindenburg's concerns proved to be unfounded. François was already complying and leading his *2. Infanterie-Division* in battle with *Gen.inf* Kondratovich's XXIII Corps at Rontzken. *1. Infanterie-Division* was redirected to Neidenburg, after it became evident that the Russian position at Soldau was merely covering their retreat. The southern exit from Neidenburg to Willenberg was still open and most of the XXIII Corps succeeded in disappearing through the woods and reaching the *Ostpreussen* border. As noon approached, François received new orders to support Scholtz's *41. InfantrieDivision*, still moving west after the struggle in Waplitz, that morning. *8. Armee War Diary* concluded in dramatic fashion at 1145 hours, "Everything depends on the *I. Armee-Korps*."[15]

An Ostgruppe Commander Confrontation

In light of the late intelligence confirming Russians at Allenstein, *GO* v. Hindenburg shifted priorities for *Ostgruppe*. *I. Reserve-Korps* and elements of *GdK* v. Mackensen's *XVII. Armee-Korps* were to stop chasing *Gen.inf* Blagoveshchensky's VI Corps and proceed west to attack the newly discovered Russian force at Allenstein. *GdK* v. Mackensen's *XVII. Armee-Korps* was to remain the optimal *8. Armee* force to block the Russian escape routes from the 28 August battleground. Communication between Mackensen's staff and *AOK 8* was still broken. *AOK 8* expected Below's staff to pass the revised *Armee-Befehl* to Mackensen. *Genltn* v. Below's chief of staff, *Oberst* Posadowsky-Wehner, was tasked with personally briefing Mackensen on change in plans. He drove 15 kilometers from Below's headquarters to *GdK* v. Mackensen's command post at Raschung with what he presumed was the *AOK 8 Operations-Befehl*. He proceeded to brief Mackensen.[16]

Posadowsky-Wehner claimed that Hindenburg directed Mackensen to coordinate with Below on retaking Allenstein.[17] The resulting discussion between Mackensen and Posadowsky-Wehner became contentious. Mackensen stated his primary objective was to pursue *Gen.inf* Blagovesh-

SORTIE DATE: 29 AUGUST 1914

AEROPLANE: A.36/14 (Taube)

PILOT: Lt. Toop

Ground track: Dtsch.Eylau-Gilgenburg-Neidenburg-Ortelsburg-Willenberg-Chorzele-Janowo-Muschaken-Neidenburg-Dtsch.Eylau.

Submitted: Dtsch.Eylau 29.8. 6:35 pm

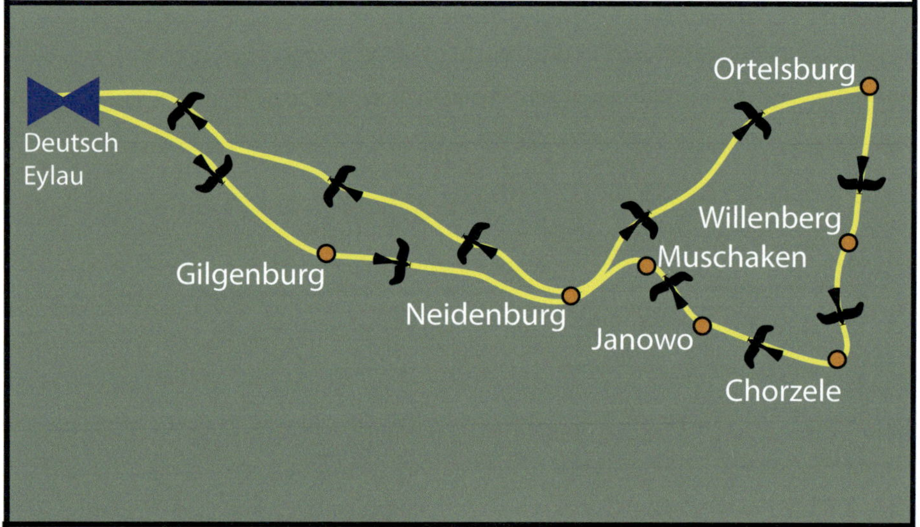

SORTIE DATE: 29 AUGUST 1914

I. Armee-Korps' FFA 14 covers the southern flanks of Samsonov's Second Army to determine exact location of picket line cutting off the Russian retreat as *8. Armee*'s envelopment proceeds.

Source: NARA RG 165

SORTIE AVIATOR REPORTS: 29 AUGUST, 1914

To Gen.Cmd. I. Armee-Korps

1. At Neu-Borowen, 5 km southwest Jedwabno, trenches, around 1-2 Batallion-Groups. Trench occupation not identified.
2. On road Ortelsburg-Jedwabno one vehicle column.- identified as food supplies - length about 6 km, moving on Worschallen. Small infantry march column about 2 btls. marching from Braunichen to Worschallen.
3. Ortelsburg free - Olschienen, Kl.Serutten burning. Road Ortelsburg-Willenberg-Chorzele free. Janowo-Chorzele free. Janowo-Muschaken. In forest area between roads Ortelsburg-O.F. Kaltenborn - Willenberg smoke ascending from many places in the woods, in my opinion not originating from forest burnings (perhaps troops in bivouacs).

—Majewski

OLt F.Flg.Abtlg.14 (FFA 14)

chensky's VI Corps. They were prevented from confirming details with *AOK 8,* as the communication network remained inoperative.[18] Posadowsky-Wehner was incorrect in stating that *AOK 8* wanted Mackensen to commit both of his divisions. Mackensen's forces were to advance on Ortelsburg while at the same time, attack Allenstein. Both locations were deviations from *XVII. Armee-Korps*' current direction.[19] Below had sent his personal chief of staff to meet with Mackensen. Mackensen knew the chief of staff and respected Posadowsky's reliability. He therefore accepted the *AOK 8 Operations-Befehl,* in question. Throughout the night, Mackensen hoped that *AOK 8* would change the *Operations-Befehl* to allow all of *XVII. Armee-Korps* to continue pursuit of VI Corps. That did not occur. He finally gave in and agreed to turn his two divisions west towards Allenstein early in the morning of 28 August.[20]

Ostgruppe Enters the Envelopment

On the eastern edge of the battlefield at 0830 hours on 28 August, an *FFA 17* observer noted a division-strength Russian infantry column at Fürstenwalde heading south, followed up 45 minutes later with a sighting of another column of equal strength heading to Friedrichshof. *FFA 17*'s reports confirmed to Mackensen, that *Gen.inf* Blagoveshchensky's VI Corps was retreating southeast, and was not a threat to ongoing operations at Hohenstein. Mackensen's *XVII. Armee-Korps* proceeded to Allenstein, aware that VI Corps would not regroup and attack his forces from the rear.[21]

8. Armee's forward command at Frögenau, received an important intelligence update early in the morning of 28 August. *8. Armee War Diary* noted that a wireless intercept message was received from Radio Station *Gouvernement Festung* Thorn, "At 0815, Army Headquarters receives information from an intercepted wireless message that the Russian XIII Army Corps is reported to be in Allenstein. XIII Army Corps lead element will arrive at 12 noon in Grieslienen to join Russian XV Corps." At 1030 hours, an *FFA 16* aircrew flying in an LVG B.I, confirmed *Festung* Thorn's intercept by reporting a long Russian column marching from Allenstein on Hohenstein.[22] At approximately the same time, a cavalry patrol reported to *Genltn* v. Below's staff, that Allenstein possessed minimal Russian troops. Below concluded Russian forces had departed to the south. *I. Reserve-Korps*' mission was to catch up and destroy them from the rear.[23]

As *XX. Armee-Korps* pursued the Russians, Third Reserve-Division infantry shot down an *RIVVF* aeroplane. *Hauptmann* Rosenthal, commander of *5.Kompagnie/Reserve-Brigade 49* recalled, "A Russian aeroplane flies above. It has already been detected by our *Leutnant of the Bicycle Kompagnie*. What's that guy above us doing here? The *Leutnant* at once commands: "Dismount! Apply a good lead! Aim well! Fire!" The flyer comes down. "Let others bury him, we have no time!"[24]

Festung Thorn's wireless intercept also clarified that *Landwehr-Division v. d. Goltz* was at risk, as it marched from Biesellen to Hohenstein. *Genmaj* v. d. Goltz knew that the Russians were at Allenstein, but he was unconcerned. As *Landwehr-Division v. d. Goltz* approached

Hohenstein, the advanced units came under fire from *Genleyt* Klyuev's XIII Corps.[25]

Based on the Thorn intercept and not the cavalry report, *AOK 8* ordered *Genltn* v. Below to change direction and advance to Zasdross – Stabigotten—attacking the Russians before entering the ongoing Hohenstein battleground. Limited communications with *I. Reserve-Korps* prompted *AOK 8* to immediately task an *FFA 16* sortie to deliver an updated *Operations Befehl* via Flieger-Meldung. The LVG B.I left Osterode and dropped the message on *I. Reserve-Korps* units at 1130 hours.[26]

GO v. Hindenburg was operating under the mistaken assumption that *XVII. Armee-Korps* was part of *Genltn* v. Below's force enroute to Allenstein. *AOK 8* was confident that Below was keeping Mackensen informed of any changes. At noon on 28 August, Below's messenger suggested that in light of the "changed situation," *XVII. Armee-Korps* should march south to cut off any Second Army efforts to escape to the east. Below suggested *XVII. Armee-Korps* turn around and resume their original route towards Ortelsburg and Passenheim.[27] Mackensen flew into a rage. Not only had he already acquiesced to Below's suggestion to cease chasing VI Corps, but Mackensen was now being asked to reverse direction, once again. The discussion became personal, when a senior General officer ordered Below to clear the main road. If Below had agreed to retrace *XVII. Armee-Korps'* march route, he would have risked his *I. Reserve-Korps* being taken for the enemy, which could have resulted in Germans accidentally firing on Germans. Still enraged, Mackensen ordered *Hauptmann* Bartenwerffer, *Generalstab* member of *XVII. Armee-Korps*, to fly to Frögenau and personally demand clarification from *AOK 8*. An *FFA 17 Fliegertruppe* pilot flew Bartenwerffer in an Albatros biplane to Frögenau. Ludendorff, distraught over the current state of *8. Armee*'s disposition of forces, was angry and on edge. It became apparent, that *Ostgruppe* was confused, behind schedule, and might not attack that day. At that precarious moment, *AOK 8* remained under the impression that *XVII. Armee-Korps* was pursuing *Gen.inf* Blagoveshchensky's VI Corps.[28]

Hauptmann Bartenwerffer was ordered to fly back and drop an order over Below's *I. Reserve-Korps'* troops to attack *Genleyt* Klyuev's XIII Corps at Grieslienen without delay. Despite the *Flieger-Meldung* being dropped at 1630 hours on *I. Reserve-Korps* troops, Below did not see the order until 2000 hours. *I Reserve-Korps* had been fighting the Russians since early afternoon. Bartenwerffer also delivered orders for Mackensen to turn off from Allenstein and expedite the closing of the road from Ortelsburg to Neidenburg. *Fliegertruppe* worked to clarify the orders, and resolve any and all of the miscommunication between *AOK 8* and *Ostgruppe*. *Genmaj* Ludendorff managed to locate a working telephone line, and gave instructions to both Mackensen and Below.[29] Armed with clear guidance, both now acted in accordance with *8. Armee*'s intentions.[30]

The Envelopment is Complete

GdA v. Scholz's *XX. Armee-Korps* combined infantry advances with devastating artillery barrages.

3 Reserve-Division and *Landwehr-Division* Unger attacked *Genleyt* Klyuev's XIII Corps and *Gen.inf* Martos' XV Corps from the west, while *Genltn* v. Below's *I. Reserve-Korps* advanced from the northeast. *AOK 8* ordered pursuit against the fleeing Russians. Both XIII Corps and XV Corps surrendered. Upon hearing of their surrender, *Gen.inf* Kondratovitch, commander of XXIII Corps, stopped defending the Neidenburg sector and retreated to the border.[31]

That afternoon, *8. Armee War Diary* recorded, "Enemy on the run to the south from Hohenstein-Waplitz. *I. Armee-Korps* will bar the way of the enemy with *I. Infanterie-Division* proceeding to Neidenburg-Muschaken. *II. Infanterie-Division* will move on to Grünfliess."[32] A few Russian units held back to slow down the German advance so that the bulk of XXIII Corps could retreat to the south.[33] Monitoring the pursuit became the new mission for the aviators.[34]

GdK v. François' *I. Armee-Korps* maneuvered in the south. *AOK 8* had not yet received confirmation from François, that his forces had cut off the Neidenburg – Willenberg road. *I. Armee-Korps* quickly formed a thin line of pickets comprised of 25 battalions stretched over 60 kilometers. This amazing feat, sealed the fate of Samsonov's Second Army. They were enveloped by *8. Armee* and trapped.[35] Despite the lack of complete certainty, *GO* v. Hindenburg and his *AOK 8* staff were confident that *Genleyt* Klyuev's XIII Corps, *Gen.inf* Martos'

Two photos showing retreat from Hohenstein involving (15th KAO) Nieuport IV.

XV Corps, and *Gen.inf* Kondratovich's XXIII Corps, were indeed, trapped. The bulk of XXIII Corps managed to escape.[36] *Genmaj* Ludendorff cabled *OHL Coblenz*, "Everything has gone well. Encircling the Russian Second Army in all probability achieved."[37] What remained, were final skirmishes with a vanquished enemy and repulse of any Russian relief actions.[38]

Fliegertruppe's Role in the Hohenstein Victory

Fliegertruppe's contribution to the Hohenstein victory was primarily due to the independent tasking policy applied by the German commanders. *AOK 8* did not specify reconnaissance assignments. Instead, they deferred each subordinate *Armee-Korps* to direct aviators to targets, instrumental in providing timely intelligence. *Fliegertruppe* support that day, quickly commenced as the fog lifted. Shortly before 0800 hours, the first sorties for battlefield reconnaissance were launched. During the Hohenstein battle, *8. Armee*'s *FFA 16* operating from Osterode, was tasked to clarify the enemy situation between Soldau and Mława, and reconnoiter the whereabouts of VI Corps and XIII Corps. Sorties by *FFA 14* (*I. Armee-Korps*) and *FFA 15* (*XX. Armee-Korps*)

GdK v. Mackensen's FFA 17 flew Albatros B.I over the eastern sectors of the Tannenberg battlefield.

flew from Deutsch-Eylau. *GdA* v. Scholtz's *XX. Armee-Korps* directed *FFA 15* to monitor the battleground around Allenstein and Hohenstein. *GdK* v. Mackensen's *XVII. Armee-Korps* tasked *FFA 17* flying from Schwöllmen, to cover the area between Allenstein and Ortelsburg.[39] Throughout the battle day, *FFA 15, FFA 16*, and *FFA 17*, flew sustained battlefield reconnaissance, particularly around Allenstein. *FFA 16*'s 1030 report confirmed Radio Station *Gouvernement Festung* Thorn's wireless intercept—that XIII Corps heading to Hohenstein from Allenstein, proved to be the most critical sighting.[40]

GdI v. François' *I. Armee-Korps* kept *FFA 14* busy, monitoring the southern approach area from Soldau and Mława. *FFA 14* in coordination with *FFA 16* and *FestFA 6* Graudenz, monitored the Russians' retreat and eliminated all doubts about Russian counter attacks coming from the south.[41] The sustained monitoring of the southern approaches allowed François to turn most of his forces away from Soldau and accomplish his original mission of re-occupying Neidenburg and barring the southern exits. That afternoon, *8. Armee* changed their mission objectives, from attacking Russians to general pursuit. *Fliegertruppe* sorties concentrated on the direction of the Russians' retreat. The *Ostpreussen* forested areas allowed fleeing troops to vanish. At 1615 hours, *FFA 14* reported a Russian column, mainly infantry, seeking cover in the woods east of the Neidenburg, at 1615 hours.[42]

An Army Command in Chaos

Since crossing the *Ostpreussen* border, *Gen.kav* Samsonov was met with conflicting guidance from the Northwest Front. A haphazard infrastructure of reconnaissance elements, combined with faulty intelligence from higher command, led to uncertainty and eventual destruction. *Gen. kav* Zhilinsky's erroneous reports that *8. Armee* was in chaos following the battle of Gumbinnen, resulted in disaster. Zhilinsky ordered Samsonov to proceed to Allenstein versus Ortelsburg. XIII Corps occupied Allenstein on 27 August. *Gen.inf* Martos' XV Corps had also been expected farther to the north, but needed rest and remained east of Hohenstein. Martos reminded Samsonov that German forces were being reinforced by the hour. A *RIVVF* pilot reported a continuous stream of German infantry moving easterly from the areas behind the Drewenz River. [43]

Confusion and frayed communications networks, plagued the Russian corps' commanders. *Genleyt* Klyuev's XIII Corps left Allenstein for Hollenstein. Klyuev lacked knowledge of *Gen.inf* Martos' XV Corps' precarious situation and the location of his forces. Samsonov wanted to order a general retreat towards Neidenburg under the cover of flank rear guards.[44] He remained committed to holding Neidenburg and ordered the Second Division to remain at Nadrau but realized later on, that his forces failed to hold the ground. The deteriorating situation left Samsonov's command staff in disarray. They subsequently generated an operations order, directing *Gen. inf* Blagoveshchensky's VI Corps to advance to Passenheim. *Gen.inf* Martos's XV Corps and *Genleyt* Klyuev's XIII Corps, were ordered to proceed to the southwest towards Gilgenburg-Lautenburg.[45]

To Samsonov's discredit in the annals of military history, he left his primary headquarters at the most inappropriate moment. He inexplicably ordered his Neidenburg headquarters to relocate south of the *Ostpreussen* border to Janow. Samsonov then left for *Gen.inf* Martos' XV Corps command post at Nadrau, 24 kilometers north of Neidenburg. In doing so, he temporarily cut off communication with *Gen.kav* Zhilinsky and Second Army's corps commanders. His only reliable communication resources were now served by dispatch riders. Samsonov's rationale for his sudden departure, remains an enigma of historical proportions, as the Second Army lost control that day.[46]

Some historians suspect that Samsonov's decision-making was primarily based upon misleading intelligence from Zhilinsky's Northwest Front headquarters, and that Samsonov and his Second Army were ultimately, tragic victims of a fractionated and disorganized intelligence network. Whether or not Samsonov might have achieved his objective of regaining control on his left flank, if supporting aerial reconnaissance data were made available to him, remains an open question. Samsonov appeared to believe that only he could save the rapidly deteriorating situation by personally taking command of *Genleyt* Klyuev's XIII Corps and *Gen.inf* Martos' XV Corps. While leaving his headquarters, he encountered British Army envoy Lieutenant-Colonel Knox and told him that *Gen.inf* Artamonov's I Corps, the 2nd Division, and *Gen.inf* Martos' XV Corps "had been forced back on his left" and were retreating south.

Knox recalled, "He had just heard that the (*Gen.inf* Blagoveshchensky's) VI Corps had been driven back yesterday afternoon in disorder on his right…He (Samsonov) concluded that he did not know what was going to happen, but even if the worst happened, it would not affect the ultimate result of the war." Samsonov's staff concluded the discussion with "he (Samsonov) was going to the XV Corps, which was suffering from hunger and heavy loss in a four-day battle, and thought he was going to collect what he could to drive the Germans back."[47]

Samsonov arrived at *Gen.inf* Martos' XV Corps headquarters at 1100 hours on 28 August. He learned from his best general that XV Corps was fought out and could no longer hold their positions against *8. Armee*. Samsonov still believed that *Genleyt* Klyuev's XIII Corps would arrive from Allenstein, but was unaware they were locked in a furious battle against I Reserve-Corps. Late on 28 August, XV Corps commenced their withdrawal. XIII Corps was decimated, trying to maneuver south to merge with XV Corps.[48] Despite the futility of the circumstances and his inevitable and tragic final moments, Samsonov remained on the battlefield with his beaten troops.

The Decisive Blow

8*. Armee War Diary* provided an update on the victory at Hohenstein with:

> "So far as now observed, the Russian I Corps is hastily retiring via Mława toward Warsaw. Russian XXIII, XV, and XIII. Army Corps cut up and scattered in the forests southeast of Hohenstein-Allenstein. Russian VI Army Corps, one division of which has been completely annihilated, is hastily retiring beyond Ortelsburg."[49]

8. Armee's *Armeebefehl* for 29 August called for all fighting to focus on routing the remaining Russian units. A new threat was expected to soon materialize, due to the arrival of *Gen.kav* v. Rennenkampf's First Army at the northern edge of the battlefield. Just as the new *Armeebefehl* for 29 August was being finalized, situation updates to *AOK 8* indicated parts of XIII Corps and XV Corps were evading through the forests south of Allenstein. *Genmaj* Ludendorff was irate after learning that the Russians were escaping.

As a result, *AOK 8* ordered *Ostgruppe's I Reserve-Korps* to prevent any further Russian moves into the forested areas in the southeastern half of the battlefield. *XVII. Armee-Korps* was ordered to completely close off the exits across *Ostpreussen*'s eastern roads.[50] *GdK* v. Mackensen's forces proceeded south and despite sandy roads causing transport to be overturned, the first elements arrived in Ortelsburg at 1500 hours on 29 August, an hour after the Russians had left.[51]

8. Armee War Diary reported "Wireless messages indicate the following: The Russian Commander in Chief [Zhilinsky] directs General Rennenkampf to support the 2d (Narew) Army. In view of this, Königsberg *Festung* has been directed to prevent the Russian Army from advancing in a westerly or southwesterly direction."[52]

That evening, *AOK 8* realized that the battle at Hohenstein had given *8. Armee* the upper hand in decimating Samsonov's Second Army. *AOK 8* then moved north and set up command location at Osterode to prepare for the next phase of the campaign—defense against Rennenkampf's First Army.[53] *GO* v. Hindenburg's direction to subordinate commanders was simple—route the remaining Russian forces in southern *Ostpreussen* and prepare against First Army's maneuver south.[54]

The *Armeebefehl* that was generated at 1730 hours on 28 August, stated for the first time, that *AOK 8*'s location was now Tannenberg. *Genmaj* Ludendorff verbally directed, "Frögenau-leave the exact time open." *Oberstleutnant* Hoffmann suggested to Ludendorff that the place of origin for the *Armeebefehl* should be the nearby village known throughout German history—Tannenberg. This location was of historical significance, as it was the site of an ancient battleground in the year 1410. Polish-Lithuanian forces at that time, had defeated the Teutonic Knights, halting their eastward expansion. Ludendorff accepted Hoffmann's suggestion and applied the Tannenberg location to the report.[55] The early historical myth that resonated with the German's victory over the Russians (in the preceding four days of one of the most significant battles in modern history) enshrined the word "Tannenberg," as an enduring emblem of German fortitude and power.

CHAPTER IX

The Final Blow

Searching for Russians

Russian leadership was now forced to contend with the annihilation of *Gen.kav* Samsonov's Second Army and strategize how to proceed. Their campaign was chaotic and vulnerable. In contrast, the Germans were actively engaged in a unified and cogent approach to their campaign. At this juncture, *AOK 8* meticulously reviewed intelligence sources for any indication that *Gen.kav* v. Rennenkampf's First Army was diverting to the south.[1] First Army's 70 kilometers proximity to the Tannenberg battle, meant reinforcements could arrive within three days.[2] Königsberg's *FestFA 5* and Airship Z-IV kept up the vigil on First Army. The evening of 28-29 August, Z-IV conducted a five-hour reconnaissance mission over the Pregel River valley. Their observer noted numerous bivouacs with camp fires as close as 40 kilometers south of Königsberg. The density of the camps and fires increased as they looked east. Z-IV dropped bombs and returned to Königsberg.[3] Königsberg's *FestFA 5* reported at 0905 hours that the First Army was still moving west. At 1430 hours, *FestFA 5* finalized their report—the First Army's southern wing had changed direction to the southwest. *8. Armee*'s *FFA 16* generated two sorties and confirmed

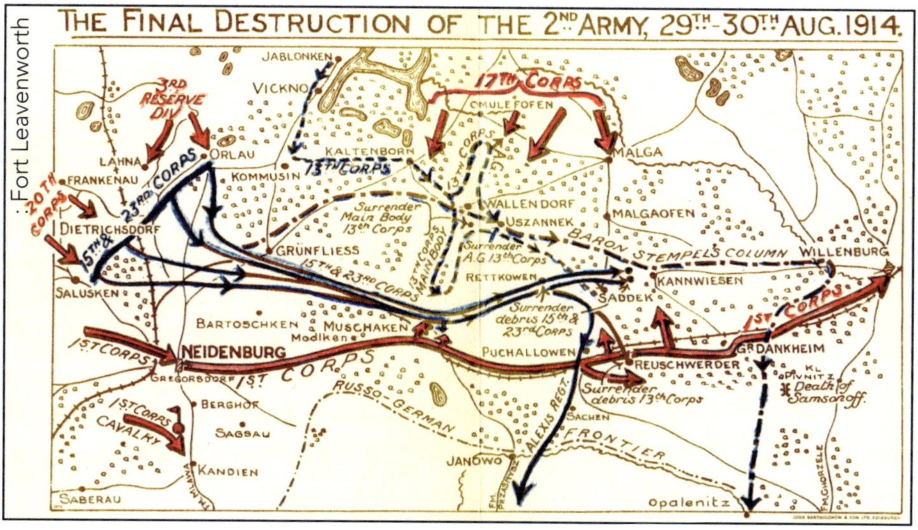

Detailed depiction of the final maneuvers of Tannenberg battle produced by the US Army's Fort Leavenworth Command and General Staff school in 1932 based on maps by Major-General Ironside. Larger map available at shootingthefront.com.

FestFA 5's reporting. First Army's right flank was still proceeding west to Königsberg, while the left flank applied elements of Russian cavalry heading south to Allenstein. Any and all efforts by Rennenkampf's First Army in support of Samsonovskaya, were implemented too late. Hindenburg was not overly concerned. The German victory at Tannenberg was affirmed, acknowledged, and consolidated.[4]

German intelligence collection searched for unexpected Russian maneuvers or reinforcements from the Soldau-Mława area.[5] Their intercepts of Russian wireless messages, revealed a strong Russian counterattack was expected; most likely to be executed by *Gen.inf* Kondratovich's XXIII Corps supported by the 3rd Guards Division.[6] On 29 August at 0710 hours, *FFA 14*'s *Ltn* von Winckler (pilot) and *Oberleutnant* Bogislav von Heyden (aerial observer) reported, "much rolling stock on Mława railroad station, closely to the northwest a camp for one brigade, a one kilometer long column, beginning ten kilometers north of Mława; near Zawadzkie one regiment and a one to two-kilometer long column on the road Mława – Neidenburg, beginning at Napierken." Initially, the report went to *I. Armee-Korps*. *GdI* v. François interpreted the observation correctly—Russian forces were preparing to attack his forces along the thinly maintained line of Soldau – Neidenburg – Willenberg. François immediately ordered more aerial reconnaissance covering all relevant land routes.[7] Reporting from other aerial platforms aided in shaping the battle. At the main battle area east and southeast of Hohenstein, the fighting continued. *GenLtn* v. Below's *I. Reserve-Korps*' Drachen-Ballon directed artillery fire and provided constant updates. At 1026 hours, from his elevated position the Drachen observer was the first to see the Russians stop fighting and retreat.[8]

FFA 15 and *FFA 16* launched sorties to monitor Russian columns that were retreating southeast. At 0700 hours, Russian infantry and artillery were spotted heading east towards Dietrichsdorf – Schwedrich, further east than was previously estimated. At 0800 hours, a sortie over Neidenburg – Willenberg recorded only a small number of Russian trains heading south. No other forces were spotted.[9] A German cavalry patrol sighted three Russian aeroplanes taking off from a landing ground south of Yanow at 1330 hours. At this point in time, there was no intelligence that corroborated unit designation; nor could intelligence identify the specific troops that the three aeroplanes supported.[10]

Fatal Results

W. Bruce Lincoln described the chaos in the Russian ranks as they fled. "The Second Army was almost deaf. As they marched through the thick forests that separated *Ostpreussen* from Russia's Polish lands, they had also become nearly blind."[11] Traumatized Russian forces were retreating through unfamiliar terrain without working telegraph or telephone. They were harassed by a vigilant and unified German enemy that had created an unambiguously gruesome battleground. *RIVVF* support failed, due to their lack of effective organization over *Ostpreussen*. No aviator had ever witnessed an encirclement or massive retreat of this magnitude. If any reporting from aerial reconnaissance had actually taken place, it never had an impact on Russian commander deliberations orchestrating their next step.[12] *RIVVF* support witnessed another loss at this grim moment. Second Army staff tasked their *21st KAO* to deliver a message packet to *Gen.inf* Kondratovich's XXIII Corps at the village of Horzhele (just south of the border). *St.Uo.* Argunov was given the mission. He took off in a Farman *HF.16* (No. 163) but was never seen again.[13]

8. Armee's encirclement of Second Army center group, was complete despite being a thin line of German infantry at the southern edge of the battlefield. There was limited Russian resistance. "Even the major Russian units fled weak German detachments and rushed eastward, or finally surrendered without weapons and food, and out of control of their officers."[14] *Gen.inf* Martos' XV Corps succumbed. *Genleyt* Klyuev's XIII Corps failed to acquire German occupied Kaltenborn, Wallendorf and Mushaken and commenced their retreat to the south, only to surrender at Uschenk. *Gen.inf* Blagoveshchensky's VI Corps continued in full retreat when it received a Samsonov order to hold on to Ortelsburg and Passenheim.[15]

8. Armee War Diary summed up 29 August with the following comments:

> At noon, the left flank of *XX. Armee-Korps* in its attack against Hohenstein, gains contact with *I. Reserve-Korps*. In consequence, a large portion of Russian troops is being driven into

SORTIE DATE: 30 AUGUST, 1914

AEROPLANE: Rumpler-Taube

PILOT: Leutnant Hesse and *Oberleutnant* Körner

GROUND TRACK: Deutsch Elyau - Neidenburg - Mlawa - Neidenburg - Deutsch Elyau

Source: NARA RG 165

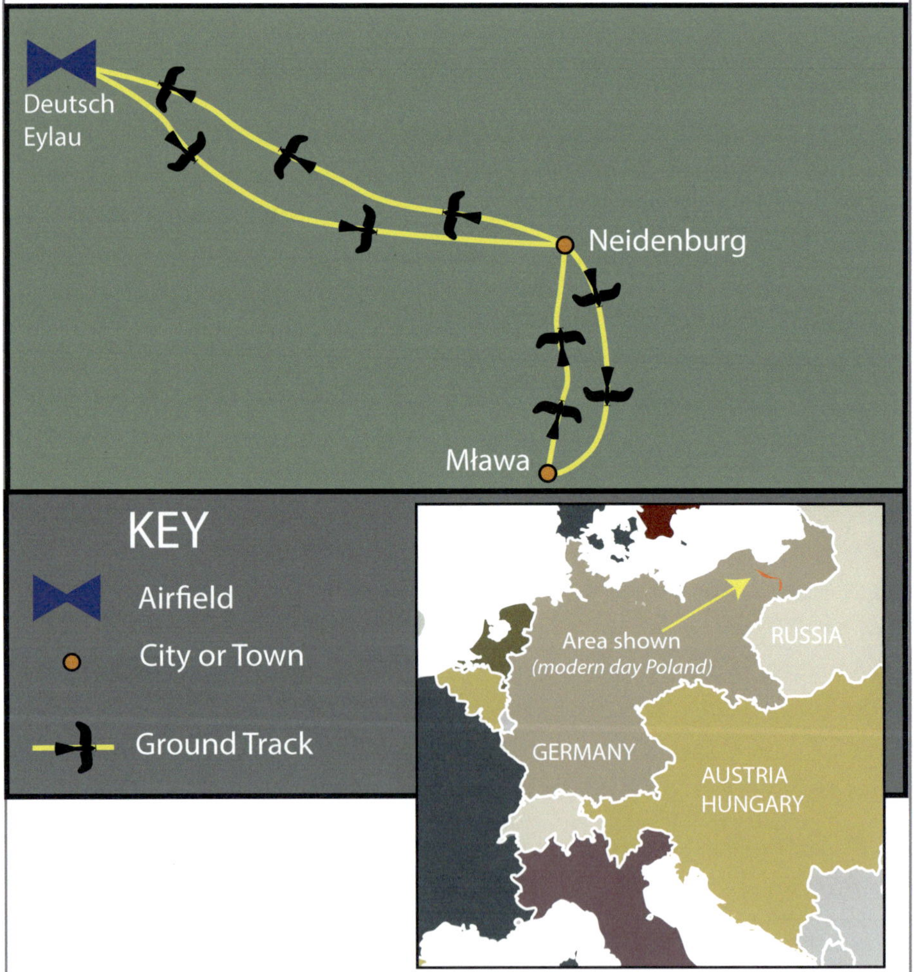

SORTIE DATE: 30 AUGUST, 1914

FFA 14 aviators, Leutnant Hesse and Oberleutnant Körner, sighted the Russian vanguard only six kilometers south of Neidenburg.

SORTIE AVIATOR REPORTS: 30 AUGUST, 1914

Message from Flieg.Abt.14. Sender: Lt Hesse, Pilot, *Oberleutnant* Körner.

Issued: dropped at Neidenburg 10:30. 10:45 am. [time of report]

Column of Infantry and Artillery marching from Mława on Neidenburg. Head of column 9:10 am near Kandien 6 km south Neidenburg. End about 1km north Mława. 1 second column marching from Stubsk on Mława. Head 8:45 am at Mława, ending at Wola. Also: 2 camps left and right of main road Mława –

Neidenburg closely south Kuklin, estimated at 1 Infantry Brigade. (Remark) Total length of the column approx. 36 km."[3]

the arms of *I. Reserve-Korps*. By evening, 41. *Infanterie-Division* secured the region of Lahna and 37. *Infanterie-Division* (*XX. Armee-Korps*) stands along the road Stabigotten-Ring-Lehnen-Hohenstein. *3. Reserve-Division* continues the pursuit between *41. Infanterie-Division* and *37. Infanterie-Division* in direction of Kurken. After a pursuit engagement at Schwedrich, *3. Reserve-Division* moves into shelter at and west of Kurken. *Landwehr-Division v. d. Goltz*, on the evening, regulates its organizations, which after the attack against Hohenstein, have become badly mixed and intermingled. The insufficient outfitting with ammunition columns and medical units, has been keenly felt. *I. Reserve-Korps* reports, that it will make a further inroad towards Hohenstein with *I. Reserve-Division* and that it had launched *36. Reserve-Division* against Wuttrienen. In view of the severe fighting for Hohenstein and the possibility of a later participation in the fighting by the Russian Wilna Army; orders are now issued to *I. Reserve-Korps* at 0630 hours, to also enter the *36. Reserve-Division* into the battle for Hohenstein and as soon as available there at, to assemble north of the forest south of Allenstein with front toward Königsberg.[16]

FLIEGERTRUPPE SAVIORS

Library of Congress

Gdl v. François poses with the four Fliegertruppe officers who saved I. Armee-Korps from the advancing Russian column on 30 August 1914. Ltn Karl Mertens (observer), Ltn Ernst Canter (pilot), Gdl v. François, Obltn Heinrich Körner (observer), Ltn Hans Hesse (pilot). Rumpler Taube in background.

That evening, Samsonov left on horseback for Janow, to resume control. Being deprived of the opportunity to take the direct road, he chose, instead, the roundabout route via Sadek. Samsonov's closest staff advised him to flee the area. His next steps proved fatal. At 0100 hours on 30 August, Samsonov disappeared. His staff heard a shot coming from the woods near the village of Willenberg. He was presumed to have committed suicide.[17] The next morning, Zhilinsky learned of his death and realized that any attempts to save the Second Army were futile. He called off all attacks in the area and directed the remainder of the Second Army to withdraw to Russian territory.[18]

Keeping First Army in Mind

A*OK 8* was certain that Zhilinsky was in the process of committing Rennenkampf's First Army south, to save what little remained of the Second Army. Prior intelligence from an intercepted wireless message dated 28 August, as well as Zeppelin Z-IV's aviation report of 29 August, noted elements of Russian cavalry heading towards Allenstein. *8. Armee* felt pressure to prepare a defense against this looming threat. That evening, German wireless intercepts learned that the First Army was to encircle Königsberg from the south. *Gouvernement Festung* Königsberg estimated four Russian army corps were 40 kilometers away crossing the Alle River.[19]

Samsonov died near the village of Willenberg. A post-war memorial to the fallen general was built to commemorate his life.

Intelligence collection also updated *8. Armee* on the residual threat of the remaining Russian forces. *Fliegertruppe* aerial reconnaissance confirmed Rennenkampf's First Army was moving east to the Norkitten-Nordenburg line and reinforcements were arriving via the Insterburg-Nordenburg line. A strong Russian infantry force was confirmed at Friedland.[20] *Gen.kav* Scheideman's II Corps proceeded to move west from the Masurian Lakes region. A wireless intercept revealed II Corps was ordered to destroy all infrastructure, in an effort to disrupt any attempt by *8. Armee* to divert back to face First Army. *AOK 8* perceived II Corps' role as further proof that First Army was not going to support and assist Second Army.[21]

The following day, *AOK 8*'s first aerial reconnaissance report was received from Festung Königsberg's *FestFA 5* providing an update on First Army. The aerial observer reported numerous bivouacs in the area. Most were abandoned. Russian march columns were reported in the sector comprising Seeburg-Rössel-Rastenburg-Angerburg-Nordenburg-Bartenstein. *AOK 8* tasked *FFA 16* to reconnoiter First Army's southern wing. *FFA 16* summarized "...one Cavalry Division in the area at and west of Roessel, near Rastenburg and Angerburg, one Infantry Division each."[22] Aerial observation confirmed that what was left of Second Army's fate was ultimately at the discretion of *8. Armee*.

The Final Maneuvers

8*. Armee* continued to envelop the remaining Russian forces in the "Tannenberg" pocket of Allenstein, Neidenburg and Ortelsburg. As the evening unfolded on 29 August, *AOK 8* prepared the *Armeebefehl*. New information confirmed a bottleneck connecting the western part of the battlefield to the east still remained open. Units from XIII Corps and XV Corps could still flee to the east. *GO v.* Hindenburg's plan focused on how to complete the annihilation of what remained of *Genleyt* Klyuev's XIII Corps, *Gen.inf* Martos XV Corps, and *Gen.inf* Kondratovich's XXIII Corps.[23]

Gen.kav Zhilinsky learned the evening of 29 August that German forces had reached Janow. He was still unaware of Second Army's fate. He ordered *Gen.inf* Sirelius' I Corps, *Gen.inf* Blagoveshchensky's VI Corps, and *Gen.inf*'s Kondratovich's XXIII Corps, to advance on Neidenburg, Willenberg and Ortelsburg. His goal was to keep Samsonovskaya's Second Army withdrawal open to the Polish Salient.[24]

SORTIE DATE: 30 AUGUST 1914

AEROPLANE: Albatros B.1 249/13

PILOT: Leutnant Scherkamp

GROUND TRACK: Osterode-Ortelsburg-Friedrichshof-Leman-Myscyniec-Fuerstenwalde-Chorzele-Janowo-Neidenburg-Osterode.

Submitted Osterode Airfield 30.8.14, 10:30 am. Recvd 10:45 am. A.O.K.

Source: NARA RG 165

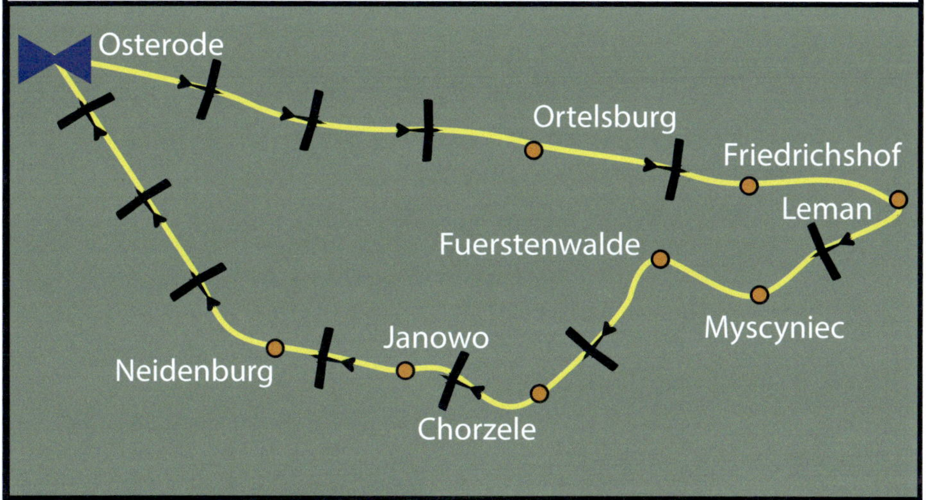

☛

SORTIE DATE: 30 AUGUST 1914

FFA 16 launched sorties to monitor Russian columns that were retreating to the Kurken Gap on roads leading out of Neidenburg. 8. Armee received an alarming report from FFA 16's Ltn Otto Scherkamp and Ltn Leo Strauch.

1. Road Ortelsburg-Willenberg 06:50 free.
2. One company head at Wawrochen (road Ortelsburg-Fürstenwalde) advancing on Ortelsburg, (s.Z.6) (06:50 am).
3. Cav. (est. one regt.) advancing on road Friedrichshof-Ortelsburg. Head Railroad/Road crossing Gr.-Jerutten (06:55).
4. Road Leman-Myszyniec vehicle column (about 50 vehicles) advancing westward. Head 07:05 near Dudy. This column later turned on Dombrowy.
5. Column of all weapons 07:45 near Lucka, head Lipowitz advancing on road Fürstenwalde-Ortelsburg. Seemingly one Inf.Div. Head see nr. 3. [above]
6. Near Mysczyniec four big cart parks, one big magazine, eight big tents. Only few infantry discernible. Troop strength assumed to be at least one division.
7. Vehicle column on road Mysczyniec-Friedrichshof, length approx 1 km. Head at Mysczyniec-Stary.
8. All other tracks and roads from Mysczyniec to all directions free of troops.
9. Near Chorzele neither troops nor vehicles. All roads free.
10. Near Gr.-Dankheim large bivouacs, layout 1 km square.
11. s.w. [southwest] of Kl.-Dankheim battery in position, oriented to n.W [northwest].
12. Near Reuchwerder Bivouac about 1 batallion.
13. On road Willenberg-Muschaken vehicle column, length 1 km, advancing east to west. Head 8:15 pm north Wichrowitz.
14. n.w. [northwest] Jägersdorf artillery positions (one detachment). Trenches west Muschaken.
15. 08:30 Russ. artillery firing to north.
16. Janowo free, also all roads and tracks from there. On Janowo market square approx. 20 vehicles.
17. 8:40 supposed own column of all weapons arrived Gregersdorf about 6 km east of Neidenburg. at 08:10.

Gen.inf Martos' XV Corps 15th KAO's Nieuport-IV aircrews were ordered to quickly retreat out of the area and joined Russian soldiers in the attempt to escape. Postcard records a burned RIVVF Nieuport IV captured at Neidenburg."NEIDENBURG after the destruction by the Russians on 22 August 1914. Russian airship that was shot down in the market"

Upon assuming command of I Corps, *Gen.inf* Leonid-Otto Ottovich Sirelius received Zhilinsky's order to immediately take all available men at his disposal and attack Neidenburg. Throughout the afternoon of 29 August, Sirelius assembled a force comprised of two regiments of the 3rd Guard Division, seven battalions from 1st Rifle Brigade, and artillery units. The surviving soldiers from the original I Corps were physically depleted, and unable to continue on. *AOK 8* failed to closely track I Corps and did not factor in the risks to *I. Armee-Korps*. Ludendorff envisioned that François would advance north against Rennenkampf's First Army—which was assumed by *AOK 8* to be advancing south.[25]

Ludendorff recalled in the early morning of 30 August, *AOK 8* received a *Flieger-Meldung* stating that a Russian army corps was marching to Neidenburg from the south and was nearing the village—putting *I. Armee-Korps* at risk. The staff overlooked the threat to *I. Armee-Korps*. They also misinterpreted a report indicating Russians at Mława were retreating to Przasnysz.[26] Further confusion was evident when Ludendorff telephoned *OHL* at 0930 hours with the news that *8. Armee* had achieved complete success. *OHL* was unaware that the Russians were attempting a counterattack on François' *I. Armee-Korps*. Ludendorff concluded in his report to *OHL*, "Everything went well. The encirclement of the Russian Second Army ... has been accomplished."[27]

The Historic Exploits of the Tannenberg Flyers – Reality versus Myth

Fliegertruppe found themselves at an inflection point in the battle of Tannenberg. This was due to the fact that the Russians still had a meaningful window of opportunity to inflict great harm on *8. Armee*. Despite *AOK 8*'s obsession with Rennenkampf's intentions, *GdI* v. François remained focused on any and all remaining threats against *I. Armee-Korps*. He suspected something was amiss, and subsequently set up his *Armee-Korps*' command post in Neidenburg's only hotel. He then ordered *FFA 14* to keep a wary eye on the Mława approach and its surroundings. *FFA 14* aviators tracked *Gen.inf* Sirelius' I Corps as it maneuvered toward Neidenburg.[28]

The morning of 30 August was a dramatic moment for *Fliegertruppe*. *GdI* v. François recalled Russian success with intelligence collection, as the first order of business that day:

> At 4:30 my striker woke me ... The most fascinating day of the Tannenberg Battle began ... Hustle and bustle in our quarters. In the small dooryard of our hotel all officers of my Command who know Russian are sitting over captured Russian documents and letters searching for intelligence. Among others they find the movement lists of the Russian I.; XVIII., II., III. (Siberian) Army Corps. Among many war maps there were also copies of our top secret war maps and plans of all Eastern Fortresses.... I went to the town square, to see the Russian artillery pieces and vehicles that had arrived overnight.[29]

Two *FFA 14* aircrews delivered their observations to François' command post at Neidenburg. The first aircrew (*Leutnan*t Ernst Canter and *Leutnan*t Karl Mertens) landed near Neidenburg at Gregersdorf. At 0835 hours, they personally reported to François and his *I. Armee-Korps* senior staff, that substantial enemy forces were leaving their bivouacs and beginning to march on Neidenburg. They were initially skeptical that a

Poruchik Alexander Iosifovich Kulvinsky flew one of the three remaining 15th KAO Nieuport IVs safely back to Russian territory.

Russian attack was threatening Neidenburg at that hour. They had no other sources to validate the report. Due to telephone communication problems with *AOK 8*'s new headquarters at Osterode, Canter and Mertens returned to their aeroplane. They took off and flew directly to Osterode. When they arrived, *AOK 8* had just received telephone confirmation from *I. Armee-Korps* on Russians advancing towards Neidenburg. *Genmaj* Ludendorff, after listening to Canter and Mertens reporting, realized they needed to fly to various command posts and alert them of the Russian threat. They were not able to take off from the field they landed on, so they drove in a motor car to Deutsch-Eylau, *FFA 14*'s primary landing ground. An available Taube was readied and they flew to the command posts with the warning of the approaching Russian threat.[30]

After Canter and Mertens left, *Gen.inf* Sirelius' I Corps continued closing in on Neidenburg. A second *FFA 14* sortie had observed the Russian advance from the south. At 0915 hours, *Leutnant* Hans Hesse and *Oberleutnant* Heinrich Körner, observed a 37-kilometer column of Russian infantry, artillery, and cavalry, advancing in the direction of Neidenburg. Körner correctly estimated the strength to be a complete Russian corps. The immediacy of the threat prompted the aviators to drop a warning to the command post at Neidenburg's town square. François looked up and saw Hesse flying his Taube in a spiral and the *Flieger-Meldung* flung overboard by Körner in dramatic fashion.[31]

GdI v. François recalled Hesse's observations in his memoirs (substance close to *8. Armee's* report [page 180]):

> Aeroplane A. 29, Lt. Hesse. Track Eylau - Soldau - Mława - Neidenburg. Message dropped 30.8., 10:25 hours am. To General Command I. A.-K. Columns of all weapons from Mława on

> Neidenburg. 9 hrs10 min at Kandien 9 km south of Neidenburg], end 1 km north Mława. A second column from Stupsk on Mława, at 8 hrs 45 min, head eastern exit Mława at 8 hrs 45 min, ending at Wola. Körner, Observer.[32]

François responded immediately, issuing verbal orders that were delivered by couriers via automobile and horseback. As *Gen.inf* Sirelius' I Corps lead column closed in on Neidenburg, François recalled, the fighting was about to commence with [Artillery] Detachment Schlimm leading the attack. He quickly dispersed an officer in a motor car to Major Schlimm's position and sent another officer to 2. *Infanterie*-Division with the order to immediately advance with all available forces against the new enemy at Gregersdorf - Sagsau. A third dispatch was sent to *Generalmajor* Paul von Mülmann, commander of *Festung* Thorn's *5. Landwehr Brigade*, to hurry up and engage via Saberau. François finished his account of the dramatic Neidenburg event: "In the meantime, the General Command had prepared the departure. When the last car left the hotel, the first rounds of Russian shrapnel burst over the town square." As *I. Armee-Korps'* communications officer in Neidenburg relayed the report to *AOK 8* by telephone, Russian artillery shells exploded on the town square, interrupting the telephone connection.[33]

GdI v. François quick response to the crisis proved timely — a unit the size of an *Armee-Korps* needed hours to shift from march order to attack. At 1100 hours, Sirelius' I Corps was preparing to attack. At 1500 hours, they were in position approximately two kilometers south of Neidenburg. Major Schlimm's detachment held until he was forced to withdraw early that evening. François had positioned almost 100 cannon and howitzers at Gregersdorf heights under Schlimm's command. The ensuing barrage created havoc in the Russian ranks. At 2100 hours, fighting ceased with the Germans reinforcing their defense positions. They could not prevent the Russians from re-occupying Neidenburg. François' staff planned a counterattack. Sirelius' I Corps retreated back to Mława before daybreak.[34]

The Russians attempted one last thrust on Ortelsburg. On 30 August, soon after the attack on Neidenburg, *8. Armee* received an alarming report from *FFA 16*'s *Ltn* Otto Scherkamp and *Ltn* Strauch. At 0650 hours, they discovered a regiment of Russian cavalry advancing from Friedrichshof to Ortelsburg. An hour later, a column of all weapons was sighted at

Gen.inf Martos, defeated commander of XV Corps, awaiting interrogation by German staff. He told his captors, "Knowing nothing of the enemy, so that I approached the Neidenburg battlefield, blindfolded."

Lucka and a division of Russian troops, sighted near Mysczyniec. By the time Scherkamp and Strauch landed at Osterode, they assumed the Russians were in Ortelsburg.[35] XVI*I. Armee-Korps' FFA 17* flew four missions west of Ortelsburg to clarify what *FFA 16* reported.[36] *GdK* v. Mackensen received *AOK 8*'s alert from Scherkamp and Strauch's report and immediately directed *35. Division* and *36. Division* to repel the threat. The Russians fired an intense barrage of artillery but lacked effective targeting. At 1930 hours, *AOK 8*'s *Armeebefehl* stated, "*XVII. Armee-Korps* defeated a Russian division near Ortelsburg and forced it back over the border."[37]

Abandoning All Hope

The remaining elements of *Genleyt* Klyuev's XIII Corps and *Gen. inf* Martos' XV Corps, surrendered. *Ostgruppe's XVII. Armee-Korps* pushed forward and closed the Passenheim-Jedwabno road cutting off the Russian escape to Willenberg. *GdI* v. Below's *I. Reserve-Korps* advanced to close the Kurken Gap. *GdA* v. Scholtz's *XX. Armee-Korps'* units pushed southeastward via the Kurken Gap to Jedwabno.[38]

By mid-day 30 August, Zhilinsky abandoned all hope to save the Second Army and ordered the relief units back. He realized it was too late.[39] The decimated I Corps and VI Corps' *KAOs* suffered a dark legacy after the campaign's defeat. *Genleyt* Klyuev's XIII Corps *13th KAO* air assets and staff were all captured by the Germans. *Gen.inf* Martos' XV Corps *15th KAO*'s Nieuport IV aircrews were ordered to quickly retreat from the area and joined Russian soldiers in an attempt to escape. One Nieuport IV under repair was burned at Neidenburg. *15*th *KAO*'s *Poruchik* Konstantin Nikolaevich Masherek and *St.Uo. Voennyi Letchik* S.Menichenko along with 22 enlisted ground crew, were captured.[40] The remaining three *15th KAO* Nieuport IVs were safely flown back to Russian territory by *Poruchik* Alexander Vasilievich Kotov, *Poruchik* Ivan Anikievich Chehutov, and *Poruchik* Alexander Iosifovich Kulvinsky.[41]

After being captured, *Gen.inf* Martos revealed that no regular cavalry had been allotted to his corps and Cossack regiments were not effective. "The first long range reconnaissance by the Orenburg Cossack regiment produced no information as it simply was not carried out." Martos also remarked to the German interrogators, "Army Headquarters did not keep us informed of the situation and neighboring corps were in the same position as myself, knowing nothing of the enemy, so that I approached the Neidenburg battlefield, blindfolded."[42]

Bad landings and take offs caused a rapid decline in the number of available aeroplanes. 15th KAO lost a Nieuport - IV before the XVth Corps surrendered.

CHAPTER X

The World Learns About Tannenberg

Tannenberg's Impact

Tannenberg proved to be one of the most critical German victories of the entire war, as it saved the Prussian heartland and averted the danger of a deeper Russian advance into industrial Silesia and towards Berlin. It was a deliverance and celebrated, as such. The battle reversed the timetable of Germany's war plan. Before Tannenberg, Germany planned for victory in the west while holding the Russians at bay. After Tannenberg, though the east no longer remained a dire threat, victory in the west remained elusive.[1]

Hindenburg cabled Kaiser Wilhelm, on 31 August:

> Most humbly I report to Your Majesty, that yesterday the ring around the bulk of the Russian Army was closed. The [Russian] XIII, XV, XXIII Corps have been annihilated. To date, we have made 60,000 prisoners, among them the Commanding Generals of the XIII and the XV Corps. The artillery pieces are still in the woods and are being collected. The total war booty cannot be judged for the time being, but it is extraordinarily high. The I and VI Corps outside the ring have also suffered considerably; they are fleeing via Mława and Myszynecz.
>
> —Hindenburg.[2]

OHL direction to *GO* v. Hindenburg and *Genmaj* Ludendorff (on 31 August) set the priority to drive the Russians out of *Ostpreussen*. Additionally, *8. Armee* was ordered to force the remaining Russian Second Army south to Warsaw.[3] *Fliegertruppe* aeroplanes and aircrews were suffering from extensive fatigue. Airframes were still intact, but standards for flight safety were lowered to allow missions to continue. Newly established *Feldfliegerabteilung* (such as *FFA 37*) assigned to *GdI* v. Below's *I. Reserve-Korps*, reported only one aeroplane available for operation during the first two weeks of September.[4]

On 31 August, *AOK 8* ordered reinforcements to advance on the recently vacated Russian occupation of Neidenburg.[5] *I. Armee-Korps'* aimed to eliminate any remaining Russian resistance and take as many prisoners of war as possible.[6] *8. Armee* now assumed

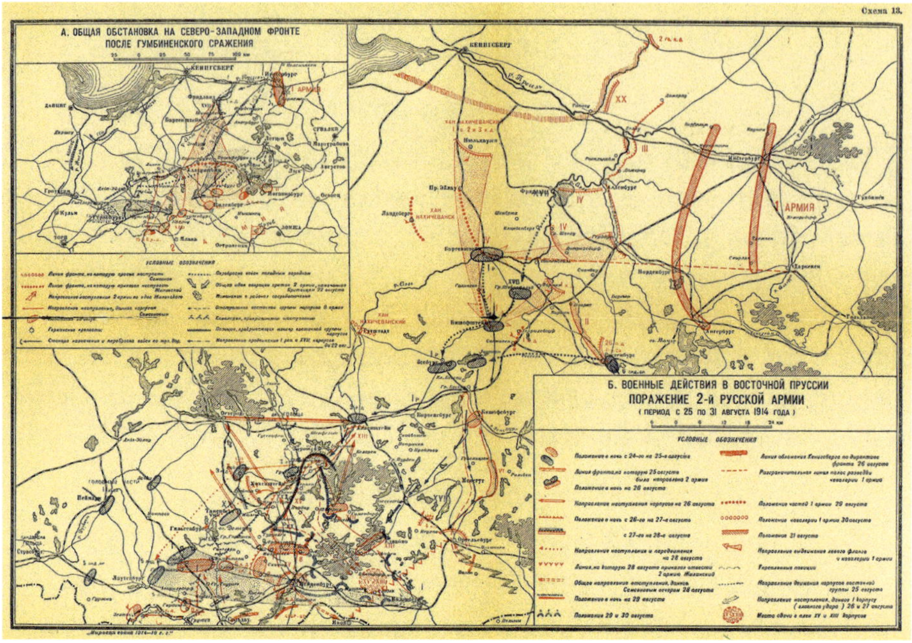

A post-war Russian Map recounts the maneuvers that took place during the Tannenberg campaign. Source: Carl J. Bobrow Larger version of this map available at shootingthefront.com.

the mission of processing Russian prisoners from *Genleyt* Klyuev's XIII Corps, *Gen.inf* Martos' XV Corps and half of *Gen.inf* Kondratovich's XXIII Corps. They were also tasked with sanitizing the battlefield of corpses and retrieving remaining weapons, ammunition, and other war materials. *8. Armee* was now faced with a unique challenge—how to effectively organize logistical requirements to transport, shelter, and feed an unexpectedly large number of prisoners of war. The initial German estimate was almost 100,000 Russians. Comprising the POW total, were 13 general officers and 350 field guns, now in German hands.[7] There was a personal reward for the German combatants of Tannenberg. *Ltn* Mahncke recalled, "A number of war chests were found among the masses of Russian weapons, equipment, war materials, and horses captured, and according to an old booty system, this money was distributed among we soldiers who had participated in the battle. I received a few Marks."[8]

GO v. Hindenburg (left) and Genmaj Ludendorff are recognized for the risks taken that ensured 8. Armee's victory over Gen.kav Samsonov's Second Army.

Gdl v. François (right) at Neidenburg on 31 August 1914, savoring the recent victory of the Tannenberg battle.

Aviation is not Mentioned

Allied leadership's assessment of Tannenberg failed to mention aviation. Russian Foreign Minister Sergey Dmitryevich Sazonov recounted to Stanley Washburn (a newspaper reporter), that Russia had lost 165,000 men in three days. Later in the war, in 1917, Mikhail Ivanovich Tereshchenko, foreign minister of the Russian Provisional government, told British ambassador Sir George Buchanan, that "France, he was sure, would never forget that Russia had sacrificed 300,000 men to save Paris." [9] Maréchal Ferdinand Foch in his post-war memoirs, paid tribute to the Russian effort with, "…we could not forget our Allies on the eastern front, the Russian Army, which, by its energetic intervention, had drawn upon itself a large portion of the enemy's forces and had thus enabled us to achieve the victory of the Marne."[10]

PRISONERS OF WAR

Russian POWs from the battles fought in Ostpreussen in August 1914. Source: Library of Congress

Newspapers gave the public the impression that the Russians were decisively winning at Allenstein. While the Second Army was being destroyed, Allied hopes were strengthened by reporting that proclaimed, "The Germans continue to retreat." It was a movement that became a front-page headline. Reporting reflected Allied intelligence. The New York Tribune on 30 August 1914, proclaimed, "Germany Rushes Troops from France to Strike at Foe in Eastern Prussia." The subsequent banner broadcast was, "Kaiser Hurries Men out of Belgium to Meet Russians."[11]

The French commander, Général Joseph Joffre, was desperately fighting at the Marne, when he personally learned of the Russians' defeat at Tannenberg. Joffre received the news through a wireless intercept of a German transmission that recounted the destruction of three Russian corps, the capture of two corps commanders, and 70,000 other prisoners. The message was simple and profound. "The Russian Second Army no longer exists."[12]

A rare media glimpse of aviation on the Eastern Front was reported by Mr. B. W. Norregaard, working for the *Flight Aero Weekly*. His descriptions captured the adventures of the French aviator, Sous lieutenant Alphonse Poiré fighting for the *RIVVF*:

> We rose to a height of about 5,000 ft. Fighting was in full swing. The captain had already made some valuable observations when the Germans, noticing my Farman machine, opened fire. A number of bullets pierced the wings of the aeroplane and ethers, struck the stays, but we flew on. It was still necessary to ascertain the exact position of certain of the enemy's forces. Then the German artillery began. Shells burst near the machine, all round. Each explosion caused the aeroplane to rock terribly. It was difficult to retain control. Pieces of shells seriously damaged two stays. This fantastic dance in the air lasted twenty minutes. The captain was wounded in the heel but continued to make observations. The machine was now much damaged, so we turned homewards and landed safely. I found ten bullet marks and two fragments of shells in the machine.[13]

AVIATION AS BOOTY

Library of Congress

For the first time in history of warfare, aviation is now part of the spoils of war. Booty includes a Nieuport IV being exhibited. Abandoned and captured Nieuport IV from XIII Corps' 13th KAO are photographed. A member of a German medical field unit displays the unique RIVVF flag shown below.

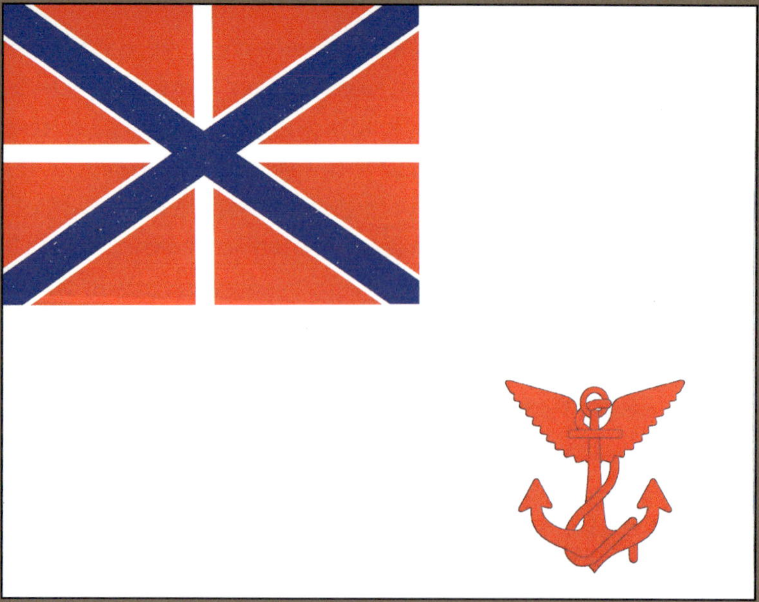

Neuigkeits

Welt-Blatt

Herausgeber: August Kirsch.

Nr. 224 — Wien, Donnerstag, den 1. Oktober 1914 — 41. Jahrgang.

Die russische Offensive in Galizien im Zusammenbrechen.

Ein denkwürdiger Armeebefehl.

Neuer großer Sieg in Frankreich bevorstehend.

Der Widerstand der Serben beginnt zu erlahmen.

Die Beute der Deutschen bei Tannenberg.

Zerlegte russische Flugmaschinen, die nach Berlin transportiert werden.

Library of Congress

Austria's Welt-Blatt acquired the photograph and shared the historic moment on the front cover that aeroplanes were now a special form of booty in warfare.

A Lasting Tribute to German Aviation

A widespread and exhaustive examination of intelligence sources, credits *Fliegertruppe* aerial reconnaissance with the triumph at Tannenberg. A total of 110 *Fliegertruppe* sorties were generated during the seven days from 24 to 31 August. *AOK 8*'s decision to allow subordinate *Armee-Korps* echelons to task *Fliegertruppe* (to cover targets), enabled them to ascertain timely enemy intent and impeding actions. Just as critical, was *8. Armee*'s *FFA 16, Festung* Königsberg's *FestFA 5,* and Airship Z-IV's collective ability, to insure effective reconnaissance over First Army.[14] In contrast, total *RIVVF* activity amounted to approximately 20 successful sorties. The absence of Russian aeroplanes flying over the battlefield, contributed to the catastrophic dearth of timely information needed to track German forces in maneuver throughout *Ostpreussen*.[15]

Major Wilhelm Siegert, senior Fliegertruppe officer, was present when Hindenburg gave praise to the aviators at Tannenberg. Siegert had a long career in aviation prior to the war. He led the formation of Brieftauben-Abeilung Ostende (BAO) on the Western Front. Later in 1916, Siegert became Inspekteur der Fliegertruppen (Inspector Air Service) stationed in Berlin.

HEADLINES

62. Jahrgang No. 412

Berliner Volks-Zeitung

Abend-Ausgabe
Sonnabend, 29. August 1914

Eine russische Niederlage in Ostpreußen.

Fünf Armeekorps und drei Kavalleriedivisionen geschlagen und über die Grenze getrieben.

The Berliner Volks-Zeitung proclaims "A Russian Defeat in *Ostpreussen*" on 29 August 1914.

Nr. 407. Morgenausgabe. Dienstag, 1. September 1914.

Königsberger Hartungsche Zeitung.

Rußlands Sedan: 70000 Gefangene.

Ganz Ostpreußen befreit und erlöst!

Königsberger Hartungsche Zeitung on 1 September 1914 compares the Russian defeat at Tannenberg to what France experienced in 1870 in the historic battle of Sedan with "Russia's Sedan: 70,000 Prisoners."

Вечернее Время

Основатель А. С. Суворинъ

ВОЙНА.

НА РУССКОМЪ ТЕАТРѢ ВОЙНЫ. Австрійцы приготовились къ отступленію за Львовъ.—На сѣверной границѣ Галиціи рѣшается участь Львова.

ФРАНЦІЯ. Французы подвигаются въ Лотарингіи.—Армія кронпринца разбита.—Франко-англійскія войска отступили къ югу.—Въ бою съ арміей ген. По погибло 50.000 Германцевъ.—На Маасѣ германскія войска понесли большія потери.

БЕЛЬГІЯ. Отступленіе германскихъ войскъ изъ Бельгіи на сѣверо-востокъ.

ГРЕЦІЯ. Греческая эскадра ушла въ направленіи Дарданеллъ.—Тысячи русскихъ путешественниковъ ѣдутъ въ Россію.

НА ДАЛЬНЕМЪ ВОСТОКѢ. Японцы заняли у Кіао-Чао три германскихъ острова.

ТУРЦІЯ. Спѣшная мобилизація турецкой арміи.—Въ турецкомъ обществѣ считаютъ, что Энверъ-паша ведетъ Турцію къ уничтоженію.

ПЕРСІЯ. Нѣмцы и Турки пытаются вовлечь Персію въ войну.—Персы остаются на сторонѣ Россіи.

Отклики войны.

Обезумѣвшая Турція. (Письмо изъ Константинополя).

Рисунки.

Больше бодрости.

Погибшіе генералы.

The Russian military's primary news for military veterans, Russian Invalid, announces the death of Gen.kav Samsonov on 1 September 1914.

New York Tribune

WEATHER

Vol. LXXIV No. 24,739 NEW YORK, SUNDAY, AUGUST 30, 1914.—SIX PARTS—FIFTY-SIX PAGES PRICE FIVE CENTS.

Germany Rushes Troops from France to Strike at Foe in Eastern Prussia; French Offensive Resumed with Success

New York Tribune reports on 30 August 1914 the impact of GO v. Prittwitz's decision after Gumbinnen requesting more German forces to hold back the Russians. On that day, 8. Armee triumphed at Tannenberg. The impact on the German advance towards Paris was soon to be felt. Source: Library of Congress

Fliegertruppe was praised for its role in defeating the Russians at Tannenberg. Hindenburg reportedly told Major Wilhelm Siegert, "*Ohne Flieger kein Tannenberg!*" ["Without flyers, no Tannenberg!"] and to have remarked to Italian news correspondent Casabino-Bando: "However, the most remarkable people are my airmen. I cannot begin to tell you the remarkable things that they have done on their reconnaissance missions." Aviation not only helped validate the signal intelligence collection on Russian force dispositions, it also played an additional role, in covering for signal intelligence success—Hindenburg's famous utterance helped divert attention away from Königsberg's signal intelligence collection.[16]

Hindenburg's revealing comments on aviation's impact at Tannenberg, has been widely quoted throughout the century. *Fliegertruppe* official history of air operations in 1914 echoed the sentiment:

> *Generaloberst* von Hindenburg, thanks to the excellent results of his air service, supported by intelligence from intercepts of the Russian command radio traffic, was always well informed about the adversary ... The flyers earned particular merits by their reports of Russian relief attempts at the end of the battle...The question of whether aerial reconnaissance had a crucial role during the operations can only be answered with "yes." As even prior to the onset of battle, the flyers, by detection of the moves of both Russian Armies, had already delivered the essential preconditions to the German commanders, that were necessary for the planning and disposition of the operation against the Narev Army.[17]

PART IV

MASURIAN LAKES CAMPAIGN

CHAPTER XI

"Our Hands Were Free to Act Against Rennenkampf."

Post-Battle Changes Underway

Tannenberg was a total victory. Hindenburg's feelings at the time were reflected in his post-war memoirs, "What secrets were hidden behind the southeast of the Masurian Lakes, we didn't know."[1] *Oberstleutnant* Hoffmann offered a more emphatic statement, "Our hands were free to act against Rennenkampf."[2] Now operating from Allenstein, *GO* v. Hindenburg was ordered to defeat *Gen.kav* v. Rennenkampf's First Army and drive the remaining Russians out of *Ostpreussen*. *AOK 8*'s estimates of *Gen.kav* Rennenkampf's First Army forces showed *Gen.inf* Smirnov's Russian XX Corps on the Deime and *Gen.inf* Epanchin's III Corps (including a division from XX Corps) in the region of Friedland-Allenburg. *Gen.art* Eris-Khan's IV Corps and 5th Rifle Brigade were assessed to be behind the Onet river near Gerdauen.[3]

It was a different German army that arrived on the scene for the next phase—the battle for *Ostpreussen*. *Garde-Reserve-Korps*, *XI. Armee-Korps* and *8. Kavallerie-Division* were transferred by *OHL* from the Western Front to reinforce *8. Armee*'s new mission. Of the three, only *XI. Armee-Korps* had aviation support with *FFA 28*.[4] *GO* v. Hindenburg and *Genmaj* Ludendorff deployed the two *Armee-Korps* and *8. Kavallerie Division* to northern *Ostpreussen*, east of Königsberg. This allowed *8. Armee* time to reposition the Tannenberg veterans to fight Rennenkampf's First Army.[5] The previous week clarified for German commanders that the optimal way to execute battle plans moving forward (particularly in finding existing gaps and maneuver space for possible envelopment) was to rely more heavily on aviation and aerial reconnaissance.[6] *Fliegertruppe*'s *FFA* serving the *Ostpreussen* sector each had six aeroplanes. An additional ten *FestFA* aeroplanes flew from Königsberg, Graudenz, and Thorn.[7] *Fliegertruppen* operations in the region received a major boost when *Etappen-Flugzeug-Park 8* was established at Dirschau to repair and replace airframes and engines. Additional aircrews for the expanding fleet were now being trained at Dirschau.[8]

SORTIE DATE: 1 SEPTEMBER, 1914

AEROPLANE: Rumpler-Taube

OBSERVER: Obltn Körner

GROUND TRACK: Deutsch-Eylau-Soldau-Mława-Ciechanow-Pragnitz-Chorzele-Willenberg-Ortelsburg-Neidenburg-Deutsch-Eylau

[Ciechanow is consistently misspelled in the report: Cichanow]

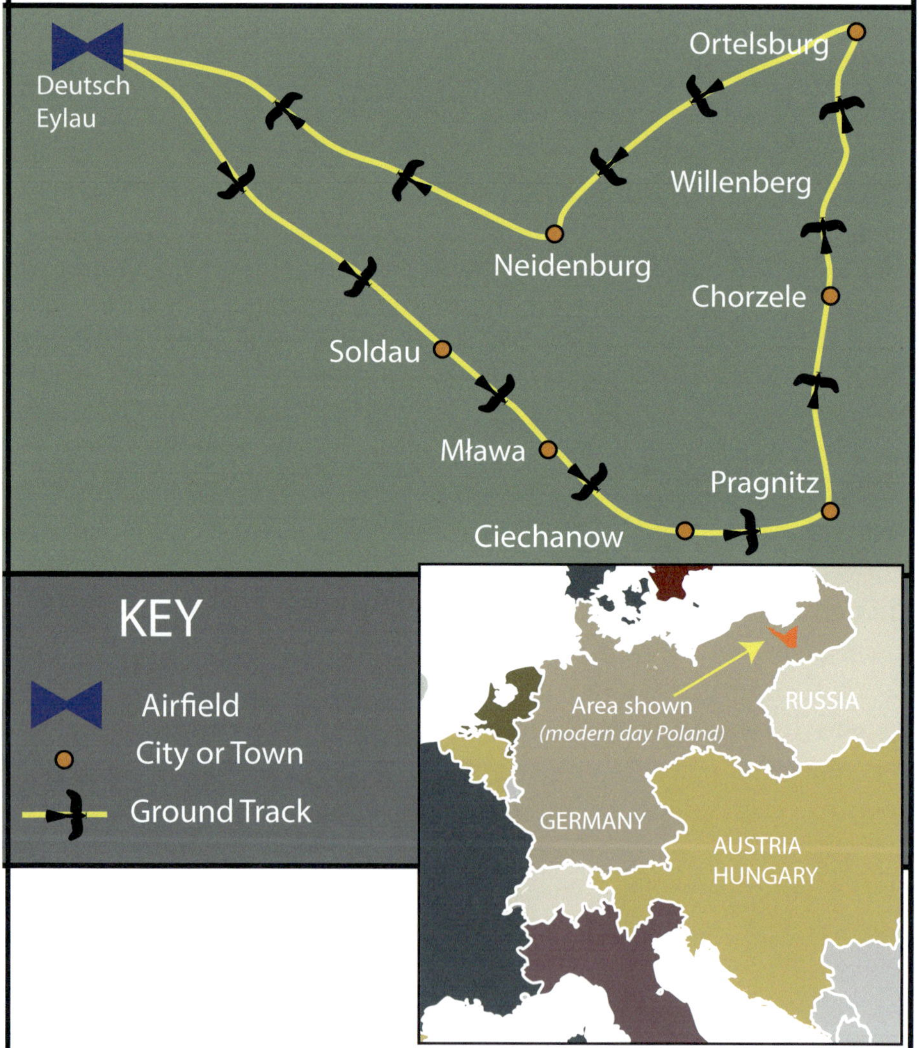

SORTIE DATE: 1 SEPTEMBER, 1914

FFA 14 conducts sortie to southern Ostpreussen to find retreating Russian forces. The next campaign is set to drive out Gen.kav v. Rennenkampf's First Army.

Source: NARA RG 165

SORTIE AVIATOR REPORTS: 1 September, 1914

1. 7:12 Railroad Station Mława empty. Road Soldau-Mława no enemy. On road Mława-Bogurzyn about 1 Brig. Inf. marching on Bogurzyn. End at Mława. 7:15 am at road Mława-Wizniewo close by Mława one camp of about one Inf. Rgt.
2. On Railroad Mława-Cichanow 1 long train, from Cychanow to Mława 1 single locomotive.
3. On road Mława-Wyszny 1 train column about 200 m long.
4. Between forest about 10 km northwest Cichanow and railroad infantry trenches.
5. North Railroad Station Cichanow 1 airfield, southwest close to Railroad Station one big camp; mainly columns and baggages, troops also about 1 Rgt to 1 Brig.
6. Northeast of Cichanow at road to Pragnitz one camp about 1 to 2 Regts.
7. On road to Pragnitz one column Bag. and Tr. (5km northeast Cichanow. At Golani 1 camp 1 - 2 Regts. Inf. On Road Golani - Pragnitz 1 column Baggage. Heading on Cichanow.
8. At Pragnitz 1 big park of baggages (8 am)
9. Road Pragnitz-Sebory-Chorsele free of enemy, dito Chorsele.

   On road to Willenberg nothing seen, to south own troops.
10. To Ortelsburg also no sightings. Dito in the forests norteast Neidenburg no enemy.

9:30 Report to Neidenburg

Fliegertruppe Aerial Reconnaissance – the Next Phase

L*tn* Mahncke recalled a major structural shift, when *Fliegertruppe* become more efficient in meeting operational demand:

> There came loud demands for the establishment of more *Fliegerabteilungen* on the Eastern Front although no one had any idea how to procure more pilots and aeroplanes. Our *Flugzeugparks* were unable to fulfill all requests and this unfortunately led to a 'go-grab-it-yourself' approach by some units who sent officers back home to appropriate everything they required. The result was confusion at our depots, promising imminent disaster. The central supply office in Berlin had only limited authority and could not stop the poaching and, in addition, flying units in the West were given preference because this front was considered more important by the Supreme Command and His Majesty. In the end, Ludendorff had to get involved. He stamped out gross inefficiency and created a new post: *Stabsoffizier der Flieger beim Oberkommando* (Staff Officer, Air) with Major Kuckein promoted to this post.[9]

Given the August experience, *Fliegertruppen*'s priorities were further defined in an effort to effectively conduct the next phase of the campaign. The area of coverage for sorties extended out to 140 kilometers from the landing ground. Target priority was First Army locations, especially rear echelons. The *FFA* commander took on the responsibility of managing aviation resources in response to higher headquarters tasking. Experienced aviators governed sortie time, routes, and frequency to support the mission.[10]

As for German perceptions of Russian aviation at this time, *8. Armee Nachrichten-Abteilung* [Intelligence Section of *8. Armee*'s *Generalstab*] reported that the limited numbers of Russian aeroplanes noted in the campaign suggested they were only trying to save their aviation assets and not put them at risk flying over German-held territory.[11]

Preparations Against First Army

Favorable September weather in the north allowed *Fliegertruppe* to fly throughout *Ostpreussen* and surveil Russian positions. Sorties over the southern *Ostpreussen* region confirmed the Russians posed no great threat. To the north, Rennenkampf's First Army appeared to be withdrawing its northern wing along the Pregel River and concentrating forces at the edge of the Masurian Lakes. The majority of aerial reconnaissance reports showed Russians withdrawing their far northern sector to position in strength the Gerdauen – Nordenburg sector. As for the remains of the Russian Second Army in the south, aviators noted elements retreating to the Narev River. What was left of *Gen.inf* Blagoveshchensky's VI Corps was observed moving south to Ostrolenka.[12] *8. Armee* generated its reconnaissance order to cover the remaining Russian-held *Ostpreussen* territory and the adjoining border. *Fliegertruppe* unit commanders were tasked to continuously monitor the following assigned sectors:

> *Gouvernment Festung* Konigsberg [*FestFA 5*] up to the Tilsit-Stalluponen line.
>
> *Garde-Reserve-Korps* [*FFA 16* in support] beyond the Allenburg-Gerdauen line to Insterburg.
>
> *I. Reserve-Korps* [*FFA 37*] to the Insterburg – Darkehmen line via Gerdauen.
>
> *XI. Armee-Korps* [*FFA 28*] to Darkehmen – Goldap line via Nordenburg and Angerburg.
>
> *XX. Armee-Korps* [*FFA 15*] to the Goldap-Filipowo-Marggrabowa line via Lotzen.
>
> *XVII. Armee-Korps* [*FFA 17*] to the Marggrabowa-Lyck-Grajewo line via Johnnisburg.
>
> *I. Armee-Korps* [*FFA 14*] to Lomza-Ostrolenka (on the Narev River) via Kolno.
>
> *8. Armee* [*FFA 16*] employed as quick reaction reconnaissance force serving on demand tasking.[13]

3 September *Armeebefehl* read, "*8. Armee* will advance in order to attack the Rennenkampf Army. Shifting of forces must be taken into account during the operation." The Germans were to proceed to jump-off positions east of Ortelsburg – Bischofsburg – Heilsburg and prepare to advance. To the north, substantial sightings were reported, especially around Angerburg and Nordenburg. *Festung* Königsberg's *FestFA* 5 generated considerable excitement when the

SORTIE DATE: 2 SEPTEMBER 1914

AEROPLANE: LVG B.I.

PILOT: Ltn Gierke

GROUND TRACK: Osterode - Bartenstein - Schippenbeil - Rastenburg - Lötzen (Intermediate landing Fortress Boyen) - Kruglanken - Angerburg - Darkehmen - Nordenburg - Gerdauen - Rastenburg - Lötzen

SOURCE: NARA RG 165

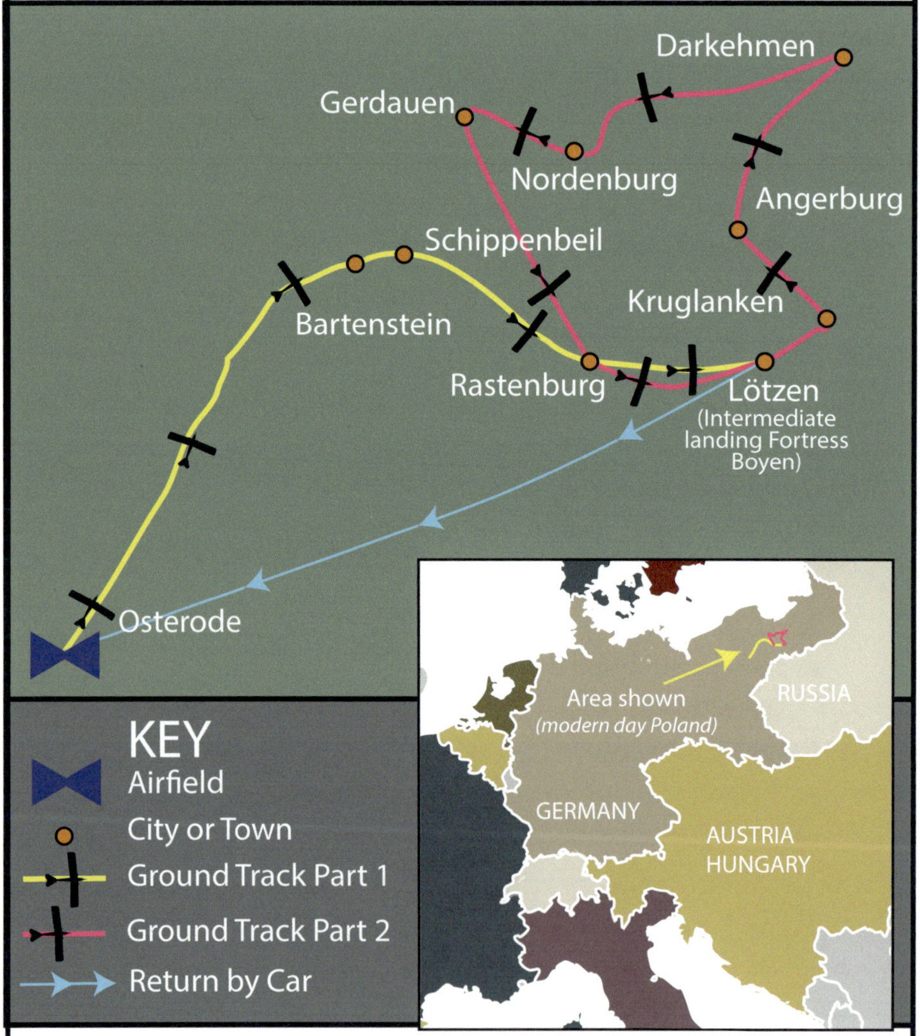

SORTIE DATE: 2 SEPTEMBER 1914

On 2 September 1914, FFA 16 covers the Masurian Lakes region to ascertain locations of First Army prior to 8. Armee moving the four Armee-Korps into position for the next offensive. Source: NARA RG 165

SORTIE AVIATOR REPORTS: 2 SEPTEMBER 1914

1. Kruglanken - Possessern - Ogonken occupied by weak infantry (est. 1 InfRgt); between Ogonken and Gruneberg infantry trenches and 2 battery cuttings.
2. Strong column marching (est. 1 InfDiv minimum) road Nordenburg - Reuschenberg apparently sesenbling 12.40 pm heading to east. In Nordenburg - Gr.Pentlack - Sawadden - Klonofken - Wickerau bivouacs and car parks still not occupied. In Darkehmen (south edge) bivouacs and car parks, apparently belonging to Columns and Trains.
3. Roads Nordenburg - Insterburg up to Szallgirren free. 12.00 - 12.35 [pm] Darkehmen - Kowarren - Nordenburg - Gerdauen, Gerdauen - Barten - Rastenburg - Lötzen.
4. Empty bivouacs on western edge Rastenburg near Alt- and Neu Rosenthal.
5. Airplane B/139 forced to return to Lötzen due to stormy weather and reduced revolutions of engine.

Upon landing propeller and wings partly damaged. Repair impossible here, airplane will be dismounted and stored in Fortress Boyen.

Pilot and observer will return to Feldfl.Abt.16 by car.

by pidgeon to Musshoff, OLtn, A.O.K. Observer

aircrew interpreted activities at the northern Deime line as Russian attack preparations. The erroneous reporting was in fact Russians establishing defensive positions. The northern *Ostpreussen* area covered by Angerburg – Nordenburg – Gerdauen – Allenburg noted Russians working on entrenchments to reinforce a strong fortified line that held the bulk of the army.[14]

Indications from the various intelligence sources, suggested the Russians were about ready to initiate a new offensive. *Fliegertruppe* reported Russians at the Angerapp River and Alle River; a warning that they were preparing to cross the Pregel and Deime rivers. All *Armee-Korps* were directed to close up as much as possible.[15] *AOK 8* envisioned an envelopment of the southern flank of the Russian forces by *I. Armee-Korps* and *XVII. Armee-Korps*. The remaining four *Armee-Korps* were to advance against the center of Rennenkampf's First Army and hold the Russians in place. *AOK 8* issued their 4 September *Armeebefehl* directing, "*8. Armee* will advance in order to attack the Rennenkampf Army. Shifting of forces must be taken into account during the operation."[16] *8. Armee* forces proceeded to their jump off positions with *Fliegertruppe* providing updates on changes to Russian positions. Principal attention was paid to the extreme northern flank of Rennenkampf's First Army along the Deime River westward between its junction with the Pregel River and Kurisches Haff (Curonian Lagoon). *8. Armee* further secured the southern flank covering the 160 kilometers of border between *Gouvernment Festung* Graudenz and the village of Chorsele. *Fliegertruppe* reported little Russian activity in southern *Ostpreussen*.[17] On the evening of 4 September, aviator reports described the Russian line between Angerburg—Gerdauen—Allenburg. The concentration of Russian forces between the Masurian Lakes and the Pregel River gave *AOK 8* an understanding of Russian resistance against an *8. Armee* attack.[18]

Inclement weather the following day, resulted in *AOK 8*'s increasing dependence on reports from aviators. The intelligence they received, never provided them the clarity they needed. Information from German residents in southeastern *Ostpreussen*, indicated large numbers of Russians—a force as great as three corps. This surprised *AOK 8* intelligence as reporting up to that time, had not observed forces in the area. The absence of aerial reconnaissance throughout the day, contributed to a limited understanding of Russian intentions. On 5 September, Second Army acquired a new leader, *Gen. kav* Scheideman, former commander of II Corps. *General ot infanterii* Vasily Flug's Tenth Army entered the Northwest Front to help fill the void from the losses suffered at Tannenberg.[19]

On 6 September, Hindenburg made the final decision to create a combat group under the command of *GdI* v. François. Their mis-

SORTIE DATE: 4 SEPTEMBER 1914

AEROPLANE: LVG B.I.

PILOT: Ltn Hagen

GROUND TRACK: Osterode – Allenburg – Nordenburg – Angerburg – Osterode

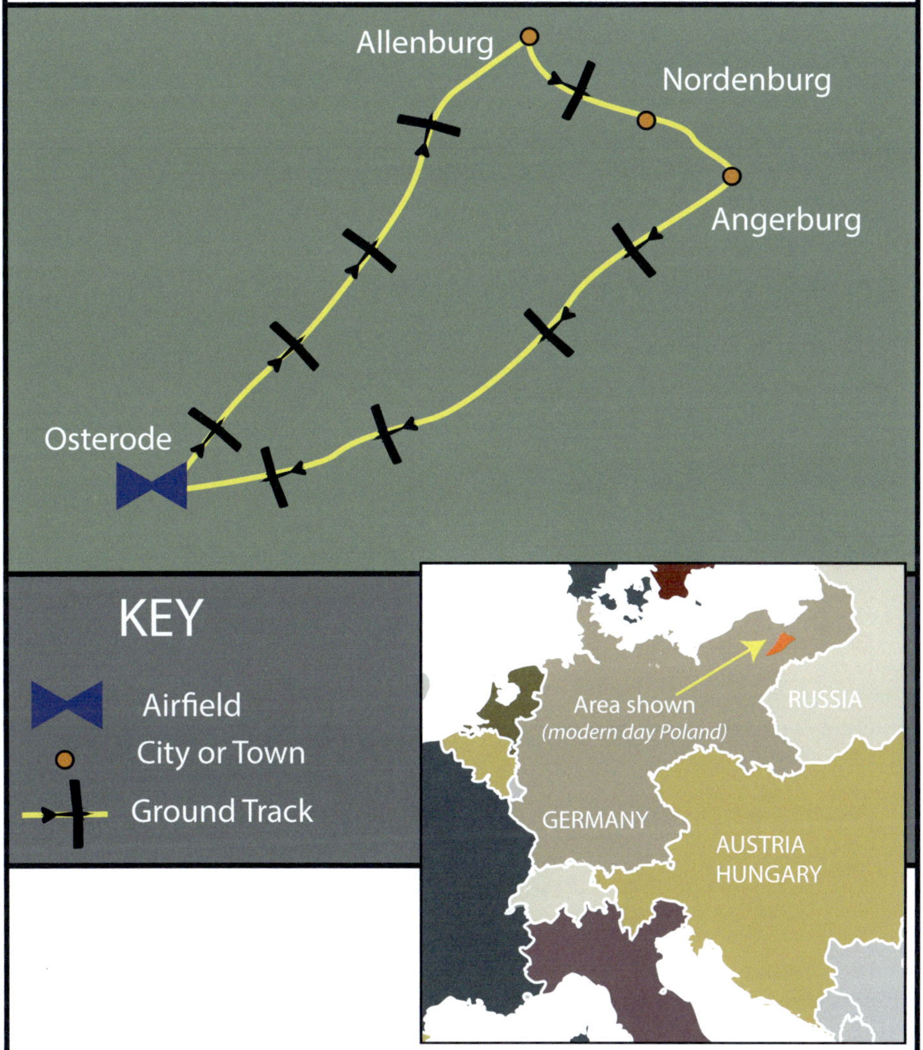

SORTIE DATE: 4 SEPTEMBER 1914

On the evening of 4 September, aviator reports described the Russian line between Angerburg—Gerdauen—Allenburg. The concentration of Russian forces between the Masurian Lakes and the Pregel River gave AOK 8 an understanding of Russian resistance against an 8. Armee attack. Source: RG 165

Feldfliegerabteilung 16

West Wandlacken around Gerdauen to Trausen strong field fortifications. Multiple lines after each other. Batteries with straight fire line on Omet (stream). On left banks on western rim of Gerdauen work in progress. There (at Gerdauen) and to east, also northest multiple isolated strongpoints.

Work near Allenburg have just begun.

Observer: OLt Richter

Flyer report about Russian positions 4/9 night (at night of Sept. 4) About 1:300,000 (scale of map sketch) Not flown to north of Allenburg

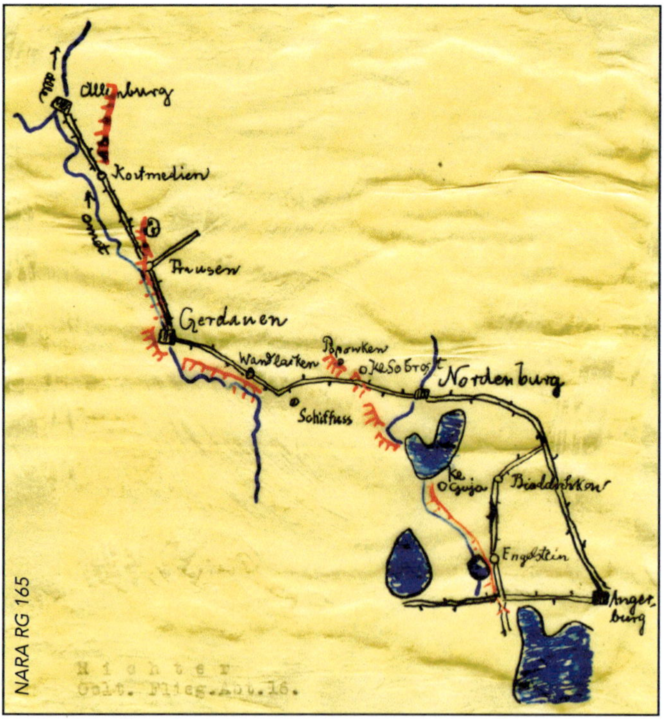

Hand drawn map by the observer Oberleutnant Richter shows enemy positions from Allenburg – Nordenburg – Angerburg.

sion was to strike at the First Army's rear echelon by executing a flanking operation through the Lötzen Gap and via Johannisburg.[20] *Fliegertruppe* now moved forward. *FFA 14* left Passenheim for Sensburg. *FFA 17* also left Passenheim but relocated to Rheinfelde. *FFA 15* departed Allenstein for Rossel. *FFA 28* also left Allenstein and now operated from Bischofstein. Finally, *Festung* Graudenz's *FestFA 6* stationed two aeroplanes at Soldau.[21]

Favorable weather enabled aerial reconnaissance sorties.

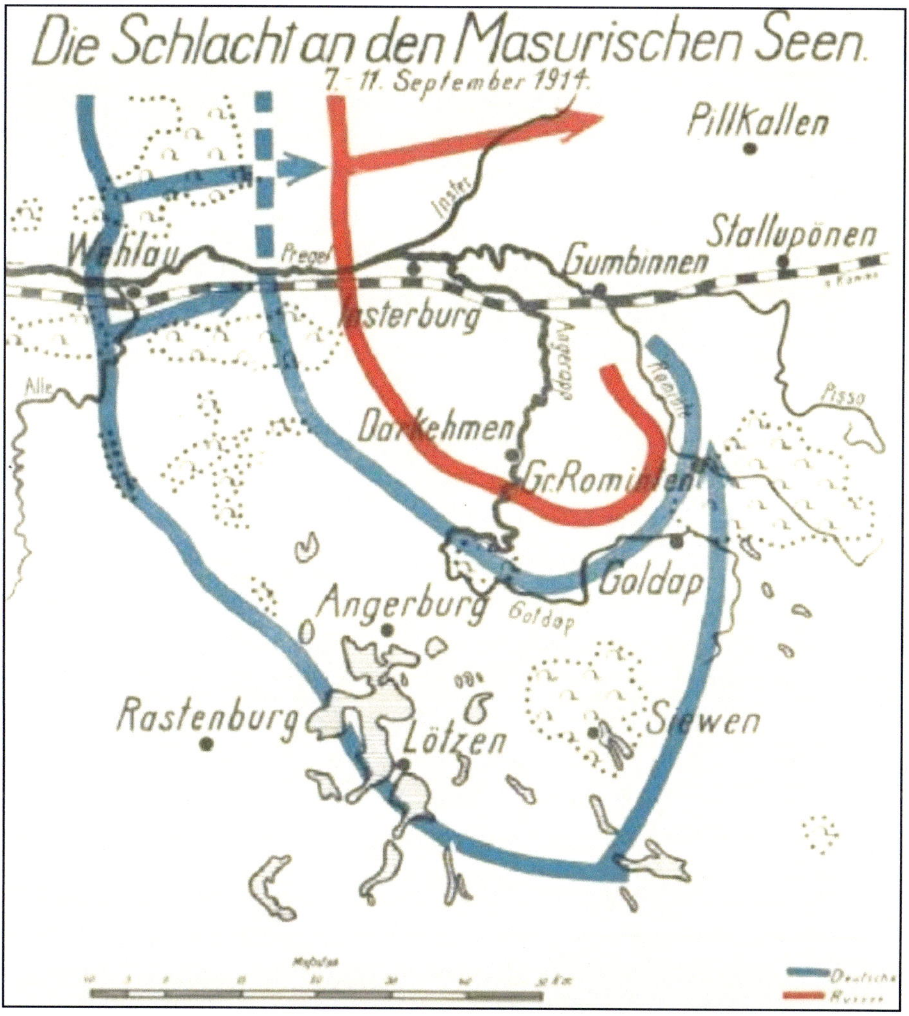

Under a newly formed combat group commanded by GdI v. François, 8. Armee commenced planning a strike at the First Army's rear echelon by executing a flanking operation through the Lötzen Gap and via Johannisburg.

Fliegertruppe reported Russians massing east of Johannisburg and at Lomsha. Longer-range sorties had no significant sightings. Unbeknownst to *AOK 8*, *General ot infanterii* Vasily Egorovich Flug's Tenth Army was approaching, with the goal of encircling *8. Armee*.[22] *RIVVF* units such as *15th KAO* (flying Nieuport IV monoplanes previously assigned to the defeated Second Army) were now ordered south to Warsaw to regroup and support Flug's Tenth Army.[23]

At Lyck in southeast *Ostpreussen*, increased activity suggested the Russians were now in the process of closing the gap between their First Army and remnants of *Gen.kav* Scheideman's Russian Second Army. *AOK 8* made immediate changes to the flanking maneuvers in progress as they received confirmation of Russian positions from *Fliegertruppe* reports. They confirmed Russian reinforcement of defensive positions along the Pregel River and Deime River. *Festung* Königsberg *FestFA* 5 estimated enemy strength in the area to be at three corps—at first glance, this appeared to be an exaggeration to *8. Armee* analysts. Later postwar assessments confirmed the aviators were correct.[24]

Fliegertruppe operations for 7 September were curtailed due to inclement weather. While aviators stood down, *Radiogross-Station* [large wireless station] Königsberg transmitted a deceptive telegram that read: "To Commanding General of the Guard Corps, Immediate attention-Urgent. Guard Corps will close up by tomorrow, close to the west of Labiau. Detrained parts of the *5. Armee-Korps* will ... [proper ciphers follow this] The Army Commander." After transmitting the false signal, *Radiogross-Station* Königsberg sent a back-channel transmission confirming, "This message was intercepted immediately by a Russian station and under 'URGENT' was transmitted." Wireless transmission deception now entered into the battleground.[25]

Festung Königsberg's *Fest FA 5* was held in high esteem by *AOK 8* for their stalwart reporting on First Army over the previous month. Now, they directly supported *AOK 8*'s main operations. *Fest FA 5* reported on 8 September that the Russian bivouacs at Insterburg were deserted. The reports confirmed Rennenkampf's First Army was establishing defensive positions to the north at Lötzen.[26] Addi-

AERIAL PHOTOGRAPHY

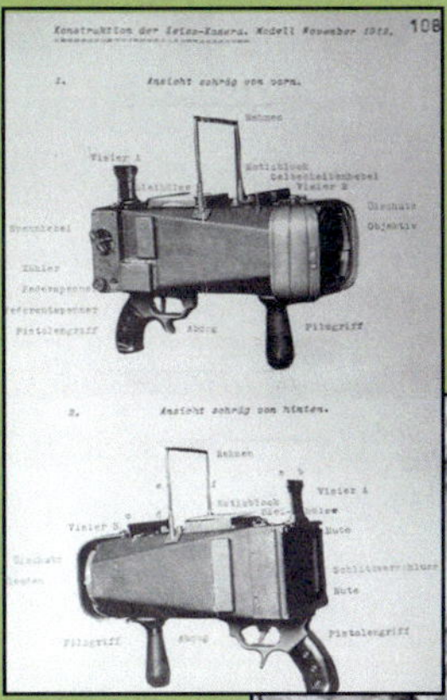

On 8 September 1914, Fliegertruppe employed the Zeiss-Kamera, Modell November 1912/14, to shoot aerial photographs of Russian field fortifications opposite Lötzen.

Below: A glimpse of the future for Fliegertruppe, working with an aerial camera.

tional *Fliegertruppe* aerial reconnaissance confirmed Russian positions extending southeast of Angerburg to the Gablich Lake. *Genltn* v. Below's *I. Reserve-Korps* cautioned they were going to advance slowly based on what they learned from the reports. At 1350 hours, aerial reconnaissance reported the main Russian position at Gerdauen had been evacuated. *AOK 8* lacked confidence in the aerial reconnaissance report and ordered *I. Reserve-Korps* to proceed cautiously.[27] Evening reports noted Rennenkampf's forces were preparing to advance, but provided no additional insight on Russian intentions. *AOK 8* speculated a Russian maneuver was in progress to attack the German's perceived weak left wing. The Russians had only two corps in the sector (XX Corps and XXVI Corps).[28]

Fliegertruppe's observations provided the crucial data for *AOK 8*'s assessment, as to Zhilinsky's perceived objectives. Zhilinsky was just alerted to the fact that two *Armee-Korps* and *8. Kavallerie Divi-*

sion had arrived in *Ostpreussen*—five days after their actual arrival.[29] Ongoing coverage of the south showed no Russian threat. The Second Army was in the process of being rebuilt with newly arrived reinforcements.[30] *8. Armee*'s main force was to strike First Army's greatest concentration between the Pregel River and north of the Masurian Lakes. Both Zhilinsky and Rennenkampf took measures to avoid a repeat of Second Army's catastrophic defeat. First Army maneuvers now strengthened their rear flanks. Aviators noted Russian units moving from Makow to Przasnysz and marching north to Sczuczyn – Grajewo. As the forces maneuvered, *GdI* v. François' *FFA 14* detected Russian forces marching towards Lyck. He responded by repositioning his right flank to hit First Army's rear echelons.[31]

8 September was a milestone for *Fliegertruppe* aerial reconnaissance on the eastern front. For the first time, aerial photography became an integral element of the reconnaissance process. *Ltn* Fink's prewar emphasis on employing aerial photography, was initiated by *XVII. Armee-Korps' FFA 17. Fliegertruppe* employed the Zeiss-Kamera, Modell November 1912/14, to shoot aerial photographs of Russian field fortifications at Lötzen. The photographs also annotated the Russian artillery positions. Timely intelligence was delivered to the German artillery headquarters via *Flieger-Meldung*.[32] The addition of aerial photography, increased the confidence in *Fliegertruppe* reports. In the coming months, aerial photography production led to standardized reporting, print development, and distribution to combatants.

The bounty of reliable and expansive *Fliegertruppe*n coverage gave Hindenburg confidence to attack the following day. *Genmaj* Ludendorff recalled the moment:

> This operation also was extraordinarily daring. … the Russian Niemen [Rennenkampf's First Army] Army … was very much stronger than *8. Armee* … The Russian fighting strength was further increased from four to six divisions, which were being assembled round Osowiec and Augustowo … Even in such a situation as this, we did not hesitate … to venture into battle. Our superior training was in our favor. Tannenberg had given us a great advantage.[33]

CHAPTER XII

Battle of the Masurian Lakes

The First Day of a New Battle

The Battle of the Masurian Lakes began on 9 September at 0330 hours. *GO* v. Hindenburg was unaware that *Gen.kav* v. Rennenkampf remained extremely cautious. Rennenkampf was consumed by the desire to avoid a repeat of Samsonov's fate at Tannenberg. He was in a controlled state of desperation. *AOK 8*'s analysts pushed *Fliegertruppe* to confirm First Army location and estimates of unit size and strength. When reports reached *AOK 8*, they clarified that Russians had commenced their withdrawal that evening, and that many front-line positions were scaled down. At 1035 hours, *Festung* Königsberg's *FestFA 5* reported a Russian division moving east from Allenburg. *XI. Armee-Korps' FFA 28*, reported Russians had given up their positions at the Deime River line and were observed marching north of the Pregel River toward Insterburg. Field fortified positions northwest of the Pregel River remained unmanned. *Fliegertruppe* reports were confirmed by ground reconnaissance patrols. At 1115 hours, they noted two large enemy columns leaving their positions southwest of Nordenburg and moving eastward.[1]

Two reports from *XVII. Armee-Korps FFA 17* confirmed Russian force concentrations northeast and east of the Masurian Lakes. Key to the report, was that aviators noted Russian forces employing deceptive cover against *Fliegertruppe* aeroplanes. Based on *FFA 17*'s reports, *XVII. Armee-Korps* changed their advance to the Gronden-Brosowken-Waldhof line.[2] Rennenkampf feared that *8. Armee* was in a maneuver to trap his forces to the south, particularly in the vicinity of the Lötzen Gap. *GdK* v. Mackensen's XVI*I. Armee-Korps* had already passed the Lötzen Gap and met severe Russian resistance eight kilometers northeast of Lötzen. Rennenkampf quickly sent his remaining reinforcements to *General ot infanterii* Alexey Evgrafovich Churin's II Corps to counter Mackensen's push into the Russian rear echelon. The fighting lasted all morning; that afternoon, the Russians left the battlefield for the east.[3] After a forced march of 115 kilometers in three days, *GdI* v. François' *I.*

On 9 September 1914, 8. Armee commences the offensive against First Army on the first day of the Masurian Lakes campaign. This map is available in a larger format at shootingthefront.com.

Armee-Korps struck II Corps at the flank and rear, causing the Russians to retire east in disorder.[4] Rennenkampf issued commands to his forces to withdraw.[5]

That afternoon, *AOK 8* was surprised by a *FFA 16* report that *Gen.kav* Scheideman's Second Army had been brought back to fighting strength and was observed advancing toward Przasnysz and

Myszinyec. Additional reports showed Russian troops (assessed as *Gen.inf* Flug's Tenth Army) approaching Ostrolenka – Szczuczyn – Grajewo – Lyck, confirming a risk to François' southern outflanking wing. *Fliegertruppe* were called upon to further substantiate Russian positions. Subsequent flights by *FFA 14, FFA 15,* and *FFA 28,* reported no further reinforcements except to field fortifications.[6] *AOK 8* diverted a brigade to the south to counter that threat. Aviator reports now showed Russian troops moving southeast. Rennenkampf had assigned them to protect his rear Russian echelon at Darkehmen.[7]

In response to Zhilinsky's order, Rennenkampf commenced a general retreat to a rendezvous point 50 kilometers east of Gumbinnen. Hindenburg and Ludendorff directed *8. Armee* to pursue, "Inflicting maximum losses with the utmost effort on the retreating

RECONNAISSANCE TARGETS 9 SEPTEMBER

On 9 September 1914, AOK 8's analysts pushed Fliegertruppe to confirm First Army location and estimates of unit size and strength.

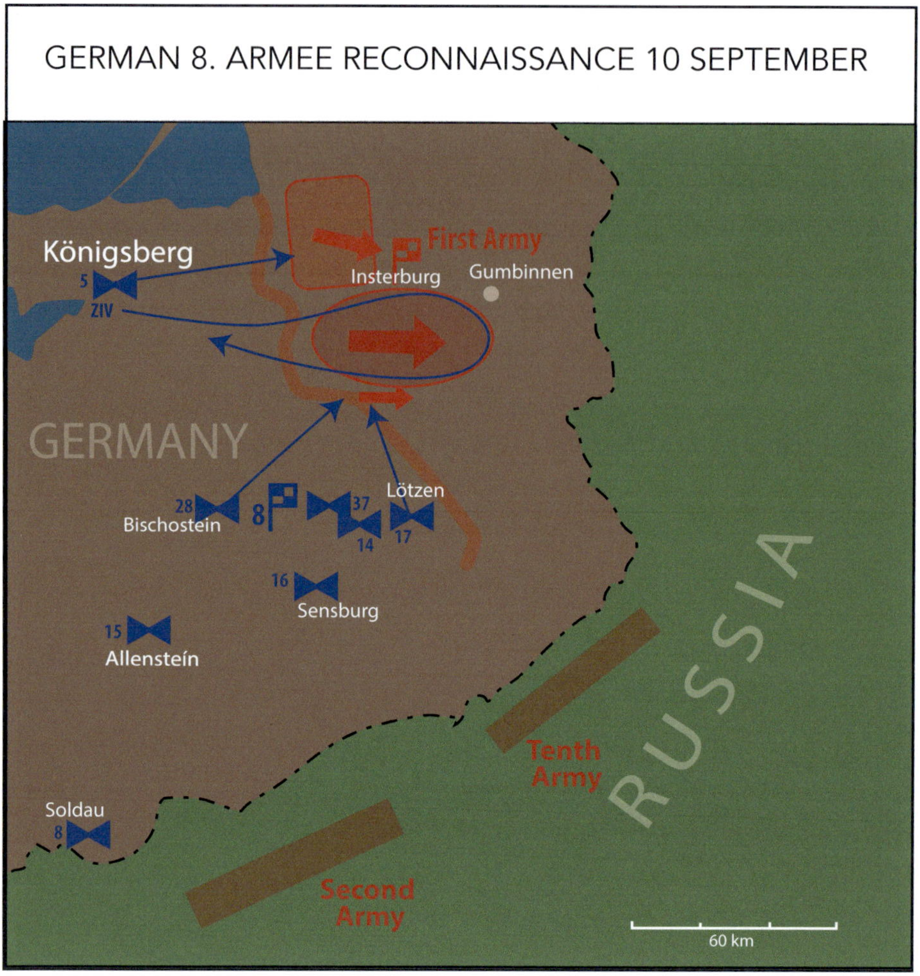

Rennenkampf commenced a general retreat to a rendezvous point 50 kilometers east of Gumbinnen. Source: Helmut Jaeger

enemy forces."[8] Zhilinsky ordered two divisions to counterattack GdA v. Scholtz's *XX. Armee-Korps* between Nordenburg and Angerburg. *XX. Armee-Korps* suffered heavy losses. Ludendorff feared that the counterattack could jeopardize the entire *8. Armee* advance. He held back *I. Armee-Korps* and *XVII. Armee-Korps*, in the event the Russians made substantial gains in territory.[9]

At this critical moment, *GdI* v. François ignored *8. Armee*'s direction to march on Wirballen with all his forces, in an effort to cut off the

On 11 September, Fliegertruppe generated an impressive 18 sorties. AOK 8 required information on the direction that First Army was heading.

Russian retreat. François diverted a division to finish off a remaining Russian rearguard force at Gawaiten, and committed only one of his divisions to advance on Wirballen. Hindenburg had to resort to persuading François to obey his command. In spite of an attempt at a reversal, François' error in splitting *I. Armee-Korps* in two, resulted in the Russians managing to escape through the resulting gap.[10]

That evening, Zeppelin Z-IV left Königsberg and surveyed the northern sector of the battlefield under clear skies and moonlight.

RUSSIAN GENERALS

General ot infanterii Nikolay Vladimirovich Ruzsky replaced Gen.kav Zhilinsky as commander of Northwest Front. Prior to his selection, he led the Third Army in the Southwest Front fighting in Galicia.

General ot infanterii Vasily Flug, Tenth Army commander until 23 September 1914. His limited time in command resulted in retaking Suwalki from 8. Armee. He had disagreements with Gen.inf Ruzsky and was transferred to Stavka.

General ot infanterii Faddey Vasilievich Sivers replaced Gen.inf Flug as Tenth Army commander on 27 September 1914. He held on to the command until 1915.

Over Insterburg, Z-IV dropped bombs on Russian positions. The Russians fired back with all available weapons striking the airship. Z-IV managed to continue the sortie and returned to Königsberg. Their 10 September report arrived at *AOK 8* at 0830 hours. Z-IV confirmed that Russian bivouacs and troop assemblies (reported previously by *Fliegertruppe*) still existed. No indication of Russian intentions was drawn from the intelligence.[11]

First Army Retreats

The Russians moved east—divisions marching in an endless column in parallel—throughout the evening. *Fliegertruppe* aerial reconnaissance confirmed that, "the principal Russian positions were only feebly occupied or not occupied at all."[12] Tasking to *Fliegertruppe* remained consistent—acquire observations on Russian positions and directions of troops retreating. *AOK 8* perceived an opportunity to pursue an attack on the Russian southern flank utilizing aerial reconnaissance. Rennenkampf took the appropriate steps to bolster his forces. The resulting battle east of the Masurian Lakes was fierce and bloody. As the struggle escalated, *Fliegertruppe* sorties confirmed that *Gen.inf* Flug's Tenth Army was not a factor. Aerial reconnaissance confirmed that the greatly depleted *Gen.kav* Scheideman's Second Army had halted its northern advance. On 11 September, *Fliegertruppe* generated an impressive 18 sorties. *AOK 8* required intelligence on where First Army was heading—either due east through Gumbinnen – Kowno via the railroad track or southeast to the Russian fortress at Olita along the Niemen River. Rennenkampf continued to retreat east to Stallupönen. Aided by superb flying weather, *FFA 14*, *FFA 17*, *FFA 28*, and *Festung* Königsberg's *FestFA 5*, did yeoman's work keeping track.[13] Along the southern *Ostpreussen* border between Mława and Lyck, *FFA 16* flying from Allenstein and *Festung* Graudenz's *FestFA 6* operating from Soldau, collectively covered the Russian's Second and Tenth Armies.[14] *8. Armee War Diary* for 12 September summed up the situation with:

> *8. Armee* troops continue to pursue in accordance with army orders. The enemy is retreating north of the Rominte Heide. Due to continuous resistance in and east of the Rominte Heide, the Cavalry Corps (two divisions) is

> being delayed. Repeatedly, it is advised of the necessity to pursue more quickly but it fails to reach the designated march-objectives, the Stallupönen-Wylkowyski road. The communication service with *I. Armee-Korps* has temporarily been discontinued, as the wireless station attached to the *Armee-Korps* was cut off from corps headquarters through the engagement near Tollming Kehmen. *XVII. Armee-Korps* has broken Russian resistance near Gawaiten east of Darkehmen and east of the Rominte.[15]

Heavy rainfall on 13 September severely hampered any attempt to conduct aerial operations.[16] The next day, *Fliegertruppe* confirmed that First Army was retreating to the northeast. One aviator reported dense march columns on the road from Wylkowyszki to Marjampol and Pilwiszki.[17] The aviators witnessed Russia's campaign in *Ostpreussen* coming to a close.

The Withdrawal from Ostpreussen

To Rennenkampf's credit, his army effectively withdrew from *Ostpreussen*. In just 50 hours, the Russians covered 85 kilometers (in challenging weather traversing roads in very poor condition). Rennenkampf protected his southern flank as *8. Armee* maneuvered to repeat an envelopment. Germany's victorious army suffered from exhaustion from weeks of constant battle. First Army managed to temporarily outdistance the Germans and established a new defensive line at the Niemen River.[18] The Germans now referred to him as Rennen von Kampf—a flight from the battlefield.[19] *AOK 8*'s focus on First Army, resulted in gaps covering *Gen.inf* Flug's Tenth Army. Sorties were flown over Augustow – Osowiec. Tenth Army was heading towards *8. Armee*'s southern wing. The Russian XXII Corps struck *3. Reserve Division* [Landwehr] at Lyck. Despite limited awareness, the Germans defeated the Russians and forced them back 30 kilometers east to Augustow. *8. Armee*'s southern flank was secured.[20] German accounts tauted Masurian Lakes as a major success, as the Russians were for the most part, driven out of *Ostpreussen*. The spoils of war with both prisoners and booty, did not measure up to Tannenberg. The German official war history,

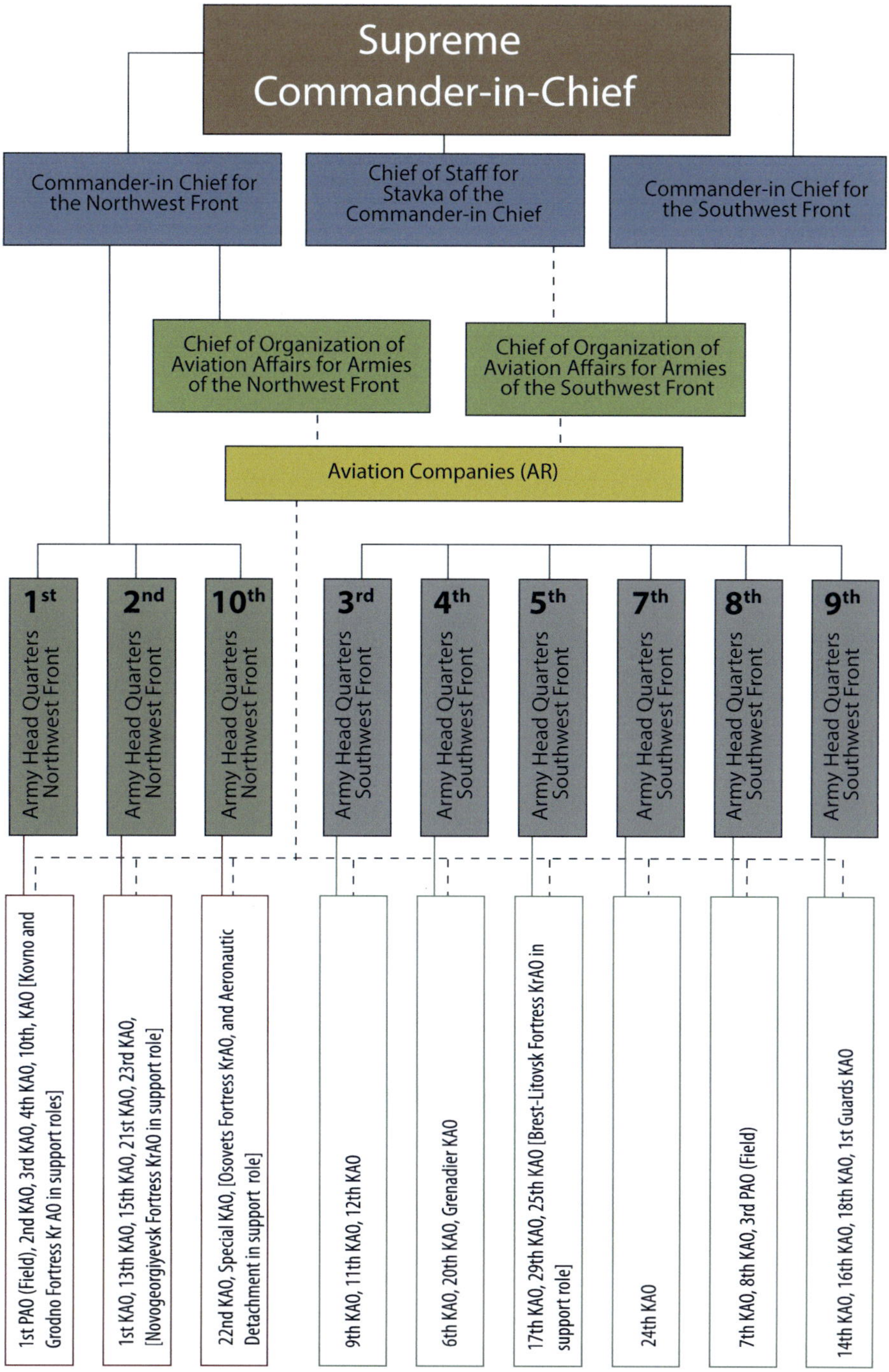

RIVVF organization changed to reflect the lessons learned from the war. Gen.inf Ruzsky initiated the formation of air groups, based on his experiences fighting in Galicia. Note the addition of Fortress KrAO now being aligned with the Russian army headquarters. Source: Carl J. Bobrow recorded taking 45,000 prisoners of war and 150 guns. Russian sources speak of a total loss of 100,000 men (70,000 killed and 30,000 prisoners of war).[21] *RIVVF*'s support to Russian forces at Masurian Lakes was minimal. German accounts of Russia's aviation role confirmed their presence on one airfield near Gumbinnen and a German aircrew sighted one *RIVVF* aeroplane.

As the battle wound down, *8. Armee* received a cable from *OHL* that directed major changes in strategy.

> Direct support for Austro-Hungary is necessary for political reasons. Operations from the area of Upper Silesia are to be considered. No more troops will be sent from the West. The amount of forces to be assigned to Silesia will be determined later. *8. Armee* will remain independent from the Austrians during joint operations.[22]

8. Armee War Diary for 15 September led off with "According to the report of an aviator, a Russian Brigade is located in Grajewo and a Division in Szczuczyn which is withdrawing toward Stawiski." *I. Armee-Korps, I. Reserve-Korps,* a cavalry corps, and main reserve forces from *Festung* Königsberg were keeping up with the pursuit of *Gen.kav* v. Rennenkampf's First Army. Critical to the shift in forces for the campaign was the assignment of the remaining *8. Armee* forces to the south to defend Silesia. *Genmaj* Ludendorff was transferred and appointed chief of staff "of an army (*9. Armee*) to be newly formed in Silesia" and under the command of *General der Artillerie* Richard von Schubert. In conclusion, the *War Diary* summed up campaigns with "The battle on the Masurian Lakes and the pursuit connected therewith, came to an end on this day."[23]

Changes Underway with Russian Command and Thinking

Critical mistakes made at Tannenberg resulted in a Russian mandate employing enciphered communications. Standard procedure now called for frequently-changed callsigns to provide appropriate cover for order of battle and unit deployment. On 14 September, *Stavka* insisted ciphers were required for all wireless transmitted military orders. Despite the change in security, it only took *k.u.k. Hauptmann* Porkorny, head of *k.u.k. Armee's Dechiffrierdienst*, five days to crack the Russian code.[24] Though improvements in secure communications had been implemented, *Gen.kav* Zhilinsky remained ignorant of the fact that *Gen.kav* v. Rennenkampf was no longer present in *Ostpreussen*. In fact, First Army had retreated to Kovno. Zhilinsky was heard to remark, "so improbable that it requires verification."[25] As Rennenkampf's First Army regrouped, Zhilinsky briefly considered ordering First Army to take the offensive, but reports of Russian casualties led him to abandon that directive.[26]

On 17 September, First Army's rearguard at Vilkoviski was sacrificed in their final battle with *8. Armee* forces.[27] *Gen.inf* Flug's Tenth Army now led Russia's fight against *8. Armee*. On 23 September, *General ot infanterii* Faddey Vasilievich Sivers replaced Flug and assumed command of Tenth Army.[28] Russian losses caused great anguish, prompting Grand Duke Nicholas to report to the Tsar, "I am inclined to think that *Gen.kav* Zhilinsky has lost his head and in general is not capable of controlling operations." Zhilinsky proceeded to blame his subordinates in his cable to *Stavka*. *Stavka* responded by removing Zhilinsky from command and replacing him with *General ot infanterii* Nikolay Vladimirovich Ruzsky, (current commander of the Third Army fighting in Galicia). Rennenkampf survived the *Stavka* purge and was able to retire without penalty the following year, after enduring a subsequent inquiry.[29]

Upon assuming command of the Northwest Front, *Gen.inf* Ruzsky initiated the formation of air groups, based on his experiences fighting in Galicia. In the first two months of combat, attrition of *RIVVF* aeroplanes spurred Russia's military leadership to rethink aviation. Many detachments had one to two operational aeroplanes left. *KAOs* were reassigned from fortresses to field armies. Aeroplanes were now an army-level resource.[30] On 27 September, *Stavka* established new manning criteria. Four aerial observers were required for each *KAO*. An *Aviatsionnaya Rota (AR)* [Aviation Company] was assigned to each army. Aviation commanders now reported directly to the chief of staff of the army they served.[31] *RIVVF* had to learn critical lessons from the past two months of combat on the Northwest Front. It's survival and credibility was dependent upon Russian military leadership's growing confidence in aviation. While operating opposite *Ostpreussen* on Russian ground, *RIVVF* was able to stabilize and subsequently contribute to building the fortified Russian force necessary to hold the advancing German armies at bay.

PART V

AVIATION SHAPES THE ONGOING BATTLES ON THE EASTERN FRONT

CHAPTER XIII

Changes in Command

9. Armee is Created

An argument between Central Powers in the east, reached a crescendo. Austro-Hungarian Archduke Friedrich contacted Kaiser Wilhelm and suggested German reinforcements (for the faltering Austro-Hungarian Galician front) be placed under *k.u.k. AOK* control. The Kaiser and *OHL* refused. To support the Kaiser's position, *GO* v. Moltke's replacement, *Generalleutnant* Erich von Falkenhayn, sent four corps as quickly as possible to Krakau. The goal was to launch an offensive operation to bolster *k.u.k. General der Infanterie* Baron Franz Conrad von Hötzendorf's retreating *k.u.k. Armee*. Subsequently, on 17 September, Kaiser Wilhelm ordered *GO* v. Hindenburg and his chief of staff, *Genmaj* Ludendorff, to assume immediate command of the newly established *9. Armee*. Hindenburg outranked Conrad.

Der Humorist

Generalleutnant von Falkenhayn replaced Generaloberst von Moltke as OHL in September 1914. 9. Armee was created during his command. He was promoted to General der Infanterie on 20 January 1915.

The decision was made by *OHL*, that Hindenburg assume control. The reason was that *OHL* only wanted Germans versus Austro-Hungarians to lead *9. Armee*. Conrad grudgingly had to give in. *GdA* v. Schubert and *Generalmajor* Paul Grünert replaced Hindenburg and Ludendorff as commander and chief of staff of the *8. Armee*.[1]

Schubert's initial command guidance to *8. Armee* stated, " ... to employ all means to destroy the Russian railroads within reach of *8. Armee* ... and to initiate sustained destruction of all Russian railroads by agents in Russia..." and ensure Germany's soil was protected from any future Russian incursion.[2] The next day, Hindenburg's *9. Armee* was tasked "... to advance toward the flanks and into the rear of the Russian Army Group which followed the Austrians." German operations were to be executed independently but in coordination with *k.u.k. AOK*. The four *Armee-Korps* comprising *9. Armee* left *Ostpreussen* and headed south to the Polish Salient.[3] On 19 September, Schubert and staff led *8. Armee* from Insterburg. On 21 September, Hindenburg and *AOK* 9 senior staff now operated from Beuthen in Upper Silesia.[4]

k.u.k. General der Infanterie Baron Franz Conrad von Hötzendorf led the Austro-Hungarian armies.

9. Armee proceeded to the Polish Salient with *XX. Armee-Korps' FFA 15* and *XVII. Armee-Korps' FFA 17*. *Fliegertruppe*'s direct support to *9. Armee*'s *AOK* 9 tasking, now involved *FFA 35*. Additional aviation support for *9. Armee* included *XI. Armee-Korps' FFA 28*, *Kavallerie-Korps* Frommel's *FFA 31*, and Division Bredow's *FFA 46*. *Feldluftschiffer-Trupp* 5 returned to *Gouvernement Festung* Posen and *Feldluftschiffer -Trupp* 1 reestablished operations with *Gouvernement Festung* Königsberg.[5]

Heads Roll at RIVVF

Within days of the final moments of the Tannenberg battle, *Zaveduyushchego organizatsiyey aviatsionnogo dela v armiyakh YugoZapadnogo fronta* [Chief of Organization of Aviation Affairs for Armies of the Southwestern Front] Grand Duke Alexander Mikhailovich, initiated a review of *RIVVF*'s perceived missteps. In a 9 September letter, Alexander cited aviators who failed to support the campaign:[6]

> General Ropp revealed many unit commanders were inconsistent. Kokayev never flew (while in) the *13th [KAO]* detachment, *21st [KAO]* Poplavko flies badly and breaks apparatuses, *23rd [KAO]*'s Prishchepov flies poorly and pilot Zhokhov flies badly... I believe that they should be expelled and more capable aviators be appointed in their place. Grand Duke Alexander Mikhailovich.[7]

Baron Kaul'bars was ignored by Alexander. Despite the perceived slight, Kaul'bars remained active, working directly with *RIVVF* units. On 20 September, while visiting Kovno [Kaunas] *KrAO*, he flew as an observer with *Shtabs-kapitan* Aleksandr Kirillovich Perepechin. Kaul'bars wanted to take advantage of the sortie to observe Russian troops in maneuver against *8. Armee*, even if it meant witnessing a retreat. Perepechin's Farman HF.22 attempted to take off in inclement weather, but the aeroplane crashed and overturned. Kaul'bars suffered several broken ribs and bruises. Poruchik Georgy Leonidovich Sheremetevsky related,

Grand Duke Alexander Mikhailovich launched an investigation of RIVVF's failures in September 1914.

> The dramatic episode with Baron Kaul'bars, confirmed the opinion at headquarters that aeroplanes were of no operational value, but the wounded hero himself did not lose his nerve in spite of his broken ribs. The insulting attitude openly shown towards us by headquarters, didn't deter us from serving with great enthusiasm in the hope of changing their unfair attitude and proving the usefulness of military

Gen.kav Baron Kaul'bars poses with Shtabs-kapitan Vladimir Nikolaevich Vargasov of the 13th KAO and an unidentified Shtabs-kapitan in front of a Nieuport IV monoplane prior to his dismissal by Grand Duke Alexander Mikhailovich.

> aviation at the front. We made all our flights from the fortress on our own initiative, without any orders or instructions about where to go or what to survey.[8]

Failed Policies Are Addressed

Grand Duke Alexander Mikhailovich maintained extensive correspondence with *General ot kavalerii* Aleksei Alekseevich Brusilov (commander of Eighth Army in the Southwestern Front) in getting Russian military leadership to rethink *RIVVF*'s role in tactical aviation. The two strongly advocated *KAO*s be realigned to serve both armies and subordinate corps. Alexander and Brusilov pointed out that the Russian army headquarters lacked aviation officers. As aeroplanes wore down, *KAO*s lacked sufficient maintenance personnel who were experienced with field operations. This resulted in plummeting numbers of ready aeroplanes to support operations.[9]

RIVVF continued to operate with a lingering bias against aviator reports. Russian aviator *Poruchik* Sheremetevsky, was ordered by *Gen.inf* Smirnov's Russian XX Corps, to ascertain the direction

Poruchik Georgy Leonidovich Sheremetevsky flew Farman HF.16 for Kovno KrAO and supported Gen.kav v. Rennenkampf's First Army during the Masurian Lakes campaign. His memoirs contained one of the most detailed accounts of RIVVF in the Great War.

the primary German force was moving in, and surmise where an attack could be expected. Sheremetevsky recalled flying reconnaissance sorties and observing German forces trying to cross the Niemen River at Srednikov. He observed, "It was such a shame that there was not an officer from the General Staff flying with me as an observer, because he could have realized the value of aerial reconnaissance."[10] He flew more sorties in the area; German locations were noted but aerial reports were disregarded. He remarked that had the Russians responded to these reports,"It would probably have changed the whole course of the campaign."[11]

The rapidly developing face of Russian operations resulted in changes for Russian "Light" (Tactical) aviation. Early on, it became clear to Russian leadership, that *RIVVF* lacked the depth to support armies in combat and their credibility suffered. *KAO* experienced great losses and operational inventory fell to one or two aeroplanes per detachment. To protect themselves, Russian army corps became more independent, in an effort to combat *8. Armee*. *RIVVF* aviators as well, were given new authority to design and execute aviation detachment missions, based on headquarters' directives. In spite of best efforts, the limited use of tactical aerial reconnaissance, proved to be ineffective. Longer-range strategical support didn't exist.[12]

On 10 October 1914, Alexander sent a report to Tsar Nicholas II and Grand Duke Nikolai, stating that five Russian pilots had been killed, ten were missing in action, and five had been injured in crashes. Ninety-four aeroplanes had been lost due to various causes, since the war commenced. Sixty-five airframes were being sent to units as replacements.[13] As a result, the policies governing Fortress Grodno's *KrAO* aeroplane assignment changed to support First Army. Fortress

I. Armee-Korps acquired Albatros B.II provided greater duration of flight and improved performance over the aging Taube.

commanders were contacted to release control of whatever *KrAO* aeroplanes they could spare, but when Rennenkampf advanced into *Ostpreussen*—not one *KrAO* aeroplane was provided. Russian military commanders realized they could have made a major difference in the field, versus remaining in a fortress far behind the front line.[14]

An Overly Independent Character

On 20 September, *8. Armee War Diary* stated that "the aim of *8. Armee* is to bind the Russian troops who have retreated in the rear of the Niemen River, and prevent their movement in the direction of Warsaw." Russian forces had suffered considerably from the defeat at Masurian Lakes and the subsequent pursuit, and *8. Armee* now commenced destruction of the Russian railways in the region.[15] On 22 September, *GO* v. Hindenburg provided *AOK* 8 with a letter requesting *8. Armee* advance and simultaneously support *9. Armee*'s offensive in the south. Hindenburg's letter also stated that should the Russians attempt to renew their invasion of *Ostpreussen*, railroads in the region were to be destroyed.[16] *GdI* v. François' *I. Armee-Korps* was subsequently assigned the task of destroying the Grodno-Wilna railroad. He reported destruction underway and requested consideration for *I.*

Armee-Korps to continue to advance further east to prevent First Army from performing a flanking maneuver from the direction of Olita and Domrowo. François was reminded his task was to destroy railroads for *AOK* 8. He replied, "*I. Armee-Korps* has been advised accordingly."[17]

On 25 September, aerial reconnaissance covered rail activity to the east of *Ostpreussen* near Grodno. The trains that were observed, failed to indicate any Russians coming or going. Russian forces were observed heading north from Sztabin. *AOK* 8 ordered destruction of the bridges leading to the village. Aviation confirmed Russian building defensive positions on the Niemen River bank between Olita and Kowno.[18] At noon on 26 September, François received a report from *FFA 14* aircrew flying an Albatros B.II, that a Russian army corps was advancing west from Olita. This prompted the Germans to cancel their plans to cross the Niemen River.[19] *8. Armee War Diary* concluded: Aviator reports provide us with no clear picture of the enemy's intentions. It appears as if a

FFA 14 symbolized Fliegertruppe's success at Tannenberg. Gdl v. François' trust in his aviators was returned by their consistent effort to acquire vital information throughout the campaign. When François assumed command of 8. Armee he now worked directly with FFA 16. FFA 14 aviators in 1914: (Left to Right) Ltn Friedrich Karl Burckhardt, Ltn Ernst Canter, Ltn Ernst Eberstein, (unknown), Ltn Kamps, Ltn Hans Wodtke, Ltn Conrad Krüger, Obltn Bogislav von Heyden, Hptm Erwin Korn (C.O.), Ltn Karl Mertens, Ltn Hans Hesse, Obltn Heinrich Körner.

transfer of the forces toward the south is underway behind the Niemen River. It has not been ascertained at which points the detraining of the Russian troops transported over the Grodno-Bialystok railroad, is taking place."[20]

On 7 November 1914, GdI v. Below replaced GdI v. François as commander of 8. Armee. Below held command until 26 May 1915.

The Germans now faced Russian fury from Augustow Wald—a battle that lasted five days. Inclement weather up until 30 September, curtailed *Fliegertruppe* sorties, forcing the Germans to resort to on-scene estimates of where the Russians were located.[21] Strategical reconnaissance continued as Z-IV flew a sortie on 24 September over Bialystok and Lomsha, followed by a flight over Warsaw, the following day.[22]

German forces faced a serious turnabout on 1 October, when a superior Russian army advanced against *8. Armee*'s thinly manned 80-kilometer front.[23] *GdI* v. François convinced *GdA* v. Schubert (now in his new role as *8. Armee* commander) that the best option was to pull back to the Angerapp River and establish a proper defense. This option was rejected by Falkenhayn's *2. OHL*. François proceeded to violate all rules of command by reporting directly to Kaiser Wilhelm (on 3 October), without including Schubert in the politically sensitive correspondence. François boasted to the Kaiser of his *I. Armee-Korps*' recent victory, and proceeded to sabotage Schubert's reputation. Kaiser Wilhelm's advocacy for François resulted in Schubert being relieved from *8. Armee* command and sent back to *2. OHL* at Charleville.[24] On 9 October, François took over command of *8. Armee*.[25] *FFA 14* aerial reconnaissance from an Albatros B.II, confirmed Russians firmly in place. The battle was fierce

FFA 15's aging Taube suffers a loss due to the airframe breaking down as the year ends.

and the Russians held their ground. At the same time, Z-IV provided limited aerial reconnaissance of the northern Niemen River and reported that no Russians were observed.[26]

Fliegertruppe aerial observation on 7 to 9 October, focused on the Russian advance. Intelligence gleaned from papers found on a Russian officer's corpse, provided further confirmation that the Russians were to attack from Warsaw.[27] The papers revealed that the Russians were still contemplating invading Germany. Four Russian armies (Second, Fourth, Fifth, and Ninth) were deploying between the San River and Novo Georgievsk [Modlin] for an attack over the Vistula River.[28]

Fliegertruppe were now pushed beyond their capacity. The initially designated six aeroplanes were reduced to three operational ones, as the campaign progressed. Vast areas of coverage now exceeded the range of aeroplanes. Liaisons between aviation units and commands were degraded by weather. As one *Fliegertruppe* aviator's diary stated, "We are only occasionally in communication with army headquarters. The roads are simply bottomless."[29]

When weather improved on 15 October, *Fliegertruppe* aerial observers reported major Russian activity and vehicles in the Grajewo-Szczuczyn sector. Russian railroad activity was extensive. Russian troops were observed advancing from Grajewo towards Lyck. François

The German generals reunite in 1924. Note Mackensen, François (in center), Hindenburg, Scholtz in the front row. General Ludendorff is noticeably absent.

determined (from aerial reconnaissance) that the Russians were leaving Grajewo-Szczuczyn.[30]

François continued to push *8. Armee* beyond its capacity. He ordered *XXV. Reserve Korps* to march a distance of 100 kilometers to Ratschki, to face Russian forces in the area. Russian heavy artillery proceeded to smash German positions. François made it clear that his forces could defeat any Russian even when the latter had artillery support.[31] Four days later, *Gen.inf* Ruzsky launched a large-scale offensive on the Vistula River against Hindenburg's *9. Armee* and *k.u.k. I Korps* on its right flank. This prompted *9. Armee* to withdraw from the area. In the northern *Ostpreussen* region, Z-IV flew its last recorded sortie in 1914 over Tauroggenon on 19 October.[32]

François' brilliant leadership enabled him to behave and lead with impunity. Despite his achievements, his violation of traditional German military protocol eventually caught up with him. His tenure as *8. Armee* commander ended on 7 November. *GdI* v. Below replaced him and commanded *8. Armee* until 26 May 1915.

François retained his rank but was transferred to the Western Front and commanded the *XLI. Reserve-Korps* for the next sixth months. While enroute west, he personally reported to Kaiser Wilhelm II and received the following counsel: "Now God be with you François, you will do well here, but you have to obey orders. You have an overly independent character!"[33]

François' support of aviation throughout the Eastern Front campaign was exemplary. The success at Tannenberg, is in part attributed to his bold movements and risk taking. Aviation fueled his confidence in implementing many crucial strategic maneuvers.

A Rocky Start for Russian Strategical Aviation

A decisive change to Russia's aviation occurred when Sikorsky's Il'ya Muromets was deployed to the Eastern Front.[34] On 13 September 1914, *Sh.Kap.* Rudnev flew the first Il'ya Muromets (No.135, Ship I) to serve the Northwest Front at Bialystok. Rudnev met with a series of problems, including strong headwinds followed by a rough landing at an airfield north of Pskov. *R-BVZ* had to send a mechanic from Petrograd to repair the aeroplane. On 24 September, he flew to Dvinsk, finally arriving at Bialystok on 6 October. Rudnev quickly discovered that Northwest Front headquarters had limited understanding of what role Il'ya Muromets could serve. Additionally, there were serious logistic and maintenance shortfalls to resolve.[35]

On 7 October, *Podporuchik* Pankratiev flew the next Il'ya Muromets (No.136 Ship II) to Bialystok. Three days later, while enroute to Bialystok, Pankratiev landed at Rezhitsa. Unfortunately, during the approach to land, Russian soldiers launched friendly fire on the aeroplane, believing it to be the enemy. Pankratiev proceeded to crash land, causing substantial damage to both airframe and one of the engines. All subsequent efforts to repair at Rezhitsa failed, so Ship II was disassembled and put on a train for Brest-Litovsk. Pankratiev took advantage of the break, and summarized the primary Il'ya Muromets issues and solutions for combat employment.[36]

Sh.Kap. Rudnev's frustration peaked when he notified *Genleyt* Peter Konstantinovich Kondzerovsky, *Stavka* duty general on 21

October, "On the unsuitability of Il'ya Muromets-type vehicles for military purposes." Rudnev listed problems experienced with Ship I. Frustration grew when *General ot infanterii* Mikhail Vasilyevich Alekseev, Southwest Front chief of staff, declared that no Il'ya Muromets units were to see front-line service until the problems were resolved. *Stavka* responded on 10 November, by ordering *GVTU* to cease Il'ya Muromets construction and deployment. *GVTU* complied and ceased construction on 3 December.[37]

Mikhail Vladimirovich Shidlovskiy, *R-BVZ* director, had personal, financial, and patriotic motives in making an appeal to Grand Duke Nikolai to support the Il'ya Muromets program.[38] As *R-BVZ* director, he stood to suffer a major financial loss if the aeroplane ceased production. Shidlovskiy argued that the Il'ya Muromets had proven (by the 1914 flight from St. Petersburg to Kiev) that it was a powerful war machine; he warned that any refusal to use the flying ships was criminal. His network also included Minister of War, Sukhomlinov. He emphasized Muromtsy problems were organizational. A centralized squadron, akin to how a naval fleet operated, could make

Il'ya Muromets Ship I, No 135 type Beh in August 1914. On the far left is Shtabs-kapitan Evgeny Vladimirovich Rudnev, in the light overcoat (4th from left). The Minister of War, Vladimir Aleksandrovich Sukhomlinov to his right is Chairman of the R-BVZ Mikhail Vladimirovich Shidlovskiy, eighth from left is Igor Ivanovich Sikorsky.

IL'YA MUROMETS AVIATORS

Shtabs-kapitan Evgeny Rudnev, flew the first Il'ya Muromets (No.135, Ship I) to serve the Northwest Front at Bialystok. The aeroplane never made it following a rough landing at an airfield north of Pskov. Rudnev's complaints to Russian commanders, nearly terminated the entire Il'ya Muromets program.

Podporuchik Alexey Vasilievich Pankratiev, flew the next Il'ya Muromets (No.136 Ship II) to Bialystok. While enroute, Pankratiev had to crash land the aeroplane due to small arms fire from Russian soldiers, believing it to be an enemy aeroplane.

Podporuchik Sergey Karlovich Modrah was initially selected as one of the RIVVF aviators chosen to fly the first line of Il'ya Muromets.

Prayer service for Il'ya Muromets Ship I, No 135, type Beh, at Korpusnoy aerodrome near Petrograd on 6 September, 1914 before flying to the Northwest front on 13 September.

The result of the forced rough landing by Shtabs-kapitan E. V. Rudnev flying Il'ya Muromets Ship I, No 135, type Beh, near Plyussa Station, 88 km. north of Pskov, on 13 September 1914. Note national colors on side of fuselage.

Forced crash landing by Podporuchik Alexey Vasilyevich Pankratiev Il'ya Muromets Ship II, No 136, type Beh, near Rezhitsa, 10 October, 1914. Evidence of the substantial damage to fuselage and engines is seen from position of the nose and the wings.

the best use of an aeroplane as unique as the Il'ya Muromets. The aeroplane's success hinged on a well-established centralized support system, providing maintenance for airframes and engines, as well as an assigned fuel and oil supply depot. Additional secondary and tertiary personnel were necessary to maintain successful operations beyond the current assignment of one aircrew to a single ship. The most significant argument offered by Shidlovskiy was Il'ya Muromets not only served the army with long-range strategical aerial reconnaissance, but also long-range aerial bombardment.[39]

General ot infanterii Mikhail Vasilyevich Alekseev, Southwest Front Chief of Staff, stated that until the problems associated with the machine were resolved no Il'ya Muromets units were to see front-line service.

Shidlovskiy's proposition was accepted. On 23 December, Grand Duke Nikolai approved Order No. 291, establishing *Upravleniye Escadra Vozdushnykh Korabley (UEVK)* [Administration of the Squadron of Flying Ships]. A chain of command was established with now *General-maior* Shidlovskiy directly subordinate to *Genleyt* Peter Konstantinovich Kondzerovsky, Adjutant General of the Supreme Commander General Staff at *Stavka*. Russia now possessed both "Heavy" aviation with Il'ya Muromets providing long-range aerial reconnaissance and "Light" aviation under Grand Duke Alexander, the nominal head of aviation and aeronautics of the Army.[40] The *Upravleniye* [Administration] was eventually dropped from the title and *Escadra vozdushnykh korabley (EVK)* [Squadron of Flying Ships] became the moniker for one of the most important operational Russian aviation units of the entire war.

General-maior Shidlovskiy reported directly to *Stavka* and supported both Northern Front (formally Northwest Front) and Southwest

General-maior Mikhail Vladimirovich Shidlovskiy, R-BVZ director. He saved the Il'ya Muromets by personally appealing to Grand Duke Nikolai. On 23 December 1914, Grand Duke Nikolai approved Order No. 291, establishing Upravleniye Escadra Vozdushnykh Korabley (UEVK) [Administration of the Squadron of Flying Ships] with Shidlovskiy in charge.

General-maior Mikhail Dmitrievich Bonch-Bruevich, Northwest Front Generál-kvartirméyster, despised what Muromtsy represented and personally prevented Shidlovskiy from receiving any meaningful assignments in early 1915.

Front.[41] As *EVK* was being established, *Genleyt* Mikhail Dmitrievich Bonch-Bruevich, Northwest Front *Generál-kvartirméyster*, stood in the way of Il'ya Muromets deployment. Shidlovskiy complained to *Stavka*, "The Chief of Staff of the Northern Front has shown a reluctance to give *EVK* any missions of military utility. As good weather is a prerequisite for flying, one can always be assured that for Bonch-Bruevich, the weather will never be suitable."[42] On 7 January 1915, *Stavka* overruled Bonch-Bruevich, and deployed Il'ya Muromets (No. 137 Ship III, 143 Ship VI and 150 Ship VII) to Yablonna. The aeroplane and crews now faced wartime conditions for the first time.[43]

CHAPTER XIV

Liberating the Homeland

Ober Ost Assumes Command

O*HL* now focused on the rapidly evolving Western Front. On 1 November, Germany's command on the Eastern Front reorganized. *2. OHL*'s newly appointed chief of staff, *Generalleutnant* von Falkenhayn (in agreement with Kaiser Wilhelm) established a new High Command for the Eastern Front named "*Oberkommando Ost*"[*Ober Ost*]. Hindenburg was appointed as *Ober Ost*. The Kaiser promoted him to the highest military rank, *Generalfeldmarschall* (*GFM*). *GdK* v. Mackensen was named *9. Armee* commander with *Generalmajor* Paul Grünert as Chief of Staff.[1]

GFM v. Hindenburg directed Mackensen's *9. Armee* which comprised eight divisions. *9. Armee* was augmented by *Genltn* v. Morgen's *I. Reserve-Korps* and *General der Infanterie* Reinhard Freiherr von Scheffer-Boyadel's *XXV. Reserve-Korps* from *8. Armee*. Hindenburg stated, "The mission of *9. Armee* is the protection of the Austro-Hungarian ... flank and at the same time, the delay and aversion of the Russian armies on their westward advance..."[2] *8. Armee* held German ground against the Tenth Army at this time. The final campaign on the Eastern Front in 1914 was launched on 11 December with *9. Armee* driving east toward Lodz in the Polish Salient.[3]

Generalfeldmarschall (GFM) Paul von Hindenburg – Ober Ost.

As 1915 approached, the Central Powers faced a

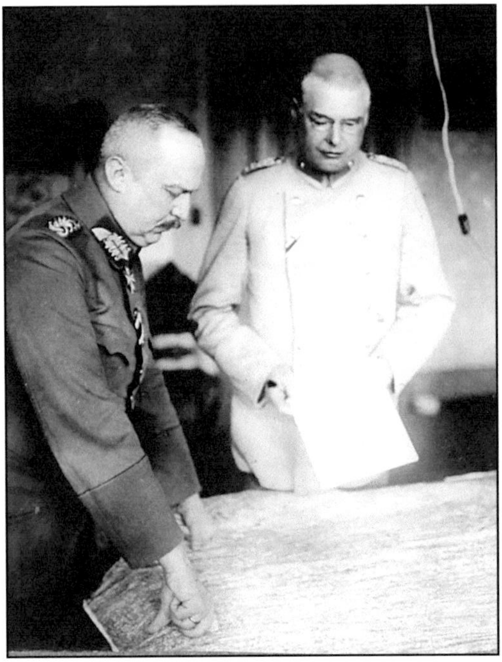

Generalleutnant Erich Ludendorff and Oberstleutnant Hoffmann provide staff leadership to Ober Ost.

serious threat. Falkenhayn feared a long-lasting war of attrition against the Western Allies. His strategy was to force Russia to conduct peace negotiations, effectively cutting short the war on the Eastern Front. *Ober Ost* commander Hindenburg strongly objected and argued that Russia could be militarily defeated. As expected, *k.u.k. GdI* Conrad also voiced strong objections against German prospects for peace-making with Russia. Falkenhayn was overruled by the Kaiser. *2. OHL* then took immediate steps to increase German military strength on the Eastern Front—with four more *Armee-Korps* assigned to *Ober Ost*. This cemented the goal of winning the war on the Eastern Front. Hindenburg stated, "could only be reached over the body of a Russia stricken to the ground." With *k.u.k. Armee* losing the Galician campaign, Falkenhayn followed Kaiser Wilhelm's directive to assist Austria-Hungary.[4]

Fliegertruppe sorties continued their aerial reconnaissance of the *Ostpreussen* border and kept *AOK 8* informed on Russian maneuvers. The establishment of *9. Armee* decreased *8. Armee*'s fighting power in *Ostpreussen*. Losing an *Armee-Korps* meant no offensive operations could be successfully launched against remaining Russian forces. *8. Armee* never lost its ability to hold fast to the German homeland that it was committed to defend. The major threat to *Ostpreussen* was *General ot infanterii* Faddey Vasilevich Sievers' Tenth Army.[5]

Russia possessed more than 100 divisions on the Eastern Front at year's end. German and *k.u.k. Armee* divisions numbered around 85. Russia initially had an advantage in determining where to mass its forces. *Stavka's* indecision allowed German *Armee* and *k.u.k. Armee* to launch their own offensives. *Generál-kvartirméyster* Danilov proposed two options: attack directly west from the Polish Salient toward Berlin or attack *Ostpreussen*. Lessons learned from the failed "Big Push" of October didn't appear to matter. *Gen.inf* Ruzsky's Northern Front envisioned driving the Germans out of territory east of the Vistula River. A 17 January 1915 *Stavka* conference approved Ruzsky's plan. A new invasion of *Ostpreussen* was scheduled for 23 February.[6]

General der Kavallerie August von Mackensen, 9. Armee commander.

Grand Duke Nikolai and *Stavka* also made changes. *Gen.inf* Sievers' Tenth Army occupied the northern sector. The Russian southern sector now fielded *General ot kavalerii* Pavel Plehve's Twelfth Army. What remained of Rennenkampf's First Army, was reconstituted and placed under the command of *General ot kavalerii* Alexander Ivanovich Litvinov. The Masurian Lakes battles culminated in one strategic outcome. Russian leadership was now convinced that any future campaign would no longer involve *Ostpreussen*. Priorities now focused on their weaker *k.u.k. Armee* opponents in Galicia and the Carpathian battleground.[7]

Changes in Organization and Aviation Leadership

On 18 December 1914, the Russian Military Council approved *Stavka*'s recommendation to form eight army flight squads.[8] A month later, *Stavka* appointed Grand Duke Alexander as Chief of Aviation and Aeronautics in the Field Army. *Aviakants* [Aviation Board Office] was established to serve as Alexander's aviation board to provide advice to the military leadership on *RIVVF* matters.[9] With

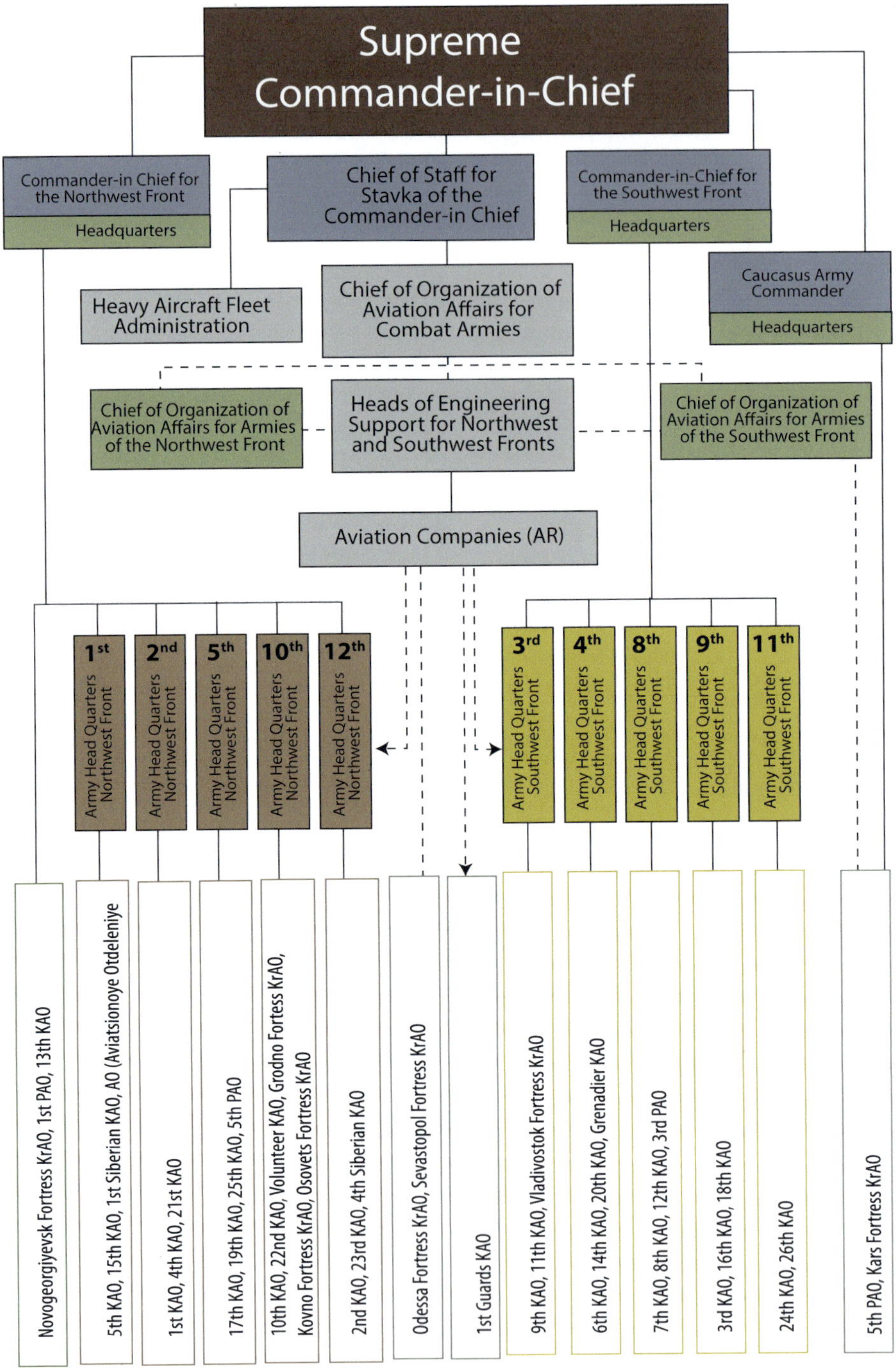

RIVVF was reorganized on 15 January 1915.
Source: Carl Bobrow

Kaul'bars' departure, Grand Duke Alexander established total control of aviation affairs for both the Northern and Southwest Fronts. On 28 February 1915, Alexander's responsibilities now included naval aviation supporting the Black Sea Fleet.[10]

Gen.kav Baron Kaul'bars was removed from the Northern Front and appointed commander of the Warsaw Military District. He was subsequently transferred (on 13 February 1915) to France and England to coordinate allied support to Russia. The allies were engaged in evaluating what type of aeroplane could best serve aviation missions. Kaul'bars returned to Russia in May 1915 and outlined French and British approaches to selecting aircraft to perform specific missions.[11]

The First Battles of 1915

Falkenhayn retained his belief that the war was to be decided on the Western Front. He attempted to strategically unify the demands of both fronts. Hindenburg and Ludendorff focused on destroying Sievers' Tenth Army with *GdI* v. Below's *8. Armee*. A new field army, *10. Armee,* was established on 26 January, under the command of *Generaloberst* Hermann von Eichhorn. *Stavka* was unaware that *10. Armee* existed until the Winter Battle of Masuria of February 1915.

The offensive commenced on 4 February, in severe snowstorms, black ice, and extremely cold temperatures. Flying was nearly impossible. *Stavka* was caught unaware when *10. Armee* attacked. They had focused on Mackensen's *9. Armee* campaign in the Polish Salient. The next day, Eichhorn's *10. Armee* attacked and pushed Sievers' Tenth Army southern flank back to Augustow. The withdrawal almost turned into a rout as *Gen.inf* Epanchin's III Corps collapsed.[12] III Corps' *3rd KAO* Nieuport IVs were destroyed by the ground crew as forces retreated.[13]

At 1400 hours, *Fliegertruppe* generated a single sortie, despite inclement weather. They reported substantial Russian forces, withdrawing. Eichhorn's southern flank was manned by *General der Infanterie* Karl Ludwig Wilhelm Hermann Litzmann's *XL. Reserve Korps.* Litzmann trusted his intelligence and ordered his divisions to intercept the Russian columns retreating from Lyck.[14] In the absence of aviation support, wireless intercepts of encrypted Russian communications made

Generaloberst Hermann v. Eichhorn, 10. Armee commander

a significant difference. The intercepts enhanced *AOK 8*'s and *AOK 10*'s planning in the final days of the Winter Battle of Masuria.[15]

Tenth Army was threatened by another German attempt at encirclement. On 14 February, they held back *8. Armee* and *10. Armee*'s advance for ten days, allowing Russian forces to retreat to Kovno and Osowiec.[16] *Fliegertruppe* flew long-range sorties deep into Russian territory. *Leutnant* Mahncke recalled the aeroplane's engines were unreliable.[17] On 21 February, *Gen.inf* Smirnov's XX Corps was trapped at Suwalki. Russian Tenth Army Commander Sievers felt it necessary to sacrifice XX Corps to allow XXVI Corps and III Siberian Corps to escape the German maneuver. The following day, Tenth Army was defeated, losing 92,000 prisoners of war, including nine general officers, 295 artillery pieces, and 170 machine guns. German losses were also substantial. *XXI. Armee-Korps* lost 6,000 men killed in action, including one general officer, two regiment commanders, and 120 officers.[18] On 22 February 1915, the last Russians were removed permanently from *Ostpreussen*. The Winter Battle of Masuria was over.

Gen.inf Epanchin, Russian III Corps commander. 3rd KAO supported his forces. During the Winter Battle of Masuria, III Corps collapsed from 10. Armee's attack.

Gen.kav Litvinov's First Army was at the northern flank of the Polish Salient, north of the Vistula River. *Ober Ost* assigned *General der Artillerie* Max von Gallwitz to lead the

General der Artillerie Max v Gallwitz, commander Armeegroup Gallwitz. Later in the Great War, Gallwitz was the primary German commander to fight the Americans at St Mihiel and Meuse-Argonne.

newly-formed *Armeegruppe* Gallwitz to attack First Army. *FFA 45* and *FFA 56* provided *Fliegertruppe* support. *Armeegruppe* Gallwitz was also tasked with preventing *Gen.kav* Plehve's Twelfth Army from assembling to the northeast of Warsaw. Gallwitz's 18 February 1915 attack was two-pronged, aimed at encircling Przasnysz from the west and east. To the west, stiff Russian infantry resistance blocked one wing of the envelopment. To the east, on 24-25 February 1915, *I. Reserve-Korps* swept past Przasnysz, then rolled in behind the town and stormed it, capturing 30,000 Russians. Inclement weather set in and stopped *Fliegertruppe* from flying. The absence of intelligence collection proved lethal. Gallwitz and staff were surprised by a 27 February Russian counteroffensive executed by *Gen.kav* Plehve's Twelfth Army, that captured 6,000 German soldiers and 58 German guns. Only an effective rearguard action kept the Russian pursuit from doing even more damage. Despite *Armeegruppe* Gallwitz's loss, Russia's intent to advance into Silesia, was thwarted.[19]

Russian Aviation is Transformed

Igor Sikorsky resolved the flaws in Il'ya Muromets design and maintenance, enabling it to effectively support wartime operations. His latest variant, Type Veh, No. 150 Ship VII "Kievskiy," met military specifications with state-of-the-art modifications to include interior ports for cameras and an internal cabin bomb rack (with a sliding door in the floor). On 6 February 1915, Sikorsky flew "Kievskiy" and personally demonstrated to skeptics that Il'ya Muromets' problems were resolved. Two days later, *Shtabs-kapitan* Georgiy Georgiyevich Gorshkov flew "Kievskiy" over Warsaw, carrying 12 passengers on board and reaching 2,500-meter altitude in 49 minutes.[20] *Fliegertruppe* retaliated when *FFA 4* (supporting *XIII. Reserve-Korps)* flew two

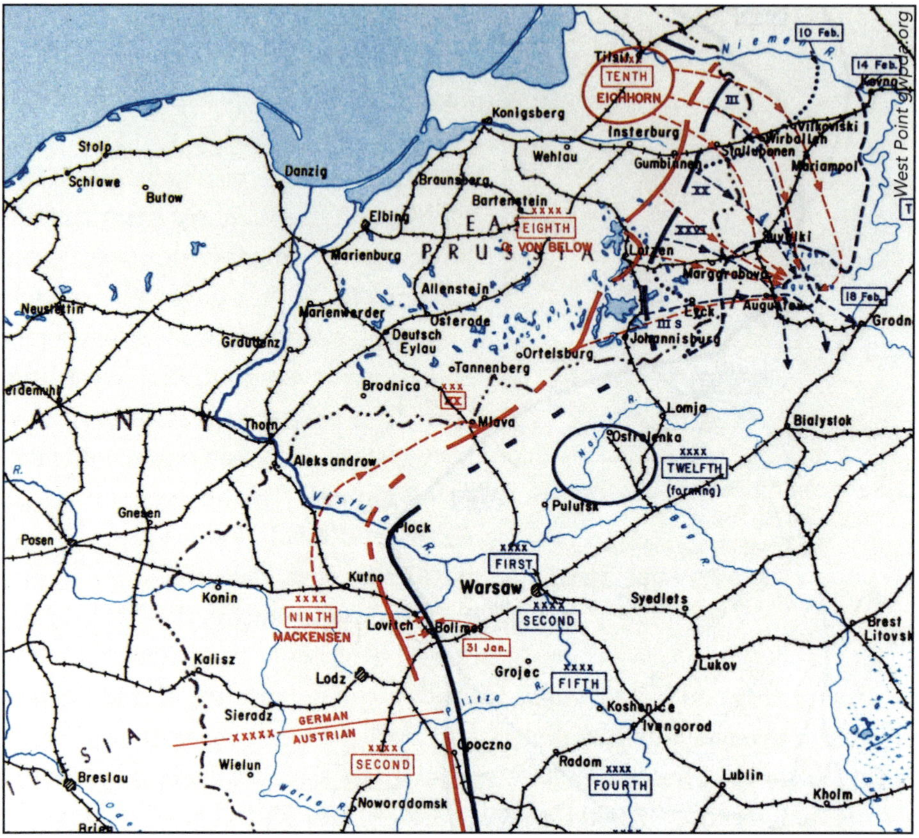

German armies now push Russian Forces out of Ostpreussen. Source: West Point

bombing raids at Yablonna, on 26/27 February and 1 March, where "Kievskiy" was stationed.[21]

The true test for Il'ya Muromets was in its effectiveness in actual combat. On 28 February 1915, Gorshkov flew a series of noteworthy aerial reconnaissance and aerial bombardment missions in "Kievskiy." Inclement weather prevented a planned aerial bombardment mission over the town of Plotsk on the Vistula River. The following day, Gorshov piloted "Kievskiy" for over two hours, dropping bombs on German lines, then engaging in aerial reconnaissance. He reported on the absence of bridges over the Vistula River. *Stavka* took notice and subsequently began to include Il'ya Muromets in their operational planning objectives.[22]

Il'ya Muromets' success, far exceeded prior *RIVVF* tactical accomplishments and capabilities. On 6 March 1915, *Sh.Kap.* Gorshkov flew "Kievskiy" on a four-hour sortie with aerial cameras, searching for German artillery near Przasnysz. Russian military leadership now possessed aerial photographs confirming that "the entire region is in a highly fortified position." A secondary mission, included dropping bombs at Willenberg. On site reporting from a Russian agent, confirmed destruction of artillery batteries, the Willenberg rail station, and enemy transport in the area.[23]

Il'ya Muromets' aerial reconnaissance sorties generated praise from *Gen.kav* Litvinov's First Army. *General-leytenant* Il'ya Zurabovich Odishelidze, Litvinov's chief of staff, stated:

On a rail car, is the disassembled Il'ya Muromets Ship III, No. 137, type Beh, Winter 1915, enroute to the UEVK squadron's Yablonna aerodrome. Source: Carl J. Bobrow

These flights proved the great advantage of Il'ya Muromets aeroplanes over other types of aeroplanes. High load-carrying capacity, convenience of reconnaissance, possibility of gaining great height (3200 meters in recent flights), long range of flight, armament on board, and great accuracy in bombing—all these constitute the great advantage over ordinary types of aeroplanes...The authenticity of reconnaissance findings (each observation can be freely photographed) also has great significance. Three combat flights fully proved what great results could be expected from Il'ya Muromets-type aeroplanes.[24]

On 9 March, the "Kievskiy" flew a three-and-a-half-hour armed aerial photographic reconnaissance mission over *Ostpreussen* to ascertain German movements opposite *Gen.kav* Litvinov's First Army. Willenberg's rail station was hit with 17 large bombs. The following day, "Kievskiy" flew an identical route and acquired aerial photographs for conducting bomb damage assessment. Upon completing the reconnaissance, an additional 45 bombs were dropped on Willenberg. The aerial photographs confirmed to First Army staff, that both the railroad station and trains at the depot were destroyed.[25] "Kievskiy" mission success changed the prevailing attitude of senior Russian commanders. Northern Front headquarters contacted *Stavka*, "We hereby request you

Aerial photograph taken over Yablonna airfield in 1915 airfield. Polkovnik Ul'yanin's designed Il'ya Muromets tent-hangers, are in place.

to inform us whether it would be feasible to make available an Il'ya Muromets type aeroplane for a unit of the Northern Front."[26]

On 18 March, *Genleyt* Odishelidze telegraphed Grand Duke Alexander, Russia's leading aviation visionary, "I believe that this type of aeroplane, by virtue of its combat qualities, deserves all kinds of encouragement."[27] Odishelidze also reported to the Northern Front command, "German pilots, captured near Przasnysz, reported they knew about our possessing large combat aeroplanes and Germany felt the absence of such an aeroplane in their own inventory."[28]

A significant sortie flown by "Kievskiy" on 31 March 1915, clearly demonstrated aviation's potential for strategical aerial reconnaissance. The movement of German and Austro-Hungarian forces to establish a major campaign in Galicia, spurred First Army to employ Il'ya Muromets. They needed detailed and verifiable information on the German's rear echelon to confirm troops in transit. In addition, First Army needed confirmation from agent reports that new forces from the Western Front

"KIEVSKIY"

Il'ya Muromets Ship VII, "Kievskiy" No. 150, type Veh. Yablonna, February 1915. The type Veh were the first Muromtsy designed specifically for military combat use.

Voyennyi Letchik Shtabs-kapitan Georgiy Georgievich Gorshkov flew "Kievskiy" and accomplished more for *RIVVF* in those initial sorties.

Shtabs-kapitan Gorshkov (commander) & Poruchik Bashko (deputy commander) in the cockpit of "Kievskiy," at Yablonna airfield. Source: Marat Khairulin

Poruchik Jāzeps Stanislavovičs Bashko, Deputy Il'ya Muromets commander.

were arriving in the region. The four-hour sortie covered 533.5 kilometers of *Ostpreussen,* to include Willenberg, Neidenburg, Strasburg, and Thorn. The Vistula River was covered by aerial photographs of Plotsk and Mława. Final coverage included battleground front lines in the area. *Gen.kav* Litvinov's First Army staff acquired 50 aerial photographs that documented the absence of German forces in *Ostpreussen.* Russian military commanders were alerted to intelligence gleaned from the sortie, indicating that Germany's forces were moving towards a new offensive in the south. *Stavka* and First Army had solid strategic information to initiate an appropriate response to the Germany's movement of forces.[29] "Kievskiy's" reporting startled First Army staff and they immediately tasked another sortie be flown the next day. That sortie further confirmed the 31 March results. This time, the Il'ya Muromets conducted aerial bombardment of key targets enroute.[30]

By April 1915, the successful wartime operations of Il'ya Muromets, influenced Russia's War Ministry to withdraw the order banning it from the field army. The Chief of the Supreme Headquarters telegraphed the Chief of the Military Technical Directorate, "The Supreme Commander ordered to ask you, in view of the clarification in the use of the air squadron, to revive the contract with R-BVZ and assist the factory in everything for an earliest execution of this order." Consequently, the original contract for 32 Il'ya Muromets was restored.[31]

The success of the Il'ya Muromets, counteracted and outweighed the dismal failures of *RIVVF* on the Northwest Front, during the first eight months of war. It rewrote the landscape of aviation history, as Russia grew to be perceived as a formidable presence in the world of strategic reconnaissance.

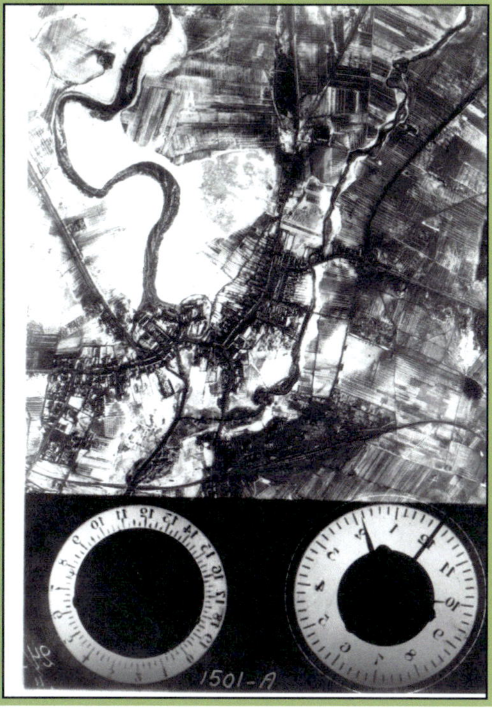

The Ul'yanin aerial camera was employed by Il'ya Muromets. An aerial photograph taken on 28 February 1915 from "Kievskiy" indicates recorded altitude and time of coverage.
Source: Carl J. Bobrow

A significant sortie flown by "Kievskiy" on 31 March 1915—the aircrew disembarking. Il'ya Muromets' early demonstration of aviation's potential for strategical aerial reconnaissance. The commander, Shtabs-kapitan Gorshkov is holding a map and the officer in the center, is carrying the Ul'yanin aerial camera case.

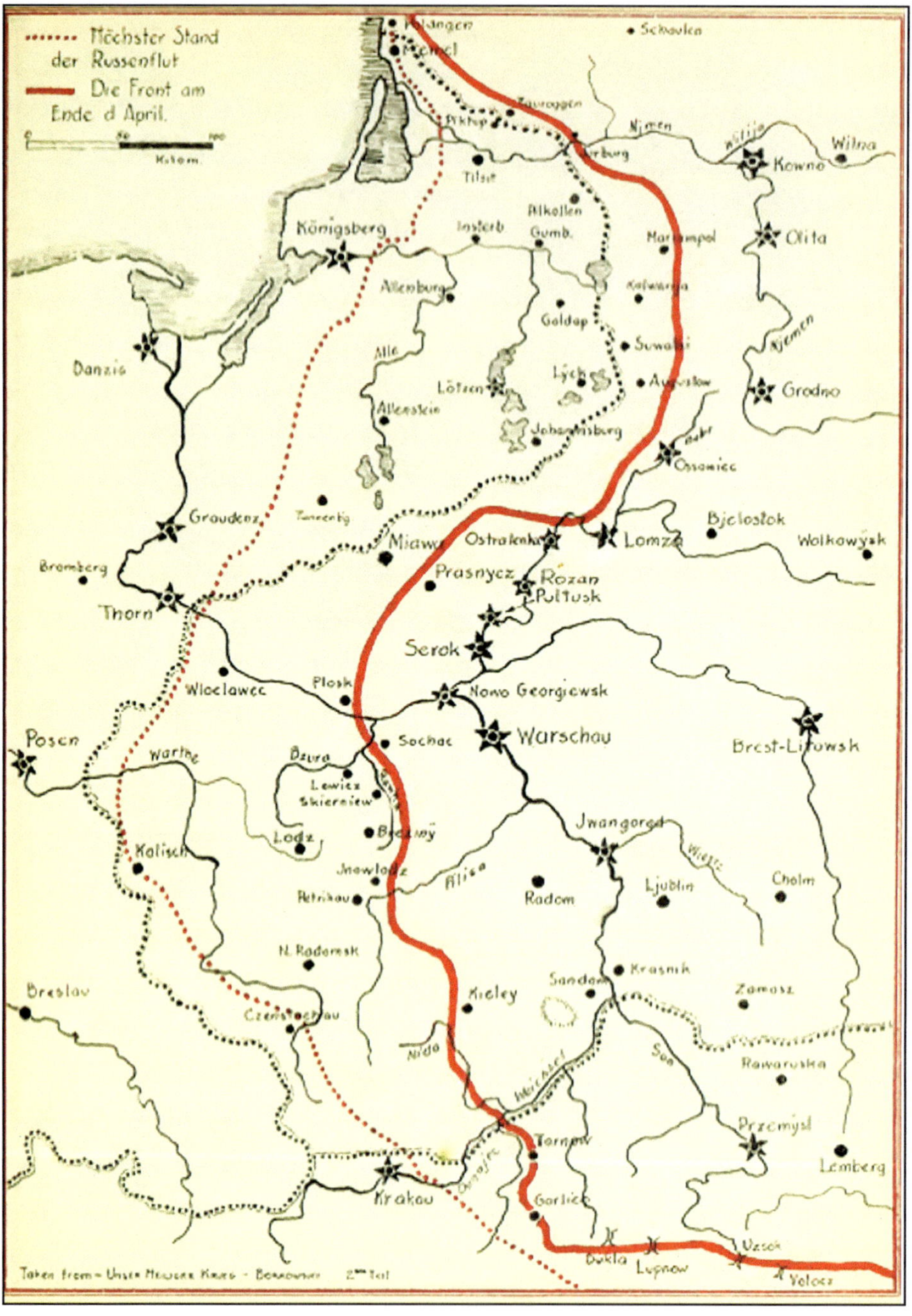

By April 1915, all Russian forces were permanently removed from Germany. The Northern Front was now fought on Russian soil.
Source: Unser Heiliger Krieg

CONCLUSION

Aviators Assume the Mantel of Reconnaissance

German aviators serving *8. Armee* entered into hallowed ground as an elite fighting force, tasked with saving their nation from a foreign invasion. Kaiser Wilhelm II's "holy duty" to protect his Prusso-German citizens from the Russians in *Ostpreussen* and Silesia, was fulfilled. *Flieger Leutnant* Alfred Mahnke reflected at year's end: "Generals and staff officers had not expected much from aeroplane reconnaissance before and at the start of the war, and believed ours and enemy's aeroplanes would be out of action after a few weeks. They were most surprised when we carried on flying; it was ideal weather, and we returned with extensive and moreover, correct reports. We almost beat the cavalry on their own turf, namely long-range reconnaissance." Aerial reconnaissance emerged as a central source of intelligence collection. Calvary lost its purpose in the Great War, as aviation took its place, transforming warfare in the 20th century.

Aerial Reconnaissance – A Retrospective

Aerial reconnaissance was aviation's primary military mission when the Great War commenced. In the decade following the Wright brother's breakthrough at Kitty Hawk, advances in aviation allowed flight to achieve the altitude and duration necessary to loiter safely over a battlefield and provide timely updates. Aeroplanes initially demonstrated military potential in minor conflicts leading up to the Great War. In the 1911 Turco-Italian War and the 1912-1913 Balkan War, they proved to the world that they could conduct reconnaissance as well as bombardment. Aviation's potential faced its greatest test with the outbreak of World War I in August 1914. The mobile battlefield dominated the way war was fought on a grand scale and the process for conducting reconnaissance evolved with each battle.

In aviation's infancy, there was a prevailing belief that airships would dominate the skies—the most famous being Germany's Zeppelin and Schütte-Lanz. The airship's ability to travel great distances and loiter over the target area was perceived as an advantage. Commanders and staffs soon learned that airship's speed, slow maneuver, and low service ceiling, were vulnerable to artillery fire. In contrast, aeroplanes surpassed airships in cost, flexible operations, and the ability to reconnoiter the battlefield nearly-unseen.

German concern over France's aeroplane advancements spurred senior military and public support for investing more resources in research and development. Military training commenced, as pilots developed airmanship skills to handle fragile aeroplanes. Aerial observer procedures were carefully cultivated and reporting priorities evolved with actual combat. Military exercises provided a glimpse of how aviators could conduct reconnaissance—a role traditionally served over the centuries, by cavalry.

Aerial reconnaissance proved to be a force multiplier in the first months of the war. Aircrews generated timely delivery of observations by either landing near the headquarters or dropping reports via containers, such as the German *Flieger-Meldung*. This real-time delivery soon became familiar to senior staff officers. Battlefield commanders now had greater capacity in locating the enemy, evaluating the terrain relevant to conducting operations, and determining intentions and strategies for upcoming battles.

The Matriculation of Intelligence

In the opening stages of the war, history was made when intelligence served a central role in defending *Ostpreussen*. The home-field advantage was critical to the German's success. Both *Fliegertruppe* and the commanders and staff they served, understood the ground that aviators surveilled. It became routine for Germany's battlefield commanders to be supported by the regular issuance of aviator reports—which both ignited and fueled Germany's victory at Tannenberg.

The noted military historian Dennis Showalter, in his acclaimed work on Tannenberg, stated that aerial reconnaissance resembled ULTRA in 1942. He acknowledged that there was abundant skepticism of the reports but at the same time, many considered it "magic." ULTRA required the highest possible security precautions—analogous to the wireless intercepts effectively employed by both German and Austro-Hungarian *k.u.k Armee* operations. Since aerial reconnaissance was open to all combatants, aviation provided the necessary cover for intercepting enemy wireless transmissions. Both mediums symbiotically defined the intelligence architecture for the modern era. Protecting the source, now governed intelligence roles and methods.

In their capacity as *8. Armee's* first commander and chief of staff, *Generaloberst* v. Prittwitz and *Generalmajor* v. Waldersee, had full knowledge of available intelligence resources. The correspondence between *Generaloberst* v. Moltke's *OHL* and *8. Armee,* revealed that Prittwitz lost sight of what was necessary to properly position *8. Armee* as the Russians approached the border. He had untested aviation capabilities and wireless intercept networks to draw upon. Instead of pushing intelligence resources to provide definitive data on Russian intentions, Prittwitz and staff became entangled with *OHL,* debating whether *8. Armee* or *OHL* should deliver the final verdict on intelligence. With his attention deflected and his priorities skewed, they lost sight of harnessing available intelligence and analyzing Russian intentions.

The initial battles at Stallupönen and Gumbinnen were far from decisive. *8. Armee's* opportunity for victory, remained a possibility. Prittwitz was reluctant to continue the attack on advancing Russian armies and elected instead, to execute his weakest option—conduct a hasty retreat to the Vistula River in an effort to save *8. Armee*. What Prittwitz (a career general who rose to the ranks by the Kaiser's patronage) lacked was operational ingenuity. He failed to seize the moment. Had he responded to critical aerial reconnaissance reports (indicating *General ot kavalerii* Rennenkampf's and *General ot kavalerii* Samsonov's armies had not filled the 70-kilometer gap between them), he could have done great damage to either or both Russian armies. Prittwitz, in discussion with *GO* v. Moltke, made it clear that he was going to withdraw troops to the rear. *AOK 8* was blindsided. The decision to retreat west, meant *Ostpreussen* was now vulnerable to invading Russians. Both Kaiser Wilhelm and *GO* v. Moltke agreed that Prittwitz had lost the mantel of command to continue the fight. After only three weeks as *8. Armee* commander, he was replaced.

General der Infanterie v. Hindenburg and *Generalmajor* Ludendorff arrived on 23 August. Ludendorff understood aviation's potential, from aeroplane performance details to logistical and maintenance requirements. He based his nightly *Armee Befehl* on intelligence collection from aviator reports and wireless intercepts and concentrated on the 70-kilometer gap that separated the two Russian armies. Rennenkampf's First Army still focused on attacking Königsberg and leaving Samsonov to face shifting German *Armee Korps*. One of the greatest strategic movement of forces in the entire Great War took place when Hindenburg and Ludendorff risked moving all corps to the south, versus retreating to the Vistula River. *8. Armee's* decision to attack the two Russian armies in *Ostpreussen* (sequentially), was in part a testimonial to their confidence in *Fliegertruppe*. Ludendorff captured the moment with the statement, "Rennenkampf's formidable host hung like a threatening thundercloud to the northeast. He need only have closed with us, and we should have been beaten."

The Great War historians gave more credit to the impact of wireless intercepts versus aerial reconnaissance. The debate over whether aviation or encryption was more critical to intelligence and whether reconnaissance served a supportive function or stood on its own, (independently as a key intelligence source) continues to this day. Since intelligence is a fusion of data, it becomes difficult to assess the percentage contribution of each intelligence source. The data remains inconclusive.

Credit went to the wireless intercepts for making the ultimate difference. The Russians failed in maintaining code books and had inept procedures distributing ciphers and keys. German cavalry and civilians cut existing land lines in occupied Russian territory and lacked the wire communication supplies to reach command headquarters. Winston Churchill amplified the prevailing 20th century mystique, by claiming that wireless transmission intercepts sealed the fate of Samsonov's Second Army. Most notably, Churchill declared, "This is the way to make war."

Generál-kvartirméyster [Quartermaster-General] Yuri Nikiforovich Danilov's *Rossiya v Mirovoy Voyne 1914-1915*, [Russia in the World War 1914-1915] post-war memoirs paid the highest tribute to *Fliegertruppe* with "As a consequence of our lack of sufficient aeroplanes, the enemy could maneuver as he wished by using a highly-developed railroad network and be certain his plans could not be discovered by us, very quickly. At the same time, with considerably better aviation facilities, the enemy was able to observe each of our steps with complete impunity." Danilov's assessment failed to hold Russian generals accountable for not aggressively utilizing available aerial resources. Long columns of German soldiers moving south (*I. Reserve-Korps* and *XVII. Armee-Korps*) were easy marks for Russian aviators and had they been flying, they would have seen them immediately. Danilov also paid tribute to the *Ostpreussen* natives who sent out signals through various means. Germans reported Russian positions "by fires, by conflagrations, by the ringing of bells, and by putting windmills into operation. Our troops were surrounded by spies on all sides, a thing which naturally restrained their activity and assisted the enemy in following all our movements."

Once aerial photography became integral to collection, the sheer volume of available information transformed both its role and application. 1915 witnessed the dawn of industrialization of an intelligence era, that is still vital and operative to the present day. The explosion of detail generated by aerial photography further revolutionized modern warfare. Maps and analysis of the battle ground were made available for daily operations. What remains unclear is why Russia failed to employ aerial photography to further their intelligence efforts, since the Ul'yanin plate camera and the Potte film camera were both successfully developed prior to the war. Our data on Russian intentions was far too limited to speculate on this absence.

Approaching the Battlefield Blindfolded

"I approached the Neidenburg battlefield, blindfolded." *Gen.inf* Martos' statement to *8. Armee* interrogators was a damning indictment of Russia's failure at Tannenberg. The case against Russian leadership was one of criminal negligence—as they never performed at the level of responsibility consistent with the command positions they held. *General ot kavalerii* Zhilinsky (commander of Northwest Front) lacked experience in fighting a major campaign and had limited understanding of aviation's potential. His arcane thinking was an institutional failure. He was acting on Russia's commitment to support France immediately upon declaration of war. His focus was not on *Rossiyskiy Imperatorskiy Voyenno-Vozdushnyy Flot (RIVVF)* aviation assets. Instead, he ordered divisions to protect Russia's territories that bordered *Ostpreussen*. His fateful decision reduced both the infantry and cavalry strength of the Russian Second Army by one third. Vital aeroplanes and crews along with over a quarter of *Gen.kav* Samsonov's artillery, were diverted to fortress garrisons.

The nine deployed *KAO's* consisted of six aeroplanes and existing supplies. As the battle front moved and shifted, *RIVVF* assets became less important than demands for men and army supplies. *KAO* operations frequently encountered breakdowns from rough field take-off and landing accidents. They were plagued by having to maintain high-maintenance, French-designed Gnome rotary engines and aeroplane components. *KAO* service and support had to cannibalize their fleet and retain fewer aeroplanes in support of the operation. Inevitably, accidents and unserviceable engines increased dramatically—prompting Russian commanders to reduce their reliance on *RIVVF*.

There are a number of factors that appear to have contributed to the Russian failure. Initially, they were inculcated into believing that their primary threat was from the Austro-Hungarian empire. Russia's planned Northwest Front offensive was only intended to complicate Germany's plans to invade France. They underestimated the German resolve to defend their homeland. Another potential factor was that Tsar Nicholas II of Russia, did not wish to conduct a major war against his cousin's (Kaiser Wilhelm II) German nation. In addition, both *Gen. kav* v. Rennenkampf's and *Gen.kav* Samsonov's serious limitations as leaders contributed to failed invasions.

Some historians suspect that Samsonov's decision-making was primarily based upon misleading intelligence from Zhilinsky's Northwest Front headquarters and that Samsonov and his Second Army were ultimately, tragic victims of a fractionated and disorganized intelligence network. Whether or not Samsonov might have achieved his objective of regaining control on his left flank, if supporting aerial reconnaissance data were made available to him, remains an open question.

Despite their long and distinguished careers in cavalry, they both lacked a clear understanding of the Germans' location throughout the campaign. Beyond the failures of Russian leadership, the issue of why the Russians were far more hesitant to embrace new aerial technology at this point in time (versus the Germans) also remains an open question.

Two Catalysts for Change

GdI v. François

GdI v. François, commander of *I. Armee-Korps*, was a maverick presence from the start of the campaign. His employment of *FFA 14* and his focused use of other *Fliegertruppe* resources, utilized aerial reconnaissance to maximum effectiveness. François' ultimate maneuver was the creation of a thinly manned picket line, designed to manipulate the Russians into believing they were totally surrounded by the enemy. His trust in aviators was most clearly exemplified at Neidenburg (on 30 August 1914), when his forces were suddenly alerted to a Russian corps approaching from the south to save Second Army from total defeat. He maneuvered them to secure positions, concentrated all artillery at his disposal, and held back the Russian advance. Without the critical aerial reconnaissance warning, Russian maneuvers on Neidenburg could have resulted in catastrophe for *I. Armee-Korps*, depriving them of a clear victory. François' unique charisma, combined with his proven leadership on the battlefield, strengthened the trust between aviators and the commands they supported. His rebellious and independent nature led to his downfall as a replacement commander of *8. Armee*, and his subsequent transfer to the Western Front as an *Armee-Korps* commander. François provided a proven template for the rest of the German commanders, that reliance upon aerial reconnaissance fortified their ability to fight and win the battle.

Il'ya Muromets

At the war's onset, *RIVVF* was mainly comprised of French-designed aeroplanes. Soon to become available but not yet part of the fleet, was the record-breaking four-engine Il'ya Muromets. After eight months of fighting, the Northwest Front finally witnessed Il'ya Muromets in wartime operations. The introduction of this revolutionary aeroplane surpassed all expectations. The inability of Russian leadership to recognize the potential of aviation was one of many reasons why the initial campaign failed. Despite the Il'ya Muromets clear advantage, bureaucratic ineptitude stifled military implementation. Sikorsky's aeroplane design successfully demonstrated long-range flight potential and durability, while providing the aircrew with ample protection against most air or ground threats. The ability to conduct in-depth reconnaissance with an aircrew of experienced military experts, meant near real-time intelligence for the field commander and Stavka. The added benefit of aerial bombardment, proved even more compelling, as commanders normalized strategic strikes against vital rear-echelon enemy targets. Had a single operating Il'ya Muromets been tasked to fly over *Ostpreussen* in the first weeks of the war, it could have re-written the history of the Great War.

Final Thoughts

Germany's victories in *Ostpreussen* in the first months proved to be their most significant military accomplishment in the Great War, fueling their confidence to conduct two simultaneous campaigns against the Triple Entente allies. Aviation's contribution was well recognized by *Generaloberst* von Hindenburg, whose Tannenberg success propelled him to the upper echelons of German military command and post-war political leadership. His poignant tribute to aviator skill at Tannenberg, has historical significance: "*Ohne Flieger kein Tannenberg!*" [Without flyers, no Tannenberg!]. Most military historians were unaware of the extent of *Fliegertruppe's* impact at Tannenberg, as the data was sparse and some suspected that Hindenburg was merely covering up for clandestine wireless intercepts in making this statement. Analysis of the *Ostpreussen* campaign of 1914, concludes *8. Armee* command made superior use of intelligence resources and effectively employed aviation to reinforce high risk decisions, leading to the envelopment and eventual surrender of Russia's Second Army.

Hindenburg's victory at Tannenberg and his subsequent rise to prominence in government and in the zeitgeist of the German soul, had consequences and repercussions. German historians often consider secondary commanders and *Fliegertruppe* to have been more instrumental in implementing battle plans and the ultimate victory at Tannenberg, than Hindenburg and Ludendorff, themselves.

The mythologizing of both Hindenburg and Ludendorff and the success of pro-Hindenburg propaganda, served to obscure other elements that factored into the battle's outcome (luck, *Fliegertruppe* excellence, and the effective interface between aviation and intelligence).

Most importantly, Hindenburg's victory at Tannenberg proved to be a moment of vast historical significance. Imagine that Hindenburg lost the battle and therefore, presumably, never acquired the office of *Reichspräsident*, with the power to appoint *Reichskanzler*, i.e., head of the government. One of his last acts as German *Reichspräsident* was the surrender of power, making Adolf Hitler *Reichskanzler*. Though the rise of the Nazis necessitated many other critical antecedents to seize, gain, and retain power, this one final fatal decision of Hindenburg's infused Hitler's initial regime (to most Germans) with a strong element of legitimacy.

As for Russia, their defeat in the first few months of the war, was complete and aviation's absence, was one of many weaknesses in their overall campaign. Alexander Solzhenitsyn, in his historical work of fiction, *August 1914*, describes the thoughts and emotions felt throughout the campaign and in the aftermath of Russia's loss. His account of the Tannenberg battle critiques Russia's challenged leadership and infrastructure (without reference to aviation). The fictional character, *Polkovnik* Georgii Kikhailovich Vorotyntsev, a staff officer assigned to General Headquarters, precisely sums up the essence of this epic moment in time:

> How could the two armies link up, when one of them was being made to advance by forced marches and the other had practically been sent on leave? How could they link up, when after the Battle of Gumbinnen, five of General Rennenkampf's cavalry divisions failed to set off in pursuit of the enemy and were not sent to the rescue of Second Army at the time when it was in crucial trouble? .... Yes, I agree that General Samsonov made mistakes, but they were tactical mistakes. The strategic mistakes must be laid at the door of Army Group Headquarters. Samsonov on his own did not have numerical superiority over the enemy, but Army Group did, and in spite of that, the battle was lost. We

> must draw our conclusions, gentlemen—otherwise what is the point of having a staff in General Headquarters, at all? We are incapable of deploying units bigger than a regiment![1]

Northwest Front's failure to employ *RIVVF* aviation assets serves as a warning. Without vigilant and relentless eyes on the enemy, commanders and staff cannot achieve mission success. History confirms that *8. Armee's* rigorous employment of *Fliegertruppe* wrote the first chapter on the application of aerial reconnaissance in the opening days of the Great War. It is a testament to its power and to the related intelligence it generated, that it became both the foundation and catalyst for victory on the modern-day battlefield.

Appendix 1

Chekhutov - The Preserver of RIVVF's Legacy

Photograph taken by Letchik vol'noopredelyayushchiysya [Pilot Volunteer] Nikolai Dmitrievich Anoshchenko of Poruchik Chekhutov and detachment officer in front of the Nieuport IV.[1]

Ivan Anikievich Chekhutov preserved the aviation legacy of *RIVVF* during the Northwest Front advance and subsequent retreat into *Ostpreussen*. He came to aviation following his initial assignment in 1910 as a *Podporuchik* [Second Lieutenant] to the *6-y Sibirskiy sapernyy batal'on* [6th Siberian Sapper Battalion]. Two years later he entered *Ofitserskaya Vozdukhoplavatel'naya Shkola* – Gatchina [Officer Aeronautical School – Gatchina] and became a *Voennyi Letchik* [Military Pilot] on 28 January 1913. He was assigned to the *15th KAO* when war commenced and took part in the Samsonovskaya operation as one of the pilots flying the two-seater Nieuport IV monoplane.

15th KAO typified *RIVVF* detachments at the start of the war. It was established in 1913, as part of the *2-ya Aviatsionnaya Rota* [2nd Aviation Company] based in the Warsaw military district at Mokotów aerodrome. *KAO* establishment included 92 enlisted men working as motor mechanics, aircraft fitters, transport crews and providing defense for the detachment. Transportation included motorcycles, Sauer-manufactured trucks and horse drawn wagons. *15th KAO* fuel allocation was rationed as the campaign started. Nieuport IV rotary engines required castor oil. Spare parts included two wings and three rotary engines.

Chekhutov commented on the Russian high command preferring cavalry over aerial reconnaissance during the campaign. Despite an impressive number of *RIVVF* sorties flown in supporting Second Army, the intelligence was not applied to the strategic initiative sought or the tactical application to ground force maneuvers. Chekhutov was one of a few *15th KAO* personnel (five officers, four NCOs and 23 enlisted escaped capture) able to escape *8. Armee*'s envelopment of XV Corps.

In 1915, Chekhutov was assigned as commander of *4-y Armeysky Aviaotryad (AAO)* [4th Army Aviation Detachment]. No information is available on his post-war life. His brother, *Poruchik* Alexander Anikievich Chekhutov was initially assigned as an observer in the *15th KAO*. He became a *Voennyi Letchik* in the *31st KAO* and the *4th AAO*. After the revolution the brother served in the Red Army Air Fleet.

In the post-war Soviet period, Chekhutov authored z vospominaniy o rabote aviatsii v Samsonovskoy operatsii [From the memories of the work of aviation in the Samsonov operation] which appeared in Vestnik Vozdushnogo Flota [Air Fleet Herald] No. 12 December 1924. It was significant for being the first detailed account of RIVVF over Ostpreussen. His article provided most of the information known today on RIVVF's aerial reconnaissance during the first weeks of the war. RIVVF flew more than 80 sorties supporting General ot kavalerie Samsonov's Second Army.

Chekhutov in his article described a number of reconnaissance flights carried out by the *KAO's* in support of *Gen.kav* Samsonov's Second Army. Intelligence collected by *RIVVF* shows important substance was made available for planning and executing operations.

1st KAO

1st KAO's five Farman HF.16 aeroplanes supported Gen.inf Artamonov's I Corps, as they moved north, in an effort to cover German forces at the southern Ostpreussen border. 1st KAO was initially tasked with conducting both a general orientation of the region and aerial reconnaissance. Farman HF.16s flew four sorties, that soon reported extensive German military activity at the border and rail activity at Soldau, with three trains heading to Deutsch-Eylau.[2]

Nieuport-IV operating from Ostpreussen during the invasion. Many Nieuport-IV were Russian built.

13th KAO

Five Nieuport IV aeroplanes supported Genleyt Klyuev's XIII Corps en route to Willenburg, Passenheim, and Allenstein. XIII Corps tasked 13th KAO with locating German forces in the Chorshele-Willenberg area. From 18 to 20 August, 13th KAO flew four reconnaissance flights on the route Chorshele – Willenberg – Jedwabno – Ortelsburg reporting:

1. Near Jedwabno there was a camp of about 1 - 2 infantry battalions.
2. A force the size of a squadron of cavalry was moving from Willenberg to Chorshele.
3. In Ortelsburg on the southern outskirts of the city – camp of about 1 infantry battalion; from Ortelsburg to Myschyniez – horse patrols.

Poruchik Ivan Anikievich Chekhutov poses in front of a 15th KAO Nieuport-IV supporting XV Corps.

15th KAO

15th KAO supported Gen.inf Martos' XV Corps heading to Neidenburg and Tannenberg. Starting on 18 August, 15th KAO generated six reconnaissance flights in two days–covering the route to Neidenburg and including Chorshele - Janow - Neidenburg:

1. Separate groups of carts were moving on roads Chorshele - Janow - Neidenburg-Chorshele – Willenberg. Chorshele – Yanov, Yanov – Neidenburg, Chorshele -Willenberg.
2. In Neidenburg there was a concentration of troops amounting to an infantry brigade with artillery.
3. Intensive movement at Neidenburg railway station.
4. On railway station Neidenburg – Hohenstein line - movement of trains in both directions.
5. In Willenberg - about 1 infantry regiment.
6. The railway Neidenburg – Willenberg was free.
7. On roads from Neidenburg to the North, there was movement of individual carts.
8. Highway Willenberg – Ortelsburg was free.
9. At the railway segment between Willenberg and Ortelsburg a single train was noticed, going on Ortelsburg.

French built HF.16 flown by RIVVF aviators assigned to 21st KAO supporting Second Army.

23rd KAO

Northeast of the Second Army and to the south of the First Army, Gen.kav Scheideman's II Corps covered the area south of the Masurian Lakes, to include Myszyniec and Ortelsburg with 23rd KAO's five Farman HF.16 aeroplanes. 23rd KAO was also assigned to support Gen.inf Blagoveshchensky's VI Corps.[3] On 18 August, 23rd KAO received orders to fly reconnaissance over Myschyniez – Ortelsburg – Johannisburg – Lomsha. From 18 to 20 August, the squad completed 4 reconnaissance flights and reported:

1. A concentration of troops up to one infantry battalion and intensive movement in the streets observed in Ortelsburg.
2. About 1 – 2 infantry battalions with 3 to 4 pieces of ordnance came to Ortelsburg from the north.
3. One cavalry squadron moved along the road from Ortelsburg to Pelty.
4. Rare train movement along the railway Ortelsburg – Johanesburg.
5. Roads running to the north and north-east of Myschyniez were free.[4]

APPENDIX 2

Clarifying Fliegertruppe Aerial Reconnaissance of 30 August 1914

The famous Körner/Hesse *Flieger-Meldung* recalled by *General der Infanterie* von François in his post war memoirs provided an important understanding of what timely aerial reconnaissance provided the morning of 30 August 1914. Another *FFA 14* mission flown by Canter/Mertens was equally significant in the sightings of the advancing Russian forces and reporting to commanders at the scene and *8. Armee* headquarters. The Canter/Mertens sortie eventually became a source of excitement for the German nation and confusion for historians of the 20th century. The Canter/Mertens sortie was never mentioned in Walter Elze's work of 1928, *Tannenberg: Das Deutsche Heer von 1914: Seine Grundzüge und deren Auswirkung im Sieg an der Ostfront*. It is not mentioned in the definitive Reichsarchiv volume *Der Weltkrieg 1914 – 1918, Band II*. The German aviation historian Peter Supf in his three-volume work *Das Buch der deutschen Fluggeschichte* published in

Pilot Leutnant Canter

Observer – Oberleutnant Körner

1935, brought the supposed sortie to life based on erroneous reporting from a single German aviation officer. *Oberstleutnant* Kuckein, asserted to have been *FFA 14's* commander when they operated out of Deutsch-Eylau, when in fact at the time, Kuckein was commander of Rear Air Park at Dirschau. *FFA 14* was commanded by Hauptmann Heinrich. The two aviators in question, Pilot *Leutnant* Canter and Observer *Leutnant* Mertens, were described in *Oberstleutnant* Kuckein's account as key to the successful aerial reconnaissance over the vital Russian reinforcement from Mława to Neidenburg.[1]

> Pilot *Leutnant* Canter and Observer *Leutnant* Mertens took off from Deutsch-Eylau at 0600 hours, following their planned mission route at 1,100 meters. Haze obscured the view. A Russian troop assembly of substantial strength was observed in the woods and marching in columns along the Mława road to Neidenburg. Russians were observed detraining at Mława Station. Observer *Leutnant* Mertens dropped bombs. The situation was deemed so serious that the aircrew decided to land next to *General der Infanterie* von François' command post. An ongoing artillery barrage negated that opportunity. Canter and Mertens turned back to Gregersdorf and landed adjacent a German battalion. The adjutant officer was immediately informed of their observations and responded. Both flyers then proceeded on bicycles to Neidenburg. While cycling enroute a motor car stopped and picked them up. At 0835 hours they arrived at *General der Infanterie* von François' headquarters and alerted him to the imminent Russian attack. The initial reaction by *I. Armee-Korps* staff was initially skeptical since they had no other information of a new threat facing them. At that time a Taube flown by *Leutnant* Körner from *Feldfliegerabteilung 14* dropped a *Flieger-Meldung* on the Neidenburg town square. *I. Armee-Korps* chief of staff, Major von Massow, quickly issued a new Korpsbefehl. *Leutnant* Canter and *Leutnant* Mertens hurried back to their Taube at Gregersdorf. They took off and flew directly to *8. Armee* headquarters at Osterode. They landed on a potato field. They were immediately received by *8. Armee* Chief of Staff *Generalmajor* Ludendorff and confirmed *Leutnant* Körner's *Flieger-Meldung* at the Neidenburg town square.

SORTIE DATE: 30 AUGUST 1914

AEROPLANE: Gotha LE.3 Taube

PILOT: Ltn Canter

GROUND TRACK: Dtsch.Eylau - Neidenburg - Mława - Janowo - Willenberg -Ortelsburg - Neidenburg - Osterode

SOURCE: NARA RG 165

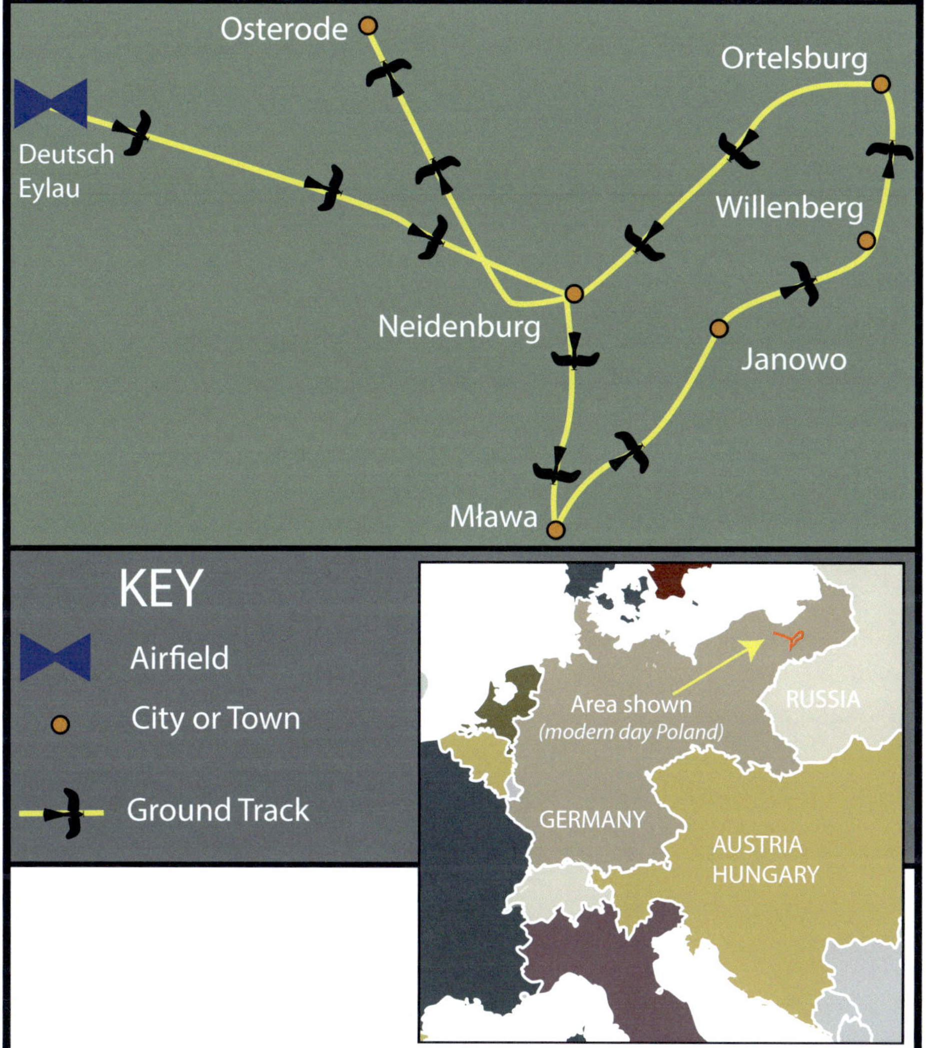

SORTIE DATE: 30 AUGUST 1914

FFA 14's Leutnant Canter and Leutnant Mertens covered a broad area to include the highway from Mława to Neidenburg. Source: NARA RG 165

1. On and near road Neidenburg - Mława At Napierken one column of vehicles heading south. Lenght about 600-1000 m. At Kuklin enemy camp, on which we dropped bombs. Strength of opponent about 1 brigade.
2. Mława and railroad stations. Troops derailing. At and around Mława big troop camps. Mława densely occupied. Strength of opponent about 1 corps. March columns forming. Direction presumably Janowo.
3. At road Kuklin - Janowo, near Koschelny and south enemy troops - mostly in camps. Strength estimated at 1 brigade.
4. Around Janowo troops (own ?), estimated at 1 brigade.
5. Around Kl.Dankheim 7 km west of Willenberg about 1 division.
6. On road Ortelsburg - Willenberg between Gross and Klein Schiemanen one Column heading south.
7. Ortelsburg free.
8. In Forest Gruenfliess and Forest Kaltenborn troops on meadows in the woods. Strength about 2 companies to 2 batallions.
9. From O.F. (Main forest warden's office) Kaltenborn one small column marching on Wallendorf (Wallenberg) (Strength not clearly discernible, about 1 regiment).

Mertens

Lt. Feldflieger Abt. 14.

Sent from Neidenburg 30.8.14 9:15 am.

[received 8. Armee] 30.8 2.45 pm.

> *Generaloberst von* Hindenburg ordered Canter and Mertens to immediately take off and drop *Flieger-Meldung* at the command posts in the sector. Unfortunately for Canter and Mertens, the Taube could not take off from the Osterode potato field. *Leutnant* Mertens then proceeded via motor car to Deutsch-Eylau main *FFA 14* landing ground and summoned another aircrew to execute the mission to deliver the message.[2]

In his memoirs, François did not recall Canter / Mertens efforts. Instead, he only described the dramatic effort by Körner / Hesse dropping the *Flieger-Meldung* in the Neidenburg square. The misunderstandings about the crucial 30 August mission report can be attributed to inaccurate recordings in German documents. There were two sorties and two mission reports put together and recorded as one. The first mission report came from Canter / Mertens and delivered to *I. Armee-Korps* command post at 0915. *GdI* v. François' senior staff doubted Canter / Mertens reports. Only after the Körner / Hesse crew, which had followed the Canter / Mertens mission, had physically dropped their *Flieger-Meldung* at approximately the same time Russian artillery shells struck the Neidenburg town square, did the veracity of Canter / Mertens oral report become understood. This quick turn of events prompted *GdI* v. François' immediate reaction. It saved *I. Armee-Korps* from catastrophe. Körner / Hesse were credited with the merits.[3]

Canter / Mertens proceeded to fly to *8. Armee*'s headquarters at Osterode after briefing François' staff. They dutifully phoned *AOK 8* at 1015 hours – just as the first rounds from Russian artillery struck Neidenburg. Their telephone call was interrupted. When Canter / Mertens arrived at Osterode, they personally briefed *GenMaj* Ludendorff. *AOK 8* responded with immediate support to *GdI* v. François'I. Armee-Korps.[4]

Appendix 3

A Russian Aviator From 15th KAO Recalls a Memorable Sortie

Poruchik Georgii Leonidovich Sheremetevsky, assigned to *15th KAO*, described in his memoirs an amazing experience escaping from almost certain capture by German forces.

On the morning of the 30th of November 1914, we flew the 30 versts from Blonie to Sochaczow and decided that we would cross the front, which was 15 versts to the west of us; Orlov would inspect the area to the north, and I to the south. We had no observers with us. The mechanics Gorshkov, Sediukov, Eliseev and Bobtsov had come from Blonie in a Renault driven by Ogloblev. The weather was foul, it was drizzling, the ground was soddened and the cloud was very low. We had very little hope of success, but we could not change the plan. This was my first operational flight in the new section. I had to prove to the officers and men that I was a flying commander and to the raise the prestige of air support, which had been undermined by my predecessors, in the eyes of Headquarters. Orlov departed first and I followed five or ten minutes later. Hundreds of rabbits ran about the field right in front of the aircraft on the take-off run, lumps of mud flew in all directions, and in spite of the fact that we had no observers on board and had removed all the tools from the aircraft, they could barely lift off the sticky earth and begin to gain height. I crossed our lines at a height of 800 metres and could climb no higher because of the cloud. Looking to the right I could see Orlov's aircraft, which was going from Sochaczow and was descending rapidly. The village of Kiernozy, one verst beyond the lines, was occupied by the Germans, and I headed toward it. In spite of the drizzle, I could see clearly that the village and surrounding area were full of troops. From above it was like an ant-heap that had been disturbed. I had not had time to have a good look when something hit me in the back and the fuel tank, which was behind me, made a noise as if it had been hit with a stick, and I felt fuel pouring over me. I at once switched off the engine and then realized that everyone below me on the ground was blazing away at me with rifles and machine guns. In the complete silence after the engine stopped it was like a drum roll or peals of thunder. Bullets were hitting the wings like heavy rain, and splinters were flying from the struts and the cockpit. My first thought was — to go back, not to be taken prisoner! I turned the aircraft through 180 degrees in a

Poruchik Georgii Leonidovich Sheremetevsky flew Farman HF.16 for Kovno KrAO and supported Gen. kav v Rennenkampf's First Army during the Masurian Lakes campaign. His memoirs contained one of the most detailed accounts of RIVVK in the Great War.

shallow gliding descent and, still hearing the crackling of enemy fire, headed towards our lines in a shallow glide, maintaining minimum airspeed. The nose of the cockpit was a quarter full of fuel and my white felt hunter's boots were completely soaked. By now I was down to 150 - 200 metres and the firing from Kiernozy had stopped. I could see the forward German trenches in front of me and knew that however shallow my glide I could not get across them. The thought of captivity again flashed through my mind. There was a ploughed field below me and I made a three-point landing on it, about 100 metres behind a German trench. Never in my life, neither before nor since, have I made such a precise landing. My mind began to work especially quickly and clearly. In a moment I jumped out of the cockpit and ran towards the trench as about ten Germans, all wearing spiked helmets and carrying rifles with fixed bayonets, jumped out of it and encircled me. When I found myself in the middle of the circle and they were only about ten paces from me, somebody gave an order and they put their rifles on the ground, stretched out their arms and advanced towards me in a ring like a round dance, evidently wanting to take me alive. I felt like a wolf at bay, and I made a great leap, punching two of them on the head, broke out of the ring between the two and dashed towards the empty trench. Running at full tilt I jumped across it and flung myself on my stomach on the no man's land on the other side. I tore off the aneroid altimeter (which I hated at that moment) strapped to my leg and, cursing my felt boots, started to run zigzagging across the ploughed ground. The Germans gave me about 200 metres start by stopping to pick up their rifles and remove the bayonets. Their bullets fell to the left and right of me, splashing me with mud. I was flying, not running, fleeing from captivity and death, and I think I would have outrun the fastest man in the world in that mad race. I worked out that I was about half a verst from our lines and felt that I could not keep up such a pace for long, especially in my high felt boots, which came up almost to my waist. Suddenly I noticed three horsemen were cutting across my path at full gallop. I pulled my Browning out of its holster as I ran, thinking they were Germans, but then I noticed that the Germans were firing at them and not at me. Two of them turned towards our lines and the third turned, galloped right up to me and seized me by the collar. I guessed his intention instinctively and jumped, landing on his horse's crupper. Without slowing, we raced towards our lines. When we were out of range of the German fire, my savior slowed down, looked back at me and said: "Your Excellency, don't forget to write a report - I am Platoon Sergeant Anton Glushchenko of the 17th Nizhnii Novgorod Dragoon Regiment." He delivered me to the headquarters of the 55th Siberian Rifle Regiment.[5]

ENDNOTES

Chapter 1

1. "Grand General Staff" in peacetime had no budgeting power; the War Minister, as final authority for allocation of funds had the last word. Airships, not aeroplanes, were the priority for *Preussisch-Deutsche's* military budget. "Grosse Generalstab" was Germany's highest military planning staff, composed mostly of Prussian General Staff officers and complemented by General Staff officers from the other German Federal States. Upon declaration of war, the *"Grosse Generalstab"* was mobilized and changed name to *"Generalstab des Feldheeres"* ["General Staff of the Army"], also named *"Oberste Heeresleitung – OHL"* [Army High Command]. *OHL's* role covered more responsibilities beyond *Generalstab*. *Generaloberst* von Moltke tenure was referred to as *1. OHL*. Karl-Dieter Seifert, *Mit Sammelgroschen zur deutschen Luftmacht* (Berlin, 2014), 98.
2. M.W. Royse, "National Aeroplane Subscription for Military Aviation." *Aerial Bombardment and the International Regulations of Warfare* (New York: Harold Vinal, 1928), 98.
3. Seifert, 186.
4. John Howard Morrow, Jr. *Building German Airpower*, 1909-1914 (Knoxville, TN: The University of Tennessee Press, 1976), 86-87.
5. Matt Bowden's comprehensive work covers *Fliegertruppe* [German Air Service]. Matt Bowden, *The Great War's Finest, An Operational History of the German Air Service, Volume 1: Western Front 1914* (Aeronaut Books, 2017), 21-22; General Ludendorff, *The General Staff and its Problems, The History of the relations between the High Command and the German Imperial Government as revealed by Official Documents, Volume 1* (New York: E.P. Dutton and Company, 1920), 32-51.
6. Chief of the General Staff of the Army, No. 12751 I. Sec, "Formation of aviation units," Ref. G.W.D. 1157/12 s. A7 of 19/8/12 in General Ludendorff, *The General Staff and its Problems*, 37-38.
7. Alfred Mahncke, *For Kaiser and Hitler, The Memoirs of Luftwaffe General Alfred Mahncke, 1910-1945*. Translated by Jochen Mahncke (Pulborough, West Sussex, UK: Tattered Flag Press, 2010), 4; Helmut Jäger, *Erkundung mit der Kamera* (München, Venorion VKA, 2007), 77.
8. German terms *Nachrichten* and *Nachrichtendienst* are not easy to translate. *Nachricht* can mean "news", "information", "message", but also "intelligence." Signals troops – the *Nachrichtentruppe* – resulted in a general shift in the use of the term *Nachrichten* from intelligence to signals. Hermann Cron, *Geschichte des Deutschen Heeres im Welt-*

kriege 1914-1918 (Berlin: Neudruck Ausgabe, 1937. Subsequently published Osnabrück, GE: Biblio Verlag, 1990), 10, 59, 224, 317.

9. Andrew H. Smoot (2017) A battle lost: re-examining the role of German radio intelligence in the Battle of Gumbinnen, *Intelligence and National Security*, 32:3, 286-299, DOI: 10.1080/02684527.2016.1270990, 287.

10. Marcus Pöhlmann, "German Intelligence at War," Journal of Intelligence History 5 (Winter 2005), 1.

11. *Kaiserlich und königlich (k.u.k.)* is the historic attribute commonly used for the Dual Monarchy of Austria-Hungary. *K.u.k. Armee* is the most definitive title for addressing the Austro-Hungarian army in World War I. *Österreich-Ungarns Letzter Krieg, 1914-1918, Volume 1 (1914)*, 192. Edited by Edmund V. Glaise-Horstenau. Vienna: Publisher of Military Science Releases. 1930. Translated by Stan Hanna, 192.

12. Reichsarchiv, ed., *Der Weltkrieg 1914-1918: Band II: Die Befreiung Ostpreussens* (Berlin: E.S. Mittler & Sohn, 1925), 45. Henceforth, *Der Weltkrieg II.*

13. Kriegswissenschaftliche Abteilung der Luftwaffe. *3. Sonderband: Mobilmachung, Aufmarsch und erster Einsatz der deutschen Luftstreitkräfte im August 1914.* (Berlin 1939). Anlage 11. Translated as Military Science Department of the Luftwaffe. 3rd Special Volume: Mobilization, Deployment and First Deployment of German Air Forces in August 1914. Henceforth *3. Sonderband.*

14. It is interesting to note *Generaloberst* von Moltke's memorandum mentions in same paragraph that the Thorn maneuvers revealed "the importance of air reconnaissance in conjunction with photography." General Ludendorff, *The General Staff and its Problems*, 46.

15. The Austro-Hungarian High Command, equivalent of the German *OHL* was abbreviated *"AOK"*, i.e. *Armee-Oberkommando*. In the German Armed Forces, an *Armee-Oberkommando* was the head command of a field army, i.e. [one level of command lower].

16. Kriegswissenschaftliche Abteilung der Luftwaffe. *8. Sonderband: Die Luftstreitkräfte im Zusammenhang mit den Operationen des Jahres 1914.* (Unpublished final draft, Berlin 1943) 443. Translated as Military Science Department of the Luftwaffe. 8 Special Volume: The Air Forces in Connection with the Operations of 1914. Henceforth *8. Sonderband*; General Ludendorff, *The General Staff and its Problems*, 40.

17. Andre Hug, (Noel C. Shirley, ed.), "Carrier Pigeon Flieger: The WWI Experiences of Dr. Andre Hug, *Cross & Cockade Journal* (US), Vol.13, No. 4, Winter 1972, 314.

18. Helmut Jäger, *Erkundung mit der Kamera*, (München: Venorion VKA, 2007), 79-80; Carl Fink, "Die Entwicklung des militärischen deutschen Luftbildwesens 1911-1918 und seine militärische wie kulturelle Bedeutung," *Wehrwissenschaftliche Rundschau*, Jahrgang 10, Heft 7 (1960), 390-392; Bowden, 17.
19. Jäger, 82, 86, 90.
20. Fink, 391-392; Jäger, 79-80.
21. Fink, 392-393.
22. Zeiss had already been involved in developing aerial cameras (30 cm [L.30] focal length and a 70 cm [L.70] focal length), that were employed on captive balloons. Fink, 390-391; Jäger, 79-80; Bowden, 17.
23. Fink, 393.
24. Fink, 393; Jäger, 79-80.
25. Bundesarchiv Militärarchiv, BAMA, PH 19/118, Freiburg.
26. *Leutnant* Hug went to war with *Feldfliegerabteilung* 26, transferred to *BAO* in December 1914, and was taken prisoner of war on 22 December 1914 near Dunkirk. *Leutnant* Fink's memories are more in-depth to the history of German aerial photography. Hug, 310-314.

Chapter 2

1. Dates in the book are as per the New Style (Gregorian) Calendar, unless otherwise noted as (o.s.) which are Julian Calendar and in the period of 1914-1917 are 13 days behind.

   Russia's newly established aeronautical team, led by Kovanko, now promoted to *Polkovnik* (Colonel), addressed all aviation relevant to military operations. The *Uchebnyy vozdukhoplavatel'nyy* park (*UVP*) [Aeronautical Training Park] was formally established by Order No. 126 of the Military Department (dated 14 July 1890); and was located outside of St. Petersburg at the Volkovo Field artillery range. In the years leading up to the Russo-Japanese War of 1904-1905, emerging aeronautical units were comprised of 65 balloons assigned to Russian fortresses at Warsaw, Novgorod, Brest, Kovno, Osovets and the Far East. The unit was reorganized as the 1st East Siberian Aeronautical Battalion in 1905. *Polkovnik* Kovanko commanded the *Vostochno-Sibirskiy polevoy vozdukhoplavatel'nyy batal'on* [East Siberian Field Aeronautical Battalion] during the Russo-Japanese War, providing both artillery adjustment support as well as intelligence updates on enemy positions. Support to the campaign was now firmly in place, contributing to the evolving role of Russian aviation. Kovanko was promoted to *General-maior* in 1906. By 1910, approximately thirty-one dirigible balloon companies had been effectively established.

Popov,V.A. Vozdukhoplavaniye i aviatsiya v Rossii do 1907 g. [State Publishing House Moscow, 1956] [Popov,V.A. Aeronautics and Aviation in Russia before 1907], 295-375; Yu.O. Druzhinin, Stranitsy Istorii Otechestvennogo Vozdukhoplavaniya (Moscow 2013) [Pages of the History of Domestic Aeronautics], 11-36; Kovanko, Alexander Matveevich // Military Encyclopedia Ed. V.F. Novitsky.

2. D.P.K. // *Russkiy invalid*. - 1908. - No. 247. S. 5 - 6.6

3. *Rossiyskiy Imperatorskiy Voyenno-Vozdushnyy Flot (RIVVF)* is Russia's Air Service name in the Great War. 4 April 1911- Russian War Council statement. Gregory Vitarbo, *Army of the Sky, Russian Military Aviation before the Great War, 1904-1914* (New York: Peter Lang, 2012), 144.

4. Sergey Pavlovich Eliseev, *Strategiya Organizovannogo Stroitel'stva Otechestvennykh Voyenno-Vozdushnykh Sil, (Moscow, 2015.) [The Strategy of The Organized Construction of The Domestic Air Force]*, 9. Carl J. Bobrow Papers.

5. *Birzhevye Vedomosti*, 30 January 1910 (Julian).

6. Vitarbo, 35-36. Gregory Vitarbo, *Army of the Sky, Russian Military Aviation Before the Great War, 1904 -1914* (New York: Peter Lang, 2012), 35-36.

7. The Gatchina operation was named *Vremenniy Aviatsionniy Otdel Ofitserskoj Vozdukhoplavatel'noj Shkoly (OBSh)* [Temporary Air Department of the Officer's Aeronautic School]. This military aerodrome was established as part of the *Uchebniy vozdukhoplavatel'niy park (UVP)* Aeronautic Training Park (Balloon Unit) for the purpose of aeroplane flight training. Over the Summer of 1910, *Poruchik* Evgeny Vladimirovich Rudnev, *Shtabs-kapitan* Georgy Georgievich Gorshkov, and *Poruchik* Ivan Lvovich Kogutov were trained by the French civilian pilot Edmond Morelle to fly aeroplanes at Gatchina. Sergey Pavlovich Eliseev, Strategiya Organizovannogo Stroitel'stva Otechestvennykh Voyenno-Vozdushnykh Sil, (Moscow, 2015. [The Strategy of The Organized Construction of The Domestic Air Force], 11, Carl Bobrow Papers; RGVIA, F.802, Op.3 D.1020 L.297-299; Vladimir Petrovich Zakharov, *Pervyy Voyennyy Aerodrom* (Moscow: Military Publishing, 1988) [The First Military Airfield] Pg 28-33.

8. Minister of War Sukhomlinov and senior staff approved the training, and Gatchina and Sevastopol aviation school pilots were invited to participate. In preparation for the event, the Sevastopol aviation school issued "Brief Guidelines for the Commanders, Troops, Pilots and Intermediaries Participating in the Flight Exercises of 1911." In turn, the Gatchina aviation school generated "Guidelines for Experimental Participation of Aviation Detachments in the Exercises Conducted by the Troops of Guard and the St. Petersburg Military District in 1911." Both

schools utilized lessons learned from French military exercises of 1910, as well as their own experiences. That summer, Gatchina and Sevastopol air units took part in exercises in the St Petersburg, Warsaw, and Kiev Military Districts. "Bol'shiye osenniye manevry s uchastiyem voyennykh letchikov", Vestnik Vozdukhoplavaniya ["Autumn Maneuvers with Participation of Military Pilots", *Bulletin of Aeronautics*] (No.13-14, 1911), 20-21; Eliseev, 13.

9. Eliseev, 11.

10. Unknown source quoted in Antulio J. Echevarria II, *After Clausewitz, German Military Thinkers Before the Great War* (Lawrence, KS: University Press of Kansas, 2000), 173.

11. Gregory Vitarbo, *Army of the Sky, Russian Military Aviation before the Great War, 1904-1914* (New York: Peter Lang, 2012), 82.

12. These units were initially located in the cities of Grodno (Vilna Military District), Kiev (Kiev Military District), Novo Georgievsk (Warsaw Military District), Kars (Caucasus Military District), Chita (Irkutsk Military District), and Spasskoe (Priamur Military District). The fortresses of Kovno, Osowiec, Sevastopol and Brest-Litovsk, also acquired aviation units. After the initial ten aviation units were in place, plans called for an additional eight units. In November 1911, when the State Duma introduced an Army Air Service Bill, the Aeronautical Department senior, *Polkovnik* Vikenty Antonovich Semkovsky stated, "Army Air Service is implemented with aeroplane units being created within the aeronautical units. These aviation detachments are expected to reach a total of 18 in 1912…by the end of 1912, 108 aeroplanes with the required number of pilots will be available in the aviation detachments." Two units (Chita and Spasskoe) were established. By June 1912, 26 pilots were trained and 32 aeroplanes fielded. Eliseev, 12.

13. Corps commanders controlled *Korpusnoy Aviatsionniy* [Corps Aviation Detachments] assigned to serve short-range aerial reconnaissance missions. Fortress headquarters commanders were responsible for *Krepostnyye Aviatsionnyye* [Fortress Detachments] to protect the fortress with short-range aerial reconnaissance support. Lastly, *Otryad Osobogo Naznacheniya* [Special Purpose Detachments] were established and among other tasks, were assigned to support major cavalry commands with aerial reconnaissance and enhance communication between cavalry groups, army or frontline elements. Voyennaya Entsiklopediya. Tom 1 (Sankt-Peterburg, 1912) [Military Encyclopedia. Vol.1 (Saint Petersburg, 1912)], 37.

14. Vitarbo, 78.

15. Philip J. Haythornthwaite, *The World War One Source Book* (London: Arms and Amour Press. 1992), 285.
16. Vitabo, 172.
17. Eliseev, 22.
18. Rossiyskiy gosudarstvennyy voyenno-istoricheskiy arkhiv - RGVIA, F.2003 Op.2 D.612.
19. Eliseev, 46.
20. Petr Dmitrievich Duz, *Istoriya Vozdukhoplavaniya I Aviatsii V Rossii (Iyul' 1914. -Oktyabr' 1917.)* [History of Aeronautics and Aviation in Russia (July 1914 - October 1917)], (1986, Moscow), 18.
21. Initially, aviation personnel were to be trained in three schools and several aero clubs. The General Staff Committee named Sevastopol Officers' Aviation School and Aviation Department of Officers' Aeronautical School at Gatchina, sufficient to meet Russia's aviation training requirements. Eliseev, 19; RGVIA. F.802. Op.1. D.1594. L.123, RGVIA. F.802. Op.1. D.1594. L.27.
22. *General-maior* Shishkevich, Head of the Aeronautical Unit of the Main Directorate of General Staff, added "…one department being in charge of aeronautics-related matters, is on the level with the department, being in charge of supplying the troops with shovels….it is a step back." *Propolkovnik* [Lieutenant Colonel] Sergey Apollonovich Nemchenko, head of the Aeronautical Unit's technical support division, followed up by declaring that the millions of rubles being spent for aviation and aeronautics were generating a large number "of different diverse-structured units and offices." He urged creation of an independent aviation and aeronautics department under direct command of the Main Directorate. Nemchenko's proposal was implemented later during the war, with the creation of the *Polevoye upravleniye aviatsii i vozdukhoplavaniya (PUAV)*, or Field Administration of Aviation and Aeronautics, to serve as the central aviation authority.Eliseev; RGVIA. F.802. Op.1. D.1594. L.123, RGVIA. F.802. Op.1. D.1594. L.27.
23. Eliseev, 19; RGVIA. F.802. Op.1. D.1594. L.123, RGVIA. F.802. Op.1. D.1594. L.27.
24. Eliseev, 13.
25. Eliseev, 29.
26. Local production of Russian aeroplanes originated from three domestic aeroplane plants: *Pervoe Rossiiskoe Tovarishchestvo Vozdukhoplavaniya (PRTV])* [Russian Association of Aeronautics, S.S. Schetinin & Co.], Aktsionevnoye obschchestvo Dux [Joint - Stock Company Dux], and *Russko-Baltiyskiy vagonnyy zavod (R-BVZ)* [Russo-Baltic Wagon

Works]. *Organizovannogo Stroitel'stva Otechestvennykh Voyenno-Vozdushnykh Sil*, Moscow, 2015. [The Strategy of The Organized Construction of The Domestic Air Force], 18. Carl Bobrow Papers.

27. At Bronnitsy (near Moscow), *5th AO* was established with three subordinated corps detachments: *Grenadier KAO, 13th KAO* and *17th KAO*. To the south at Odessa, *6th AO* had four detachments: *7th KAO, 8th KAO, 24th KAO, 25th KAO. 1st AO* included *5th KAO, 16th KAO* and *22nd KAO*. Finally, Russia's far eastern aviation comprised *4th Siberian Fortress AO* at Spasskoe. Three fortress detachments supported their respective fortress operations at Grodno, Sevastopol and Vladivostok. RGVIA F.2008. Op.2. D.643. L.170; Eliseev, 31.

28. Some sources say 263 Russian aeroplanes were available at the start of the war. The single *PAO* was assigned to *General ot artillerii* Nikolai Iudovich Ivanov's Southwest Front headquarters. *KAO* units arrived at forward landing grounds in accordance with the plan for deployment of armies. *KrAO* remained with their fortresses. Aviation company and subordinate corps detachments were in place when war was declared. *4th Aviatsionnaya Rota (AR)* [4th Aviation Company] formed at Lida to support the Vilna Military District had seven corps detachments: *2nd KAO, 3rd KAO, 4th KAO, 6th KAO, 10th KAO, 20th KAO* and *21st KAO. 4th AR* demonstrated Russia's dilemma in integrating aviation resources, compatible with prevailing army priorities. Vilna Military District's chief rendered assistance to complete the detachment's formation and initial training, as increased interaction between aviators and commanders demonstrated the unit's mission was receiving great support. When Russia invaded *Ostpreussen*, the *4th AR* supported both the Russian First Army and the Russian Second Army. Georgii Leonidovich Sheremetevsky, "White Nights" *The '14 – '18 Journal*. Edited and translated by Margaret Travers. (Australian Society of World War I Aero Historians, 2001), 18.

29. Sheremetevsky, "White Nights," 17; Duz, 18.

30. RGVIA, F.802 Op.4 D.2892; RGVIA, F.352 Op.2 D.5 L.115.

31. Echevarria, 140.

32. *Podyesaul* Vyatcheslav Mikhailovich Tkachev was one of the best known Russian aviators in World War 1. His accomplishments are described in Book 2. Vitarbo explained the Russian introduction with aviation as "a façade of modernity, a Potemkin village in the sky." The sole exception was the "Sikorsky's."Vitabo, 177.

33. Sean McMeekin, *The Russian Origins of the First World War* (Cambridge, MA: The Belknap Press of Harvard University Press, 2011), 79.

34. McMeekin, 81.

35. *Gen.kav* Zhilinsky was the former Stavka chief of staff. Paul Robinson, *Grand Duke, Nikolai Nikolaevich, Supreme Commander of the Russian Army* (DeKalb, IL: Northern Illinois University Press. 2014), 143.

36. David R. Stone, *The Russian Army in the Great War, The Eastern Front, 1914-1917* (Lawrence, KS: University of Kansas Press, 2015), 57-58.

37. RGVIA. F.2003. Op.1. D 1224; Eliseev, 29.

38. Yvatcheslav Mikhailovich Tkachev, *Kryl'ya Rossii. Vospominaniya o proshlom russkoy voyennoy aviatsii 1910-1917* gg. SPb: GITS Novoye kul'turnoye prostranstvo, 2007, 102 [Wings of Russia. Memories of the past of Russian military aviation 1910-1917, St. Petersburg: State Research Center 2007]

39. *Aeroplan i yego boyevoye znacheniye. K dokladu poruchika Mikos v Ofitserskom garnizonnom sobranii 27 fevralya* 1912g. - Kiyev, 1912. S. 17. [Airplane and Its Combat Application: For the Report of Lieutenant Mikos in the Officers' Garrison Assembly on 27 February 1912 - Kiev, 1912; page 17.]

40. *Polozheniye o polevom upravlenii voysk v voyennoye vremya.* - SPb., 1914. [Provisions on Field Administration of Troops at Warfare. Saint Petersburg, 1914.]

41. Farman MF.11 and Nieuport IV's were powered by the Gnome rotary engine. The Gnome rotary engine required frequent cleaning of piston rings, due to castor oil being the primary lubricant. Inevitably, accidents and poor landings increased dramatically. Germany's *Fliegertruppe* also faced similar maintenance challenges throughout the campaign, but possessed a home field advantage. They had already established airparks as well as a functioning logistic tail, which ultimately made a critical difference in aviation's role in August 1914 on the Eastern Front. Eliseev, 29.

42. RGVIA F.2008. Op.1. D.657. L.43.

43. Vyatcheslav Mikhailovich Tkachev, *Kryl'ya Rossii. Vospominaniya o proshlom russkoy voyennoy aviatsii 1910-1917* gg. SPb: GITS Novoye kul'turnoye prostranstvo, 2007, 70 [Wings of Russia. Memories of the past of Russian military aviation 1910-1917, St. Petersburg: State Research Center 2007].

44. Alex Marshall, "Russian Military Intelligence, 1905-1917: The Untold Story behind Tsarist Russian in the First World War," *War in History, (4)* 2004, 395-396.

45. Marshall, 406.

46. Hermann von François, *Marneschlacht und Tannenberg: Betrachtungen zur deutschen Kriegsfuehrung der ersten*
Henceforth François, *Marneschlacht.*

47. Sikorsky, an engineering graduate of Kiev Polytechnic Institute, applied his intuitive genius in developing the world's first multi-engine enclosed cabin aeroplane. In addition to Sikorsky, the crew consisted of *Starshi Leytenant* Georgy Ivanovich Lavrov of the Imperial Russian Navy, *Sh.Kap.* Christopher Fridrikhovich Prussis of the Imperial Russian Army, and the *R-BVZ's* trusted mechanic, Vladimir Sergeyevich Panasyuk.

48. *Stars.Lient* and Lavrov's participation in the epic flight, helped spur navy interest. Their intent in adapting Sikorsky's design, was to modify and adapt the airframe into a giant float-plane for naval operations. Boris Petrovich Dudorov, Personal papers and Manuscript, "Voyenno-morskiye VVS" [The Naval Air Force], Carl J. Bobrow Papers; Boris Vyacheslavovich Drashpil, Personal papers and Manuscript, "Istoriya aviatsii v rossiyskom imperatorskom flote" [History of Aviation in the Russian Imperial Navy], Carl Bobrow Papers; Marat Khayrulin *"Il'ya Muromets," Gordost' russkoy aviatsii.* 2010 Marat Khairulin [*"Ilya Muromets,"* Pride of Russian aviation.] 23.

49. *Leitenant* Yankevich, *Rukovodstvo Po Vozdushnoy Fotografii*, Petrograd, 1917, 87-96; [Handbook of Aerial Photography. Petrograd 1917, 87-96].

50. Boris Viktorovich Krasnopevtsev, *Osnovnyye Sobytiya V Istorii Fotogrammetrii V 19 i 20 Vekakh*. Moscow, 2008. [Main Events in the History of Photogrammetry in the 19th and 20th Centuries], 4.

51. Yankevich, 81, 83-85, 87; Krasnopevtsev, 4; Duz, 42, 44.

52. "Ogoniok" ("Spark"), March 19, 1911 No.12, 16.

53. On 26 October 1911, at a ceremony presided over by Tsar Nicholas II, Gelgar received the *Voennyi Letchik* [Military Pilot] earned from the *Sevastopol'skaya ofitserskaya shkola aviatsii* [Sevastopol Officer Aviation School]. While at Sevastopol, Gelgar conclusively demonstrated aerial photography's potential to locate submerged submarines by employing a camera that photographed through a hatch, constructed on the cabin floor of a Blériot XI. RGVIA F.2008. Op.1. D.362.

NOTE: Russian has two words (дивизия and дивизион) which are usually translated "division." дивизия is what we usually think of as a division with brigade- or regimental-sized subunits. In 1914, a дивизион was a cavalry or artillery unit with two subunits (squadrons for cavalry, batteries for artillery). At the present time, дивизия is used for all artillery units (all types) and we usually translated it as battalion. An alternative is to transliterate the term (divizion) and provide an explanation the first time it appears. Discussion with Andrew Smoot.

CHAPTER 3

1. David R. Stone, 55.
2. Alexander Watson, *Ring of Steel, Germany and Austria-Hungary in World War I* (New York: Basic Books, 2014), 161.
3. German military identified Russian Armies by the ground occupied in relation to the most significant feature on the topographical map (ie., Nieman Army, Narev Army). John Keegan. *The First World War* (New York: Alfred A. Knopf, 1999), 140.
4. Bruce W. Lincoln, *Passage Through Armageddon, The Russians in War and Revolution*. (New York: Oxford University Press, 1994), 64.
5. Georgii Leonidovich Sheremetevsky, "Memoirs of a Russian Aviator, The Recollections of Georgii Leonidovich Sheremetevsky," Edited and translated by Margaret Travers. *Cross & Cockade International* 36/2, 2005, 97.
6. Marshall, 401; Norman Stone, *The Eastern Front, 1914-1917* (Abingdon, Oxon, UK: Book Club Edition, 1975), 59.
7. *Der Weltkrieg II*, 304-306; Watson, 161.
8. Major-General Ironside's post-war account of Tannenberg makes it clear that Russian cavalry failed in successfully accomplishing their main task: collecting accurate information on enemy's dispositions. Edmund Ironside, Major-General Sir, *Tannenberg, The First Thirty Days in East Prussia* (Edinburgh and London: William Blackwood & Sons Ltd., 1928), 142.
9. Yuri Nikiforovich Danilov, *Rossiya v Mirovoy Voyne*, Berlin, 1924 [*Russia in the World War*], 7.
10. *Österreich-Ungarns Letzter Krieg, 1914-1918, Volume 1* (1914), 209-210.
11. Keegan, 143-144.
12. Jamie H. Cockfield, *With Snow on Their Boots, The Tragic Odyssey of the Russian Expeditionary Force in France During World War I* (New York: St Martin's Press, 1998), 62.
13. Captain James Coghlin's Command and General Staff study of the Tannenberg battle revealed from Colonel Conger, U.S. Army Attache to Germany, that the German military had constructed defensive barriers in the Lötzen region many years before the war. The purpose was to prevent communication between two Russian armies invading *Ostpreussen*. The principal barrier was at Lötzen, closing the gap between two of the Masurian Lakes. A barrier between Willenberg and Ortelsburg consisted of cleared lanes about 50 meters wide through the thick forest. Blockhouses armed with machine guns were established at intervals. This thwarted Russian patrols trying to

link the two armies. "Notes on the Battle of Tannenberg, G-2 Report, No. 8578, CONFIDENTIAL, Colonel A. L. Conger, GSC, U.S. Military Attache, Berlin, 4 May 1927 in Captain James J. Coghlan, *Operations of the I German Corps at the Battle of Tannenberg, 26-31 August, 1914* (Fort Leavenworth, Kansas: The Command and General Staff School, 1931-1932), 4-5; *Der Weltkrieg II*, 67-68.

14. Ivan Anikievich Chekhutov, "Iz vospominaniy o rabote aviatsii v Sam sonovskoy operatsii," *Vestnik Vozdushnogo Flota No. 12* (dekabr') 1924 g. [Ivan Anikievich Chekhutov, "From the memories of the aircraft in operation Samson," Journal of the Air Force, No. 12. 1924, 2].

15. RGVIA F.2048 Op.2 D.115.Аэродром Лида — сто лет полетов / Д. Киенко, В. Сливкин (Lida airdrome - One Hundred Years of Flight / D. Kienko, V. Slivkin Smolensk), 2013, 25-26.

16. Chekhutov, 2.

17. Chekhutov, 5., Viktor Kulikov, *Iz Istorii Vozdushnykh Voyn Za Otechestvo! Khronika boyevykh deystviy 15-go aviaotryada Russkoy armi na frontakh Pervoy mirovoy voyny,* (Avia-khobbi, 2007); [The History of Air Wars for the Fatherland! Chronicle of the Fighting of the 15th Squadron of the Russian Army on the Fronts of the First World War, Air Hobbies, 2007].

18. *Samsonovskaya* (Самсоновская). It is first introduced as the name of Second Army's operation in East Prussia (as in Самсоновская Операция). However, its use appears to expand to being shorthand (at least as much as a 13-letter word can be short hand for anything) for Second Army, Second Army's headquarters, and maybe even for Samsonov himself. Discussions between author and Andy Smoot.

19. The Hughes' alphabetic transceiver was a printing telegraph, unlike the Morse code of dots and dashes, it contained a printing wheel and a keyboard of 26 keys containing alphabet and punctuation characters at both the transmitter and receiving units. Lincoln, 56-57; David Kahn, *The Codebreakers, The Story of Secret Writing* (New York: The Macmillan Company, 1967), 622; Norman Stone, 50-51.

20. Kahn, 622.

21. Novo Georgievsk was the primary Russian fortress northwest of Warsaw. *German 8th Army War Diary*, Folder V, 2 August to 1 October, 1914, Annex Folder 1, 25 (RG 165 NARA subsequent reference is *German 8th Army War Diary*, English Translation, 2-3.

22. Walter Elze, *Tannenberg: Das Deutsche Heer von 1914, seine Grundzüge und deren Auswirkung im Sieg an der Ostfront* (Breslau, GE: Hirt, 1928), 201, [Tannenberg: The German Army of 1914, It's Basic Features and Their Impact on the Victory of the Eastern Front].

23. The Russian First Army was tentatively named Wilna Armee to reflect its home station on the Wilna River. Elze, 209-210, (Document 14).

24. Elze, 201.

25. Zeppelin Z-V was also employed to drop propaganda leaflets. Mahncke, 27.

26. *German 8th Army War Diary*, English Translation, 5.

27. Annex I, *German 8th Army War Diary*, English Translation, 13. Dennis E. Showalter, *Tannenberg, Clash of Empires* (Hamden, CT: Archon Books, 1991), 152-153.

28. Grosser Generalstab, *Die Schlachten und Gefechte des Großen Krieges 1914-1918* (Berlin: Verlag von Hermann Sack, 1919), 7. Henceforth *Schlachtenkalender*; *German 8th Army War Diary*, English Translation; Georg Paul Neumann, *Die deutschen Luftstreitkräfte im Weltkrieg* [The German Air Force in World War] (Berlin: E. S. Mittler und Sohn, 1920), 362.

29. Annex I, *German 8th Army War Diary*, English Translation, 14.

30. *8. Armee-Oberkommando (AOK 8)* is German military terminology and abbreviations. Annex I, *German 8th Army War Diary*, English Translation, 25.

31. Smoot, 292; John R. Cuneo, *The Air Weapon 1914-1916, II* (Harrisburg, PA: Military Service Publishing Company, 1947), 112. Henceforth Cuneo, II.

32. Smoot, 292.

33. Report covering 8 August on page 25. 9 August report on page 22-23. Annex I, *German 8th Army War Diary*, English Translation, 22-23.

34. Annex I, *German 8th Army War Diary*, English Translation, 30.

35. *Genmaj v.* Waldersee quote in Cuneo Notes, 404.

36. *3. Sonderband*, 84; Annex I, *German 8th Army War Diary*, English Translation, 30.

37. *Schlachtenkalender*, 7; *3. Sonderband*, 86.

38. Annex I, *German 8th Army War Diary*, English Translation, 33.

39. *German 8th Army War Diary*, English Translation, 9; Cuneo, 403; Elze, 98.

40. *German 8th Army War Diary*, English Translation, 10.

41. *German 8th Army War Diary*, English Translation, 13.

42. 1st Russian Cavalry Division, 2nd Russian Cavalry Division, 3rd Russian Cavalry Division, and 5th Rifle Brigade. Discussion in Notes. Cuneo, 404; Smoot, 288.

43. Annex I, *German 8th Army War Diary*, English Translation, 48.

44. *Leutnant* Mertens in eighteen days made history for *Fliegertruppe* over Neidenburg. Annex I, *German* 8th *Army War Diary*, English Translation, 59-60.
45. *Leutnant* Rother was the pilot for the sortie that flew with *Leutnant* Mertens. Annex I, *German* 8th *Army War Diary*, English Translation, 66.
46. Both Hesse and Körner played an important role on 30 August over Neidenburg. Annex I, *German* 8th *Army War Diary*, English Translation, 67.
47. *I. Armee-Korps* reported on 11 August that no Russian aeroplanes had been sighted. Elze, 207-209; *Der Weltkrieg II*, 64-67.
48. François, *Marneschlacht*, 148-150.
49. *3. Sonderband*, 87-88.
50. Annex I, *German 8th Army War Diary*, English Translation, 77.
51. *Leutnant* Eberstein was the pilot. Annex I, *German 8th Army War Diary*, English Translation, 96-97.
52. *3. Sonderband*, 85; François, *Marneschlacht, 164.*
53. *German* 8th *Army War Diary*, English Translation, 16-17; Annex I, *German* 8th *Army War Diary*, English Translation, 97-105; *3. Sonderband*, 88.
54. Eliseev, 47.
55. *General-leytenant* Oranovskii was Quartermaster General during the Russo-Japanese war managing HUMINT in the region. Kulikov, "For the Fatherland!" WW1.; Chekhutov, 9-12.
56. Kulikov, "For the Fatherland!", 2; Chekhutov, 9-12.
57. Chekhutov, 9.
58. *Der Weltkrieg II*, 62.
59. Annex I, *German* 8th *Army War Diary*, English Translation, 134.
60. *German* 8th *Army War Diary*, English Translation, 20.
61. *German* 8th *Army War Diary*, English Translation, 21.
62. The reports say the *Leutnant* "threw the automobile in the lake." Annex I, *German* 8th *Army War Diary*, English Translation, 127.
63. *Genmaj* Frantz delivered the documents. Annex I, *German* 8th *Army War Diary*, English Translation, 153-158.
64. Annex I, *German* 8th *Army War Diary*, English Translation, 159.
65. *Genmaj* v. Waldersee quote in Notes. Cuneo, 404; Elze, 98.
66. Smoot, 288.
67. Annex I, *German* 8th *Army War Diary*, English Translation, 153.
68. *German* 8th *Army War Diary*, English Translation, 23.

Chapter 4

1. *3. Sonderband*, 88; Smoot, 289.
2. Wyschtyniez is east of Rominte Heide. Annex I, *German 8th Army War Diary*, English Translation, 168-169.
3. Annex I, *German 8th Army War Diary*, English Translation, 169-170.
4. Sonderband, 90-91; *The West Point Atlas of War*. Edited by Brigadier General Vincent J. Esposito. New York: Tess Press. 1997, 32. Henceforth *West Point Atlas*.
5. *Der Weltkrieg II*, 63.
6. Annex I, *German 8th Army War Diary*, English Translation, 163.
7. Annex I, *German 8th Army War Diary*, English Translation, 163.
8. Annex I, *German 8th Army War Diary*, English Translation, 165-167.
9. Annex I, *German 8th Army War Diary*, English Translation, 172.
10. Elze, 197; *Der Weltkrieg II*, 55-56; Robert Forzyk, Combat – 1914-15 *German Infantryman versus Russian Infantryman*, (Oxford, UK: Osprey Publishing, 2015), 38.
11. A unique perspective of chain of command existed. Each *Kommandierender General* (Korps Commander) had direct access to His Majesty, Emperor Wilhelm II. *Der Weltkrieg II*, 76; Elze, *Tannenberg*, 100-102.
12. Francois, *Marneschlacht*, 148-150.
13. *German 8th Army War Diary*, English Translation, 26.
14. *German 8th Army War Diary*, English Translation, 27.
15. Elze, 197; *Der Weltkrieg II*, 55-56.
16. Annex I, *German 8th Army War Diary*, English Translation, 172.
17. Annex I, *German 8th Army War Diary*, English Translation, 173.
18. Annex I, *German 8th Army War Diary*, English Translation, 184-186.
19. Annex I, *German 8th Army War Diary*, English Translation, 186.
20. Report provided in Annex I does not state headquarters or time of report. Annex I, *German 8th Army War Diary*, English Translation, 183.
21. *Der Weltkrieg II*, 69-78.
22. Annex I, *German 8th Army War Diary*, English Translation, 176-177.
23. *German 8th Army War Diary*, English Translation, 31.
24. *German 8th Army War Diary*, English Translation, 32.
25. *German 8th Army War Diary*, English Translation, 30-31.
26. Cuneo, II, 116.
27. *3. Sonderband*, 90-91.

28. *3. Sonderband*, 90.
29. *German 8th Army War Diary*, English Translation, 36.
30. *3. Sonderband*, 90-91.
31. Smoot, 291.
32. Smoot cites 1620 hours, Cuneo cites 1640 hours. Smoot, 291; Cuneo, 113-114.
33. *3. Sonderband*, 91.
34. Cuneo, II, 114.
35. Smoot, 292.
36. Cuneo, II, 117.
37. David Stone, 65-66.
38. David Stone, 65.
39. *Gen.kav* Samsonov had greater challenges to meet Northwest Front objectives. *Gen.kav* Rennenkampf prior to the war was the Vilna Military District commander, comprising the majority of First Army. Rennenkampf's direction was based on a long-arranged plan. *Gen. kav* Samsonov was Governor of Turkestan when war was declared. When he arrived to command Second Army he did not know the staff or the subordinate commanders. A healthy Samsonov might have had the strength to challenge *Gen.kav* Zhilinsky of the strict time schedule mandated for Second Army's advance into *Ostpreussen*. Ironside, 118-120; Chekhutov, 13.
40. *3. Sonderband*, 92.
41. Northwest of Myszyniec. *3. Sonderband*, 91.
42. Kulikov, "For the Fatherland!"
43. *German 8th Army War Diary*, English Translation, 39, 42.
44. Cuneo, II, 114.
45. Robert Forczyk, *Combat: German Infantryman versus Russian Infantryman* (Oxford: Osprey Publishing Ltd, 2015), 42.
46. Cuneo, II, 114.
47. David Stone, 65-66.
48. Cuneo, II, 114.
49. Russian 28th Division and 29th Division were also cited being in defensive positions. Smoot, 292.
50. *German 8th Army War Diary*, English Translation, 36-38, 42.
51. Cuneo, II, 114-115.
52. *German 8th Army War Diary*, English Translation, 39, 42.

53. *Der Weltkrieg II*, 95 f.
54. M. Schwarte, *Der Grosse Krieg 1914-1918*, (Leipzig: Johann Ambrosius Barth, 1922), 551.
55. *German 8th Army War Diary*, English Translation, 39, 42.
56. *Der Weltkrieg II*, 117; Elze, 119; Paul von Hindenburg, Field Marshal, *The Great War* (London: Greenhill Books, 2006), [Original – *Aus meinem Leben*, Leipzig, 1934], 69. [Hindenburg did not write his memoirs himself. His ghostwriter was Merz v. Quimheim. (*Pöhlmann, Kriegsgeschichte und Kriegspolitik*, 60)].
57. *Der Weltkrieg II*, 122.
58. Chekhutov, 13.
59. Cuneo, II, 117.
60. Smoot, 292.
61. *German 8th Army War Diary*, Annexes, Folder 1a, 54; Elze, 226; Cuneo, II, 116; Prit Buttar, *Collision of Empires: The War on the Eastern Front in 1914* (Oxford, UK: Osprey Publishing, 2014), 135; Smoot, 292.
62. Cuneo does not reference the source of the quote. Cuneo, II, 116; Elard von Löwenstern, Baron, *Der Frontflieger* (Berlin: Bernard & Graefe, 1937), 35.
63. *3. Sonderband*, 94; Cuneo, 116; Smoot, 292.
64. Smoot, 292.
65. *German 8th Army War Diary*, English Translation, 40.
66. *Der Weltkrieg II*, 67.
67. Cuneo, 117.
68. Cuneo, 117.
69. David Stone, 66-67.
70. Elze, *Tannenberg*, 242; Buttar, *Collision of Empires*, 136.
71. Cuneo, II, 117-118.
72. *West Point Atlas*, 32-34.
73. *3. Sonderband*, 95.
74. *German 8th Army War Diary*, English Translation, 40-41.
75. *Schlachtenkalender*, 9.
76. Winston S. Churchill, *The Unknown War* (New York: Charles Scribner's Sons, 1931), 188.
77. *German 8th Army War Diary*, English Translation, 43.
78. Chekhutov, 13-14.
79. *Der Weltkrieg II*, 120-121.

80. RGVIA F.6071. Op.1. D.2, 4, 5, 6, 7, 8 L.3; Donat Aduevich Makijonek was awarded the Cross of St. George 4th class for the action. Victor P. Kulikov, *Russian Aces of World War 1* (Osprey Publishing, 2013), 58; Kulikov, *Pol'skii Orel v Russkom Nebe – Letchik-Istrebitel' Donat Makienok.* AviaMaster 6/2000. [Kulikov, Polish Eagle in the Russian Sky - Fighter Pilot Donat Makijonek. AviaMaster 6/2000.]

81. *Der Weltkrieg II*, 246.

82. *Der Weltkrieg II*, 111; Keegan, 145; Churchill, 177-178; *West Point Atlas*, 34.

Chapter 5

1. *German 8th Army War Diary*, English Translation, 49.

2. Erich Ludendorff, *Ludendorff's Own Story, August 1914 – November 1918* (New York and London: Harper & Brothers Publishers, 1919), 55. Henceforth, Ludendorff.

3. Traditional view held *AOK 8 Generalstab* conceived the maneuver. Schäfer in Deutches Offizierblatt (1926/17), 730-734, made the case from the records review that *GO* v Prittwitz envisioned only defensive options and held no objectives to conduct a major attack on the Russian Second Army approaching from the south. A subsequent view from General Kabisch in Deutsches Offizierblatt (1927/17), 721-725, related *GO* v. Prittwitz had no plan for an all-out attack.

4. Generalfeldmarschall von Hindenburg, [Original – *Aus meinem Leben*. Leipzig GE, Hirzel Verlag 1934), 69 f. [Field Marshal von Hindenburg, *The Great War* (London: Greenhill Books, 2006), 70.

5. Max Hoffmann. *War Diaries and Other Papers, Volumes I and II*, Translated from the German by Eric Sutton, Martin Secker, London, 1929 (*War of Lost Opportunities and The Truth About Tannenberg*).

6. Elze, Tannenberg, 281 (Document 142).

7. *Der Weltkrieg II*, 114, 120.

8. *8. Sonderband*, 414-415.

9. *Fliegertruppe* assignments that supported Hindenburg's initial command included *8. Armee's FFA 16's* L.V.G. biplanes and *XX. Armee-Korps' FFA 15's* Jeannin "Stahltaube" monoplanes. Both *Feldfliegerabteilungen* operated from Osterode and covered southern *Ostpreussen. GdK* v. Mackensen's *XVII. Armee-Korps' FFA 17's* Albatros B.I's flew out of Schwöllmen, 50 kilometers to the north of Osterode. *Festung Boyen/Lötzen (FestFA) 7* had two aeroplanes at Allenstein supporting 3. Reserve Division. Further south at *Gouvernment Festung* Graudenz, *FestFA 6* monitored for Russian forces south of Soldau. Additionally, *8. Armee's* rear air park at Dirschau south of Danzig

also was available for tasking. In northern *Ostpreussen*, *Gouvernment Festung* Königsberg's *FestFA 5* operated from Königsberg flying Taube sorties over *Gen.kav* v. Rennenkampf's First Army. Zeppelin Z-V also supported the campaign flying from Posen. *8. Sonderband*, 414-415.

10. Lincoln, 71; *8. Sonderband*, 414-415.
11. Discussion with Andrew Smoot; Tsikhovich, Appendix 2.
12. *Der Weltkrieg II*, 123-125.
13. David Stone, 72.
14. David Stone, 62-63.
15. Chekhutov 14-15.
16. Keegan, 146-147.
17. Evans, 121; John Sweetman, *Tannenberg 1914* (London: Cassell & Co. 2002), 118-119.
18. RGVIA F.6071. Op.1. D.2, 4, 5, 6, 7, 8 L.3; Kulikov, V*iktor P., Pol'skii Orel v Russkom Nebe – Letchik-Istrebitel' Donat Makienok*. AviaMaster 6/2000. [Kulikov, Victor P., *Polish Eagle in the Russian Sky - Fighter Pilot Donat Makijonek*. AviaMaster 6/2000.]
19. *Der Weltkrieg II*, 124-126.; Elze, 122-123; *German 8th Army War Diary*, English Translation, 53.
20. *German 8th Army War Diary*, English Translation, 52.
21. Elze, Tannenberg, 289 (Documents 161, 162); 8th Army War Diary, Annex Folder 1a, 104, item 11-12; discussions with Andrew Smoot.
22. *8. Sonderband*, 414.
23. Discussion with Andrew Smoot; Showalter, 227.
24. *German 8th Army War Diary*, English Translation, 54.
25. Annex I, *German 8th Army War Diary*, English Translation, 109.
26. Tage Carlswärd, Karl-Albert Mügge, *Operationen und Nachrichten-verbindungen, Voggenreiter im Osten* (Potsdam, 1939) 92-93; 3 Sonder-band, 91.
27. Marshall, 407; Danilov (*Rossiya v Mirovoy Voyne*) quoted in Lincoln, 57; David Stone, 71-72; Showalter, 217.
28. Kahn, 622.
29. Elze, Tannenberg, 289, (Documents 161, 162); *Der Weltkrieg II*, 136-137.
30. Elze, Tannenberg 289, (Document 161)
31. *Der Weltkrieg II*, 136-137.
32. Elze, Tannenberg, 289 f. (Document 162).
33. Former *Oberstleutnant* Theobald Schäfer, postwar Reichsarchiv Ar-

chivist, had at his disposal all of the main records of the Tannenberg battle when he wrote "Legendenbildung über Tannenberg" [Tannenberg Urban Legends] *Deutsches Offiziersblatt*, 1923.

34. Ludendorff, 57-59.

35. A station was under construction at Graudenz. The building and antenna were ready, but the radio apparatus had not yet been assembled. Discussion with Andrew Smoot, Carlsward, 13-14.

36. Theobald von Schäfer, *Tannenberg, Schlachten des Weltkriegs*, Bd. 19 (Oldenburg GE: Gerhard Stalling 1927), 46-47. Henceforth Schäfer, *Tannenberg*.

37. Discussion with Andrew Smoot.

38. In addition to Professor Deubner, *k.u.k. Hauptman* Hermann Pokorny was another leading Central Powers cryptologist in World War I. Book 2 details his accomplishments. Kunibert Randevig, "Die Deutsche Funkaufklärung in der Schlacht von Tannenberg", *Wissen und Wehr*, (Berlin 1932/5), 128 -141; Kahn, 629.

39. Cuneo, II, 120.

40. *8. Sonderband*, 420.

41. Cuneo, II, 122.

42. *8. Sonderband*, 418-420.

43. Cuneo, II, 122.

44. *8. Sonderband*, 418-420.

45. Löwenstern, 35; Cuneo, II, 120.

46. The report most likely refers to Russian XIII Corps. [Elard Baron von Löwenstern, *Der Frontflieger* (Berlin: Bernard & Graefe, 1937), 35.

47. *8. Sonderband*, 413-415.

48. Elze, *Tannenberg*, 285, (Document, nr. 152).

49. Chekhutov, 16-17.

50. *z vospominaniy o rabote aviatsii v Samsonovskoy operatsii, L.Chekhutov, Vestnik Vozdushnogo Flota No. 12 (dekabr')* 1924 g [From the Memories of the Aircraft Application in Samsonov's Operation, L.Chekhutov, Air Force Journal No. 12.1924, 24-28].

51. A. A. Noskoff, General, *"Der Russische Nachrichtendienst vor dem Gorlice Durchbruch," Militär-Wochenblatt Nr. 42* (Berlin: 1935), 1670 f, [The Russian Intelligence Service before the Gorlice Breakthrough].

52. Elze quotes eight Russian messages on 25 August 1914. The messages revealed Russian intentions that day—assuring *8. Armee* that *General ot kavalerii* Rennenkampf's Russian First Army was not to turn south to immediately support *General* Samsonov. Elze, *Tannenberg*,

289 ff. (Documents 161-168). A. Noskoff, General, *Militaer-Wochenblatt Nr. 42* (Berlin: 1935). [The Russian Intelligence Service before the Gorlice Breakthrough].

53. Danilov, 8.

54. Lincoln, 57 (quoting Danilov, *Rossiya v Mirovoy Voyne*, Berlin, 1924 [Russia in the World War], 191-192); Rostunov, *Russkii front*, 100-101); Norman Stone, 52.

55. Elze, 124-125; *Der Weltkrieg II*, 117-118.

56. Cuneo, II, 120-121.

Chapter 6

1. Three German operations comprise the battle of Tannenberg. Usdau, Bischofsburg, and Hohenstein.

2. *General der Infanterie* von Hindenburg was promoted to *Generaloberst* on 26 August 1914. Hindenburg, 70-72; *Der Weltkrieg II*, 117-118.

3. *AOK 8* generated the 26 August *Armeebefehl* the evening prior. *German 8th Army War Diary*, English Translation, 57.

4. *Der Weltkrieg II*, 171.

5. Hindenburg, 73 f.; *Der Weltkrieg II*, 142-143.

6. Major-General Sir Alfred Knox, *With the Russian Army 1914-1917, Vol. I* (London: Hutchinson & Co. 1921), 67.

7. David Stone, 66-67.

8. Churchill, 196-198; Lincoln, 72; Showalter, 217; Keegan, 140.

9. Lincoln, 73; *Der Weltkrieg II*, 135; Noskoff, *Tannenberg*, 24.

10. Churchill, 198.

11. Samsonov order for 26 August in Captain James J. Coghlan, *Operations of the I German Corps at the Battle of Tannenberg*, 26-31 August, 1914 (Fort Leavenworth, Kansas: The Command and General Staff School, 1931-1932), 3; Ironside, 142.

12. David Stone, 72.

13. *German 8th Army War Diary*, English Translation, 58-60.

14. Hermann von François, *Tannenberg. Das Cannae des Weltkriegs in Wort und Bild* (Berlin: Verlag Deutscher Jägerbund, 1926), 206; 8. Sonderband, 420; Elze, 127-129; *German 8th Army War Diary*, English Translation, 58-60.

15. *8. Sonderband*, 421.

16. *Der Weltkrieg II*, 148.

17. Neidenburg was occupied by XV Corps. Soldau was occupied by I Corps. Löwenstern, 37.
18. *Der Weltkrieg II*, 154.
19. *Generalleutnant* v. Morgen promoted on 18 August 1914. *8. Sonderband*, 422; *Der Weltkrieg II*, 155.
20. *8. Sonderband*, 418 ff, 423-424; Elze, 301.
21. *Der Weltkrieg II*, 134; Elze, 301.
22. Norman Stone, 64; Showalter, 240.
23. Schäfer, Tannenberg, 58; *Der Weltkrieg II*, 154-155; Elze, 129
24. *8. Sonderband*, 421; *Der Weltkrieg II*, 149-151.
25. Löwenstern, 33.
26. Löwenstern, 37.
27. Buttar, *Collision of Empires*, 136.
28. Geoffrey C. Evans, *Tannenberg 1410:1914* (Harrisburg, PA: Stackpole Books, 1971), 113.
29. Sweetman, 115-116.
30. Chekhutov, 18.
31. *8. Sonderband*, 424.
32. *8. Sonderband*, 416.
33. *Der Weltkrieg II*, 134.
34. *8. Sonderband*, 424; *Der Weltkrieg II*, 171-173; Cuneo, II, 123-124.
35. *German 8th Army War Diary*, English Translation, 58-60.
36. Richard N. Armstrong, "Tactical Triumph at Tannenberg" *Military History*, August 1997, Vol 14, No. 3, 62.
37. *8. Sonderband*, 424; *Der Weltkrieg II*, 171-173; Cuneo, II, 123-124.
38. *8. Sonderband*, 425; *Der Weltkrieg II*, 176.
39. *Der Weltkrieg II*, 174-176; Sweetman, 119-120.
40. Ludendorff, Own, 58.
41. *8. Sonderband*, 425.
42. *8. Sonderband*, 425, 433.

Chapter 7

1. Elze, 305 (Document 201); *Der Weltkrieg II*, 156.
2. *Der Weltkrieg II*, 157.

3. The position of General Hell and *8. Armee* staff officers that served at Tannenberg recorded by Colonel Conger conflicted with Ludendorff's opinions as stated in his accounts of the battle. Coghlan, 4.
4. Hindenburg, 74.
5. *8. Sonderband*, 425.
6. Knox, 69; Buttar, Collision of Empires, 173.
7. Sonderband, 427-430; *Der Weltkrieg II*, 154.
8. *8. Armee War Diary* states 1. Armee-Korps took Usdau at 0500 on 27 August. François, 212, 214; Schäfer, Tannenberg. 70-77; West Point Atlas, 38. *German 8th Army War Diary*, English Translation, 62.
9. François, 33.
10. *8. Sonderband*, 427-429.
11. Elze, 314 (Document 226).
12. *8. Armee War Diary* cited, "Aviators have ascertained that Russian columns are retreating from Soldau to the south as far as Illowo (I. Armee-Korps)." 8 Sonderband, 428. *German 8th Army War Diary*, English Translation, 75.
13. *8. Sonderband*, 422-423.
14. *German 8th Army War Diary*, English Translation, 66.
15. Schäfer, *Tannenberg*, 70.
16. *8. Sonderband*, 427-429; Cuneo, II, 122-123
17. Buttar, *Collision of Empires*, 171.
18. *8. Sonderband*, 427; Armstrong, 62-63.
19. *Der Weltkrieg II*, 177-178.
20. Russian archival records show that he was captured by the Germans. RGVIA F.493 Op.3 D.156 L.5.
21. Unknown source quoted in Schäfer, Tannenberg, 115.
22. Knox, 70; Sweetman, 125.
23. *Schlachtenkalender*, 21.
24. Zeppelin Z-V. Knox, 73.
25. Hauptmann Grüner's fate was controversial. Sources suggested he was killed in the crash. *8. Sonderband*, 436.
26. *German 8th Army War Diary*, English Translation, 75.
27. R. L. Rimell, *Zeppelins at War! 1914-1915* (Berkhamsted, Hertfordshire, UK: Albatros Productions, Ltd, 2014), 15; Ernst A. Lehmann, *Auf Luftpatrouille und Weltfahrt* [On Air Patrol and World Trip] (Berlin: Wegweiser-Verlag. 1936), 74-75.

28. *Schlachtenkalender*, 7; Neumann, 362; Lehmann, 74-75.
29. Chekhutov, 18-19.
30. *Der Weltkrieg II*, 370.
31. *Der Weltkrieg II*, 160-162.; West Point Atlas, 38.
32. *8. Sonderband*, 433-434.

Chapter 8

1. Postwar analysis proved *GdA* v. Scholtz was correct. Schäfer, *Tannenberg*, 116-118.
2. *Der Weltkrieg II*, 182.
3. The remainder of the *AOK 8 Armee-Befehl* stated: "The Landwehr-Division advances at about 0500 hours over Menchengut and attack toward Hohenstein. Reinforced *I. Reserve-Korps* and parts of the *XVII. Armee-Korps* will attack the enemy near and at Allenstein. Should the enemy march to the south, the German forces will, with the right flanks, deflect toward Stabigotten. *XVII. Armee-Korps* will also pursue in direction of Willenberg. *I. Armee-Korps* has today repulsed the enemy at Usdau and forced him to retire in a southerly direction toward Soldau. The *I. Armee-Korps* covers the rear of the embracement." *German 8th Army War Diary*, English Translation, 68.
4. Hindenburg, 75.
5. German postwar military enthusiasts saw the battle of Hohenstein as a tribute to the outstanding strategic skills of Hindenburg and Ludendorff. In reality, Hohenstein was chaotic, decided by the fortune of war in favor of 8. Armee. It became the stuff of legend that shaped the remainder of the 20th century. Ludendorff, 64.
6. Elze, 308 (document 212); Showalter, 263-264.
7. Ludendorff, 64.
8. *Der Weltkrieg II*, 218; *8. Sonderband*, 433.
9. *German 8th Army War Diary*, English Translation, 68.
10. *8. Sonderband*, 433.
11. *Der Weltkrieg II*, 184.
12. Schäfer, Tannenberg, 157.
13. *German 8th Army War Diary*, English Translation, 69.
14. Major v. Massow, *First* Generalstabsoffizier (Ia), *I. Armee-Korps*, was tasked and personally delivered the message. François, *Marneschlacht*, 218; Elze, 139; Schäfer, *Tannenberg*, 153.

15. Schäfer, Tannenberg, 157; *German 8th Army War Diary*, English Translation, 70.
16. *Der Weltkrieg II*, 198; Schäfer, *Tannenberg, 170.*
17. *Der Weltkrieg II*, 198-200.
18. *Der Weltkrieg II*, 198-199; Elze, 136-138; Schäfer, Tannenberg, 170.
19. Showalter, 264.
20. *Der Weltkrieg II*, 199; Elze, 136-138; Showalter, 265, 286.
21. *8. Sonderband*, 431.
22. *German 8th Army War Diary*, English Translation, 73; Showalter, 287.
23. Showalter, 287.
24. Showalter reported an *RIVVF* aeroplane was also sighted flying over I Reserve-Korps while enroute to Allenstein. Showalter, 286-287; Schäfer, Tannenberg, 139-140.
25. *Der Weltkrieg II*, 189.
26. Sonderband 8, 430; Schäfer, Tannenberg, 171; Showalter, 287.
27. Elze, 140.
28. Schäfer, Tannenberg, 172.
29. *Der Weltkrieg II*, 200 ff; Goodspeed, 96.
30. *Der Weltkrieg II*, 198-200.
31. Schäfer, Tannenberg, 157.
32. *German 8th Army War Diary*, English Translation, 70.
33. Schäfer, Tannenberg, 157.
34. *8. Sonderband*, 430.
35. *West Point Atlas*, 38.
36. Elze, 141.
37. *Der Weltkrieg II*, 202-205; Elze, 321 (document 249).
38. *Der Weltkrieg II*, 194.
39. *8. Sonderband*, 430.
40. *8. Sonderband*, 430 f.
41. *8. Sonderband*, 431 f.
42. *8. Sonderband*, 431 f.
43. Evans, 138.
44. Noskoff, *Tannenberg*, 68-70.
45. *Der Weltkrieg II*, 178-180.
46. *Der Weltkrieg II*, 209; Noskoff, *Tannenberg*, 56.

47. Knox, 73-74.
48. Major-General Ironside commented that Gen.inf Martos' XV Corps had borne the brunt of the fighting during the Tannenberg campaign. They were the only troops which had shown themselves capable of opposing the Germans, and their success appears to have been due to the energetic personality of their commander, General Martos. (Ironside, 181, 189). *Der Welkrieg II*, 178-180; Noskoff, *Tannenberg*, 68-70; Showalter, 297-298.
49. *German 8th Army War Diary*, English Translation, 76-77. Elze, 321 (Document 248).
50. Elze, 321-323.
51. Showalter, 290-291.
52. *German 8th Army War Diary*, English Translation, 79.
53. *Der Weltkrieg II*, 204-205; Elze, 136-137.
54. *Der Weltkrieg II*, 194.
55. As Dennis Showalter put it, "The stain of an ancient defeat would be blotted out by a modern victory; the Teutonic Knights would be avenged." Showalter, 292.

Chapter 9

1. Elze, 326 (Document 259).
2. *Der Weltkrieg II*, 204-205.
3. *8. Sonderband*, 433.
4. *8. Sonderband*, 435.
5. *8. Sonderband*, 430.
6. *Der Weltkrieg II*, 161.
7. Sonderband, 435.
8. Sonderband, 436.
9. *8. Sonderband*, 435.
10. RIVVF unit was unknown. Schäfer, Tannenberg 196; Chekhutov, passim.
11. Lincoln was a noted scholar on Russian history. Lincoln, 72.
12. Evans, 144.
13. RGVIA F.493 Op.3 D.156 L.6.
14. *Der Weltkrieg II*, 228.
15. Evans, 133.

16. *General der Infanterie* von Below was promoted on 30 August 1914. *German 8th Army War Diary*, English Translation, 78.
17. *Der Weltkrieg II*, 237; Noskoff, *Tannenberg*, 122-127.
18. *8. Sonderband*, 441.
19. *Der Weltkrieg II*, 217-218.
20. *German 8th Army War Diary*, English Translation, 90-91.
21. Armstrong, 64.
22. *8. Sonderband*, 437-438.
23. Elze, 321-323.
24. *8. Sonderband*, 441.
25. Ironside, 189-190.
26. *8. Sonderband*, 436.
27. *Der Weltkrieg II* 207; Elze, 326
28. See Appendix 2. Exploits of the "Tannenberg Flieger" [Tannenberg Flyers] was told so often to German wartime and postwar public that it became a true myth, claimed by *Fliegertruppe* as evidence of their battle deciding impact. It became an urban legend. As such it was prone to dramatizing and embellishing, even distortions. The most accurate portrayal of this legend is acquired from contemporary and reliable German language sources. Elze, 145, 335-336; *8. Sonderband*, 440.
29. François, Tannenberg, 228-229.
30. Supf, 258-260; *8. Sonderband*, 430.
31. *Der Weltkrieg II*, 220, 223; Elze, 335 (285); *8. Sonderband*, 430.
32. See Appendix 2. Analysis of the reporting shows the most accurate depiction of the timely information was detailed by François. François, *Tannenberg*, 228-229.
33. François, *Tannenberg*, 228-230; *Der Weltkrieg II*, 220, 223; Elze, 335 (Document 285).
34. *Der Weltkrieg II*, 223-224.
35. Elze, 336 (Document 286).
36. *8. Sonderband*, 436.
37. *German 8th Army War Diary*, English Translation, 83; *Der Weltkrieg II*, 227; Cuneo, II, 128.
38. *8. Armee War Diary* succinctly commented, "20,000 men, 100 officers among which commanding general XV Corps with staff are made prisoners." *German 8th Army War Diary*, English Translation, 81; *8. Sonderband*, 434.
39. *Der Weltkrieg II*, 235-237; *8. Sonderband*, 441.

40. RGVIA F.493. Op.3 D.156. L.5.
41. Chekhutov, 21; Kulikov, "For the Fatherland!", 2.
42. *General ot infanterii* Martos' interrogation from Germans after the battle of Tannenberg. Evans, 106-107.

Chapter 10

1. Keegan, 149-150.
2. Elze, 343, 346.
3. *Der Weltkrieg II*, 268.
4. *8. Sonderband*, 449-250.
5. Schäfer, Tannenberg, 210-211.
6. *Der Weltkrieg II*, 223-225.
7. *Der Weltkrieg II*, 229-230.
8. Mahncke, 31.
9. Cockfield, 13-14.
10. Ferdinand Foch, Maréchal, *The Memoirs of Marshal Foch*. Translated by Col. T. Bentley Mott. (Garden City, NY: Doubleday, Doran and Company, Incorporated, 1931), 191-192.
11. "Germany Rushes Troops from France to Strike at Foe in Eastern Prussia." *New York Tribune*, 1 September 1914.
12. Barbara Tuchman, *The Guns of August* (New York: The Macmillan Company, 1962), 385-386; Douglas Porch, *The French Secret Services, From the Dreyfus Affair to the Gulf War* (New York: Farrar, Straus and Giroux, 1995) , 74.
13. This rare description of an actual sortie over German lines was a message from Petrograd published by *Flight* dated 11 September 1914. *Flight Aviation Journal*. 18 September 1914.
14. *8. Sonderband*, 443.
15. *Fliegertruppe* operated at altitudes up to 1,800 meters averaging 2-3 hours per sortie. *8. Sonderband*, 444.
16. Officers who witnessed Hindenburg's statement recalled hearing, "Without our flyers, we would not have been able to do some of the things we did." Cuneo, II, 128; Jäger, 98.
17. *8. Sonderband*, 444-445.

Chapter 11

1. Hindenburg, 78.
2. *Oberstleutnant* Hoffmann in Churchill, 215 .
3. *German 8th Army War Diary*, English Translation, 94.
4. *8. Sonderband*, 450.
5. *German 8th Army War Diary*, English Translation, 96.
6. *8. Sonderband*, 450.
7. *8. Sonderband*, 448.
8. *8. Sonderband*, 449.
9. See Appendix 1 for discussion of Major Kuckein's erroneous title regarding the report on the Canter / Mertens sortie. Mahnke, 32.
10. *8. Sonderband*, 454.
11. *8. Sonderband*, 454.
12. *German 8th Army War Diary*, English Translation, 96.
13. *8. Sonderband*, 452.
14. *8. Sonderband*, 452-454.
15. *German 8th Army War Diary*, English Translation, 96-97.
16. *8. Sonderband*, 452.
17. *Der Weltkrieg II*, 273.
18. *8. Sonderband*, 454; *West Point Atlas*, 40.
19. *Der Weltkrieg II*, 275.
20. *German 8th Army War Diary*, English Translation, 103.
21. *8. Sonderband*, 455.
22. *8. Sonderband*, 456.
23. Kulikov, "For the Fatherland!"
24. *8. Sonderband*, 456-458.
25. *German 8th Army War Diary*, English Translation, 106.
26. *8. Sonderband*, 461 - 462.
27. *German 8th Army War Diary*, English Translation, 107-109.
28. *Der Weltkrieg II*, 282-284.
29. Armstrong, 64.
30. *8. Sonderband*, 460.
31. *8. Sonderband*, 461 - 462.
32. *8. Sonderband*, 460.
33. Ludendorff, 75.

Chapter 12

1. *8. Sonderband*, 468; *Der Weltkrieg II*, 291-292.
2. *German 8th Army War Diary*, English Translation, 117-118.
3. *Der Weltkrieg II*, 286; Armstrong, 64.
4. *Der Weltkrieg II*, 291.
5. *West Point Atlas*, 40.
6. *8. Sonderband*, 464-465.
7. *8. Sonderband*,464-465.
8. Ludendorff, 50; Buttar, *Collision of Empires*, 236.
9. *West Point Atlas*, 42.
10. Reichsarchiv, ed., *Der Weltkrieg 1914-1918: Band V: Die Militärischen Operationen zu Lande* (Berlin: E.S. Mittler & Sohn, 1929), 299. Henceforth, Der Weltkrieg V.
11. *8. Sonderband*, 467; *German 8th Army War Diary*, English Translation, 116.
12. Sweetman, 176.
13. *German 8th Army War Diary*, English Translation, 119.
14. *8. Sonderband*, 473.
15. *German 8th Army War Diary*, English Translation, 127.
16. *8. Sonderband*, 477.
17. *8. Sonderband*, 477; *German 8th Army War Diary*, English Translation, 132.
18. *Der Weltkrieg II*, 299; *8. Sonderband*, 474-476.
19. Lincoln, 77.
20. *8. Sonderband*, 474-480; *West Point Atlas*, 40.
21. *Der Weltkrieg II*, 316-317.
22. Der Weltkrieg V, 407.
23. *German 8th Army War Diary*, English Translation, 136-139.
24. *Dechiffrierdienst* was *k.u.k. Armee's* wireless intelligence service led by *Hauptmann* Porkorny. The Russian cipher system was a numerical polyalphabetic that enciphered several letters in succession in a single cipher alphabet. Kahn, 628.
25. Lincoln, 77.
26. David Stone, 79.
27. Churchill, 221-224.
28. *West Point Atlas*, 42.

29. Rennenkampf was executed by Bolshiviks in 1918. Armstrong, 64; *West Point Atlas*, 42; Dowling, 20.
30. RGVIA. F.2003. Op.1. D 1224. L.215.
31. RGVIA F.2008 Op1 D.612.

Chapter 13

1. *Generalleutnant* Erich von Falkenhayn assumed *OHL* leadership – now *2. OHL*. Baron Franz Conrad von Hötzendorf is subsequently refered to as Conrad. *German 8th Army War Diary*, English Translation, 139; *Der Weltkrieg II*, 409; *Der Weltkrieg 1914-1918: Band V, Die Märischen Operationen zu Lande,* (Berlin: GE: Mittler & Sohn, 1929), 501. *8. Sonderband*, 481.
2. Initiated on 16 September 1914. Der Weltkrieg V, 521.
3. *Shooting the Front Eastern Operations, Book 2* discusses this campaign in detail. Der Weltkrieg V, 411.
4. Hindenberg served as overall commander of *8. Armee* and *9. Armee. German 8th Army War Diary*, English Translation, 140-141.
5. The aging Schütte-Lanz Airship SL-2 was transferred to the Western Front. *8. Sonderband*, 487-488.
6. His letter was addressed to *Kapitan 1-go ranga* (Naval 1st rank Captain) Nikolay Fedorovich Fogel. Fogel was an adjutant to Grand Duke Alexander Mikhailovich and a member of the "Special Committee for Strengthening the Navy on Voluntary Donations." Fogel went on to become *Upravlyayushchim delami Otdela Vozdushnogo Flota* Chief Executive Officer of the Air Fleet Division).
7. Eliseev makes a point that Mikhailovich identified those aviators that did not agree with him when he identified failures at Tannenberg. RGVIA. F. 2019 . Op. 1 . D. 541 . L.286.
8. Perepechin was flying a Farman HF.22. Sheremetevsky, "White Nights," 19.
9. Brusilov commented on correspondence in his written Memoirs. Eliseev, 48-50.
10. Verst is a Russian measurement equivalent to 1.07 kilometers. Sheremetevsky, "White Nights," 20.
11. Sheremetevsky, "White Nights," 51-52.
12. P. Kreyson, "From the Reminiscences of the Combat Activity Carried out by the 23-rd Corps Aviation Detachment during the Imperialistic War". *Air Fleet Proceedings* (No. 8, 1930), 7-8.; RGVIA 16 Ib. F.2008. Op.1. D.612. L.9.

13. Eliseev, 48.
14. *Grodno Krepostnyye Aviatsionnyye Otryad (KrAO)* [Grodno Fortress Aviation Detachment]. Sheremetevsky, "Memoirs of a Russian Aviator," 97.
15. *German 8th Army War Diary*, English Translation, 143.
16. *General ot infanterii* Thadeus von Sievers assumed command of Russian Tenth Army on 23 September 1914. *German 8th Army War Diary*, English Translation, 148-149.
17. *German 8th Army War Diary*, English Translation, 151.
18. *German 8th Army War Diary*, English Translation, 161.
19. *German 8th Army War Diary*, English Translation, 164; *8. Sonderband*, 482; Der Weltkrieg V, 511-512.
20. *German 8th Army War Diary*, English Translation, 165.
21. *8. Sonderband*, 483; Der Weltkrieg V, 512, 524.
22. *Schlachtenkalender*, 43.
23. *8. Sonderband*, 483.
24. *GdA* v. Schubert eventually took command of *7. Armee* on the Western Front and was promoted to *Generaloberst* on January 1917. Der Weltkrieg V, 516, 526-527.
25. Der Weltkrieg V, 529.
26. *8. Sonderband*, 484-485.
27. Ludendorff, 67.
28. *Österreich-Ungarns Letzter Krieg, 1914-1918, Volume 1* (1914), 453.
29. Cuneo, II, 136.
30. *8. Sonderband*, 486.
31. *Der Weltkrieg V*, 534-538.
32. Almost four months later Z-IV accomplished their next recorded sortie over Tauroggen. *Schlachtenkalender*, 55.
33. François, *Marneschlacht*, 177.
34. Prior to the war, on 2 May 1914, *Glavnoye voyenno-tekhnicheskoye upravleniye (GVTU)* [Main Military Technical Directorat] concluded a contract with R-BVZ to construct 10 Il'ya Muromets aeroplanes. *Polk* Ul'yanin, commander of Gatchina led the assignment process for Il'ya Muromets aviators. *Sh.Kap*. Georgiy Georgievich Gorshkov, *Sh.Kap.* Evgeny Vladimirovich Rudnev, *Podporuchik* Alexey Vasilyevich Pankratiev, *Sh.Kap.* Sergey Mikhailovich Brodovich and *Podpor.* Sergey Karlovich Modrah, all who were staff at Gatchina, were chosen to lead the first four ships and assigned to receive flight in-

struction directly from the aircraft's designer, Igor I. Sikorsky. RGVIA F.2008 Op.1 D.487; RGVIA F.2008 Op.2 D.614; RGVIA F.2019 Op.1 D.542; RGVIA F.2003 Op.1 D.1511;
RGVIA F.2003 Op.2 D.616; RGVIA F.2003 Op.2 D.617; RGVIA F.2003 Op.2 D.711.

35. RGVIA F.2003 Op.1 D.1511; RGVIA F.2144 Op.2 D.135.

36. RGVIA F.2003 Op.1 D.1511; RGVIA F.2144 Op.2 D.135.

37. RGVIA. F.2003. Op.2. D.613; RGVIA F.493 Op.2 D.4,. Kapitan et. al.

38. Finne, 71-72; Kapitan, et.al., 10-11.

39. Finne, 71; Sheremetevsky, "White Nights," 20.

40. RGVIA F.2003 Op.2 D.617; RGVIA F.2003 Op.2 D.614; RGVIA F.2019 Op.1 D.542 h.1; RGVIA F.493 Op.11 D.10; RGVIA F.2003 Op.2 D.616; RGVIA F.2003 Op.1 D.711.

41. By order of the Supreme Commander on 20 April 1915, No. 291, the Administration of the Squadron of Flying Ships was renamed the Squadron of Flying Ships. RGVIA F.2003 Op.2 D.614,

42. RGVIA F.2003 Op.2 D.622; RGVIA, F.2003 Op.2 D..625; Tim Wilson, "Eskadra Vozdushnikh Korablei (EVK) and the Brusilov Offensive: June-August 1916," Over the Front, 2003, 121.

43. Il'ya Muromets No. 137 Ship III, 143 Ship VI and 150 Ship VII were deployed to Yablonna, 16 kilometers from Warsaw and 40 kilometers from the front line. The base met *EVK's* criteria for operations with a good landing ground for the Muromtsy, and rail connections serving logistic and personnel requirements. Kapitan et. al.

Chapter 14

1. *Ober Ost* is short for *Oberbefehlshaber der gesamten Deutschen Streitkräfte im Osten*, German for "Supreme Commander of All German Forces in the East." Reichsarchiv, ed., *Der Weltkrieg 1914-1918: Band VI, Die militärischen Operationen zu Lande, Der Herbstfeldzug 1914, Der Abschluß der Operationen im Westen und Osten* (Berlin: E.S. Mittler & Sohn, 1929), 45. Henceforth, Der Weltkrieg VI.

2. Der Weltkrieg VI, 42.

3. Richard L. DiNardo, Breakthrough, The Gorlice-Tarnow Campaign, 1915 (Santa Barbara, CA: Praeger. 2010), 14.

4. *XXI, XXXVII, XXXIX* and *XL Armee-Korps* assigned to *Ober Ost. Generalfeldmarschall* von Hindenburg and *Generalleutnant* Ludendorff promoted on 27 Nov 14. David Stone, 128-129.

5. Peter Graf Kielmansegg, *Deutschland und der Erste Weltkrieg* (Frankfurt, GE: Akademische Verlagsgesellschaft Athenaion, 1968), 70.

6. David Stone, 126-127.
7. Kielmannsegg, 80.
8. Each flight squad was assigned six Voisin aeroplanes with a dedicated repair shop. Aviation support to the armies achieved greater mobility thanks to engineering transport made available. *KAOs* were now moving more frequently serving Northwest Front armies and corps. *23rd KAO* redeployed eight times between January and March 1915. Eliseev, 58.
9. RGVIA. F.2003. Op. 2. D.14.
10. RGVIA F.2008. Op.1. D.39; RGVIA F.2003, Op.D.616; RGVIA F.2003, Op.D.236.
11. RGVIA F.2000 Op.3 D.301; RGVIA F.2003 Op.1 D.1677; Eliseev, 58.
12. *FFA 52* supported *10. Armee*. David Stone, 132-134; Reichsarchiv, Edited by. *Der Weltkrieg 1914-1918, Band VII, Die militärischen Operationen zu Lande -Die Ereignisse im Winter und Frühjahr* (Berlin, GE: Mittler & Sohn, 1931), 172; Ludendorff, 145-147.
13. Tkachev, *Vospominaniya o proshlom russkoy voyennoy aviatsii 1910-1917.*
14. *FFA 49* supported *XL Reserve Korps*. West Point, 52; Prit Buttar, *Germany Ascendant*, (New York: Osprey Press, 2015), 96-98.
15. *K.u.k. Hauptmann* Pokorny and *Hauptmann* Deubner wireless teams solved a new Russian cipher code known as RSK. Kahn, 631.
16. Buttar, *Germany Ascendant*, 84; Mahncke, 33.
17. Mahncke, 34.
18. David Stone, 132-134; Der Weltkrieg VII, 172 – 238; Ludendorff, 145-147; *West Point Atlas*, 52.
19. David Stone, 135; Lincoln, 120, Norman Stone, 119.
20. Type Veh Muromtsy is the approved title for Il'ya Muromets. Kievskiy" honorific was applied because the very same engines from the Il'ya Muromets of Sikorsky's epic roundtrip from St Petersburg to Kiev were installed on this prototype Il'ya Muromets Type Veh airframe. While this impressive sortie took place, a Russian aviator in a Morane Saulnier approached thinking the four-engined aeroplane was German. Fortunately, there was no exchange of fire and Gorshkov landed "Kievskiy" safely. "Finne, *Sikorsky*, 72-73.
21. *Schlachtenkalender*, 87.
22. A few hours after Gorshkov returned from the 1 March sortie, *Fliegertruppe* conducted an aerial bombardment strike on Yablonna airfield.RGVIA F.493 Op.3 D.54., Finne, *Sikorsky*, 75-77; Khairulin, "*Il'ya Muromets*," 50-54.

23. The agent reported, "The Station building and the warehouse, six freighter cars and commandant's car were destroyed and the commandant was injured; in the town several houses were destroyed, two officers, seventeen lower ranks and seven horses were killed, panic reigns in the town and residents hide in cellars in clear weather." RGVIA. F.2008. Op.2. D.1276., RGVIA. F.2019. Op.1. D.543., RGVIA. F.2019. Op.1. D.542. Duz, 357.

24. The sortie covered Surovo-Krukovo-Zaremby-Willenberg-Khorzhele. RGVIA. F.2008. Op.1. D.1276., RGVIA. F.2019. Op.1. D.522.; Duz, 356

25. *"Kievskiy"* conducted an aerial photographic reconnaissance mission to ascertain German troop movements in the Mława area. The 17 large bombs ranged in weight from 16.38 to 32.76 kg each. RGVIA Fond493 Op.3 0.54, Finne K.N. Igor Sikorsky, the Russian Years. Washington, 1987, 75-77. Khairulin M. "Il'ya Muromets". *Pride of Russian Aviation*, Moscow, 2010, 50-54.

26. RGVIA Fond493 Op.3 EDeloHR 54.

27. RGVIA. F.2008. Op.2. D.1276., RGVIA. F.2019. Op.1. D.543., RGVIA. F.2019. Op.1. D.542.; Duz, 358.

28. German media coverage stated that the Russians possessed special aeroplanes that caused great damage and were not vulnerable to antiaircraft artillery fire. RGVIA. F.2008. Op.1. D.1276.; Duz, 86.

29. *"Kievskiy"* employed both Ul'yanin and Potte aerial cameras at an altitude up to 3600 meters. *"Kievskiy's"* crew included Gorshkov, *Por.* Jāzeps Stanislavovičs Bashko (deputy commander), *Sh.Kap.* Alexander Alexandrovich Naumov (artillery officer) and *Sh.Kap.* Aleksey Aleksandrovich von Goertz, an artillery (intelligence) officer on Litvinov's First Army staff. The rail lines of communication between Willenberg, Neidenburg, Soldau, Lautenburg, Strasburg, and Thorn were filmed. Anatoly Mikhailovich Sergienko, Aircraft squadron "Ilya Muromets": the history of creation and combat use in documents and memoirs: [in 3 volumes] Belgorod, 2014. Ostaschenko A.A.; RGVIA F.493 Op.3 D.54.; I.S. Bashko "V Nebe nad Vragom" Riga, 2017. [Jāzeps S. Bashko, *The Sky Over the Enemy* (Riga, 2017), 58-62.]

30. *"Kievskiy"* Ilya Muromets flew the 1 April sortie. EVK 3 Volumes by A.M. Sergienko.

31. RGVIA F.493 Op.11 D.10.; Duz, 360.

Conclusion

1. Solzhenitsyn, Alexander. *August 1914*. Trans. Michael Glenny. New York: Farrar, Straus and Giroux, 1971. 618-619.

Appendix 1

1. Dmitrievich Anoshchenko [1894-1974] was a Russian pioneer of aviation and aeronautics. His talents included being an inventor, cinematographer, director and teacher. He was awarded the Order of St. George and five other military awards for his service. In 1914 he served as a volunteer pilot in *15th KAO*, photo-documenting operations and flying sorties during the Tannenberg campaign. In 1916, Anoshchenko commanded the Fifth Army Aeronautical Detachment on the Northern Front.
2. Chekhutov, 9.
3. Chekhutov, 12-13.
4. Chekhutov, 12.

Appendix 2

1. Peter Supf cites the mission as a critical part of aviation's role at Tannenberg. Peter Supf. *Das Buch der deutschen Fluggeschichte.* Band II. (Berlin: Vorkriegszeit - Kriegszeit - Nachkriegszeit. 1935), 258; *8. Sonderband*, 438, 440.
2. Supf, 260. *8. Sonderband*, 438, 440.
3. Supf, 259. *8. Sonderband*, 438, 440.
4. Supf, 260. *8. Sonderband*, 438, 440.

Appendix 3

1. Sheremetevsky, "White Nights," 24-25.

BIBLIOGRAPHY

ARCHIVAL AND MANUSCRIPT SOURCES

Bayerisches Hauptstaatsarchiv, Abteilung IV, Kriegsarchiv. München, Germany (GE). Abbr.: BY-HStA-KA.

Kriegsgliederung und Kräfteverteilung der Russischen Fliegertruppe. 15. April 1917. BY-HStA-KA.

Der Kommandierende General der Luftstreitkräfte. *Das Taktische Lichtbilderbuch*. II. Teil. *Die Deckung gegen Fliegersicht*. Nr. 2500 / 17 Lb. Juni 1917. BY-HStA-KA.

Kriegswissenschaftliche Abteilung der Luftwaffe. 3. Sonderband: *Mobilmachung, Aufmarsch und erster Einsatz der deutschen Luftstreitkräfte im August 1914*. Berlin, GE, 1939.

Kriegswissenschaftliche Abteilung der Luftwaffe. 8. Sonderband: *Die Luftstreitkräfte im Zusammenhang mit den Operationen des Jahres 1914*. (Unpubl. final draft), Berlin, GE, 1943.

Stärkenachweisungen [Tables of Manning and Equipment].
BY-HStA-KA.

Bundesarchiv Militärarchiv, Freiburg, GE. Abbr.: BAMA.
PH 19 / 118.

National Archives And Records Administration, College Park, MD, Army War College Historical Section. *German 8th Army War Diary, Folder V, 2 August to 1 October, 1914*. (with Annex Folders I, Ia, Ib, Ic, and Id. (Trans.) Record Group 165, Entry 320, Boxes 64-66.

Rossiyskiy Gosudarstvennyy Voyennyy Arkhiv - Russian State Military Archive. Moscow Russia (RU). Abbr.: RGVA.

Rossiyskiy Gosudarstvennyy Voyenno-Istoricheskiy Arkhiv. Moscow, RU. Abbr.: RGVIA.

Rossiyskiy Gosudarstvennyy Arkhiv Voyenno-Morskogo Flota - Russian State Archive of the Navy. Moscow, RU. Abbr.: RGAVMF.

Rossiyskaya Gosudarstvennaya Biblioteka - Russian State Library. Moscow, RU. Abbr.: RGB.

CONTEMPORARY PUBLISHED SOURCES

Bashko, Jāzeps S. *The Sky Over the Enemy*. Riga, Estonia, 2017.

Baur, Fritz. *Wir Flieger. Der Krieg im Fliegerlichtbild 1914 - 1918*. Wien, Austria: Eigenverlag, 1930, und Österreichische Druck u. Verlagsges., 1930.

Bismarck, Armin v. *Das Fliegerbuch. Flugabenteuer an allen Fronten*. Berlin, GE: Ullstein, 1918.

Bruchmüller, Georg. *Die Deutsche Artillerie in den Durchbruchschlachten des Weltkrieges*. Berlin, GE: Mittler und Sohn, 1922.

Churchill, Winston S. *The Unknown War*. New York: Charles Scribner's Sons., 1931.

Coghlan, James J., Captain. *Operations of the I German Corps at the Battle of Tannenberg, 26-31 August, 1914*. Papers of The Command and General Staff School, Fort Leavenworth, KS: (1931-1932).

Cron, Hermann. *Geschichte des Deutschen Heeres im Weltkriege 1914 – 1918*. Berlin, GE: Neudruck 1937. Osnabrück, GE: Biblio Verlag, 1990.

Chekhutov, Ivan Anikievich. *Iz Vospominaniy o Rabote Aviatsii v Samsonovskoy Operatsii*. Vestnik Vozdushnogo Flota, No. 12 (1924) str. 24-28.

Danilov, Yuri Nikiforovich. *Rossiya v Mirovoy Voyne*, Berlin, GE, 1924. [Russia in the World War].

Deubner, Christian. *Die deutsche Geheimwaffe war die Funkaufklärung*. https://www.welt.de/geschichte/article131900487/Die-deutsche-Geheimwaffe-war-die-Funkaufklaerung.html

"Der russische Nachrichtendienst vor dem Gorlice-Durchbruch," Militär Wochenblatt Vol. 119, No. 42 (May 1935), 1671.

Eltester, Oberstlt. *Nachrichtentruppen in der Schlacht bei Tannenberg. Funkstationen," Deutsche Nachrichtentruppen. Die F-Flagge* (June 1938), 103-104.

Elze, Walter. *Tannenberg. Das Deutsche Heer von 1914. Seine Grundzüge und deren Auswirkung im Sieg an der Ostfront*. Breslau: Ferdinand Hirt, 1928.

Fink, Carl. *Die Entwicklung des militärischen deutschen Luftbildwesens 1911-1918 und seine militärische wie kulturelle Bedeutung*. Wehrwissenschaftliche Rundschau, Jg. 10, Heft 7 (1960) 390-399.

Fink, Carl. *Einiges von der Entwicklung unserer Flugzeugkameras*. Illustrierte Flug-Welt, (Juni 1920).

Fink, Carl. *Das Reihenbild als Kartenersatz*. Illustrierte Flug-Welt, (Mai 1920).

Finne, K. N. *Igor Sikorsky, The Russian Years*. Carl J. Bobrow, Ed., and Von Hardesty. Washington, D.C.: Smithsonian Institution Press (1987). Transl. from Finne, K. N. *Russkiye vozdushnyye bogatyri I. I. Sikorskogo* Belgrade, 1930.

François, Hermann. *Marneschlacht und Tannenberg. Betrachtungen zur deutschen Kriegsführung der ersten sechs Kriegswochen*. Berlin, GE: August Scherl GmbH, 1920.

François, Hermann. *Tannenberg. Das Cannae des Weltkrieges in Wort und Bild*, Berlin, GE: Verlag Deutscher Jaegerbund, 1926.

Foch, Marshal. *The Memoirs of Marshal Foch*. Transl. by Col. T. Bentley Mott. Garden City, New York: Doubleday, Doran and Company, Inc., 1931.

Golovine, Nicholas N. *The Russian Campaign of 1914: The Beginning of the War and Operations in East Prussia*, translated from Russian by Captain A.G.S. Muntz, The Command and General Staff School Press, Fort Leavenworth, Kansas, 1933.

Grosser Generalstab, *Die Schlachten und Gefechte des Großen Krieges 1914-1918*. Berlin, GE: Verlag Hermann Sack, 1919.

Handbook of the Russian Army 1914. General Staff, War Office, 1914. Nashville, TN: The Battery Press. 1996.

Hindenburg, Field Marshal Paul von. *The Great War*. London: Greenhill Books, 2006. Original *Aus meinem Leben*, Leipzig: Hirzel, 1934.

Hoffmann, Max. *War Diaries and Other Papers, Volumes I and II* [War of Lost Opportunities and The Truth About Tannenberg], Transl. from German by Eric Sutton, London: Martin Secker, 1929.

Hug, Andre. *Carrier Pigeon Flieger. The WWI Experiences of Dr. Andre Hug*. Cross & Cockade Journal, Vol. 13, No. 4 (Winter 1972) 289-321.

Humphries, Mark Osborne and John Maker, Editors. *1915, Germany's Western Front, Translations From the German Official History of the Great War, Vol. II.* Waterloo, Ont.: Wilfrid Laurier University Press, 2010.

Ironside, Sir Edmund, Major-General: Tannenberg. *The First Thirty Days in East Prussia*. Edinburgh and London: William Blackwood & Sons Ltd. 1928.

Ives, Herbert E. *Airplane Photography*. Philadelphia: J.B. Lippincott Co. 1920.

Knox, Major-General Sir Alfred. *With the Russian Army 1914-1917*. London: Hutchinson and Company, 1921.

Kozlow, N. "*A Study of the Military-Technical Supply of the Russian Army in the World War*." Moscow, RU: Government Military Publications Division, 1926. Transl. by Charles Berman, Washington, D.C.: The Army War College, 1931.

Krasnopevtsev, Boris V. "*Osnovnyye Sobytiya V Istorii Fotogrammetrii V 19 I 20 Vekakh,*" Moscow, RU: State University, Geodesy and Cartography (May 2008).

Kreyson, Pavel Martynovich. "*Vospominaniy O Boyevoy Deyatel'nosti 23 Korpusnogo Aviatsionnogo Otryada V Imperialisticheskuyu Voynu,*" Moscow, RU: Vestnik Vozdushnogo Flota No. 8-9 (1927).

"Kriegsgeschichtliche Einzelschriften der Luftwaffe, Band 3: "Mobilmachung, Aufmarsch und erster Einsatz der deutschen Luftstreitkräfte im August 1914." Berlin, GE: E. S. Mittler & Sohn, 1939.

Langsdorff, Werner. *Flieger am Feind*. Gütersloh, GE: Bertelsmann, 1934.

Lehmann, Ernst A. *Auf Luftpatrouille und Weltfahrt*. Berlin, GE: Wegweiser-Verlag, 1936.

Löwenstern, Baron Elard v. *Der Frontflieger*. Berlin, GE; Bernard & Graefe, 1937.

Ludendorff, Erich. *Meine Kriegserinnerungen*, 1914-1918. (German Edition). Berlin, GE: Mittler & Sohn, 1919.

Ludendorff, Erich. *Ludendorff's Own Story, August 1914 – November 1918*. (English Edition). New York and London: Harper & Brothers Publishers, 1919.

Ludendorff, Erich. *The General Staff and its Problems, The History of the relations between the High Command and the German Imperial Government as revealed by Official Documents, Volume 1*. E.P. Dutton and Company, New York, 1920.

Luft- und Erdbilder der russischen Stellungen im Rigaer und Jakobstädter – Brückenkopf (Herbst 1917).

"Luftbildwesen vor fünfundzwanzig Jahren." Luftwelt, Bd. 4 No. 11 (1937).

Mackensen. *Briefe und Aufzeichnungen*. (W. Förster, Hg.) Leipzig, GE: Bibliographisches Institut, 1938.

Mackenthun, Walter. *Das Militär-Flugzeug. Erklärung und Verwendung der Apparate. Das Verhalten der Truppe*. Berlin, GE, 1911.

Mahncke, Alfred. *For Kaiser and Hitler, The Memoirs of Luftwaffe General Alfred Mahncke, 1910-1945*. Translated by Jochen Mahncke. Pulborough West Sussex, U.K.: Tattered Flag Press, 2010.

Neumann, Georg Paul. *Die deutschen Luftstreitkräfte im Weltkrieg*. Berlin, GE: Mittler & Sohn, 1920.

Nikolsky Sergey Nikolaevich; Nikolsky Mikhail Nikolaevich: *Bombardirovshchiki "Il'ya Muromets" v boyu: vozdushnyye linkory Rossiyskoy imperii* [Bombers "Ilya Muromets" in Battle: Air Battleships of the Russian Empire], Yauza, 2008.

Nikolsky Sergey Nikolaevich; Nikolsky Mikhail Nikolaevich: *Muromtsy v Boyu. Podvigi Russkikh Aviatorov* [Muromets in Battle. Feats of Russian Aviators], Eksmo 2010.

Nikolsky, Sergey Nikolaevich; Khairulin, Marat; Pankratiev, Natalia. *Khronika Vozdushnogo Korablya "Il'ya Muromets II"* [Chronicle of the Airship "Ilya Muromets II"] Russian Knights, 2019

Novikoff, A. A.: *"Der russische Nachrichtendienst vor dem Gorlice-Durchbruch"* Militär-Wochenblatt Bd. 119, Nr. 42. Berlin, GE, (1935) 1670.

Noskoff: *Der Mann der Tannenberg verlor*. Berlin, GE: Vorhut-Verlag Otto Schlegel, 1938.

Österreich-Ungarns Letzter Krieg, 1914-1918, Volume 1 (1914), Edited by Edmund v. Glaise-Horstenau. Transl. by Stan Hanna. Wien, Austria: Publisher of Military Science Releases. 1930.

Österreich-Ungarns Letzter Krieg, 1914-1918, Volume 2 (1915), Edited by Edmund v. Glaise-Horstenau, Transl. by Stan Hanna. Wien, Austria: Publisher of Military Science Releases. 1931.

Randevig, Kunibert. *Die deutsche Funkaufklärung in der Schlacht bei Tannenberg*. Wehr und Wissen, Vol. 13 (1932).

Reichsarchiv, Edited by. *Der Weltkrieg 1914-1918, Band II: Die militärischen Operationen zu Lande: Die Befreiung Ostpreussens*, Berlin, GE: Mittler & Sohn, 1925.

Reichsarchiv, Edited by. *Der Weltkrieg 1914-1918: Band V, Die Militärischen Operationen zu Lande*. Berlin, GE: Mittler & Sohn, 1929.

Reichsarchiv, Edited by. *Der Weltkrieg 1914-1918, Band VI, Die militärischen Operationen zu Lande - Der Herbstfeldzug 1914 - Der Abschluß der Operationen im Westen und Osten*. Berlin, GE: Mittler & Sohn, 1929.

Reichsarchiv, Edited by. *Der Weltkrieg 1914-1918, Band VII, Die militärischen Operationen zu Lande - Die Ereignisse im Winter und Frühjahr*. Berlin, GE: Mittler & Sohn, 1931.

Ritter, Hans. *Der Luftkrieg*. Berlin, GE, 1926.

Schäfer, Theobald: *Tannenberg. Schlachten des Weltkriegs*, Band 19. Berlin, GE: Reichsarchiv, 1927.

Schröder, Hans. *An Airman Remembers*. Translated by Claud W. Sykes. London: John Hamilton, Ltd., 1936.

Sheremetevsky, Georgii Leonidovich:*"White Nights." The '14 – '18 Journal*. Edited and transl. by Margaret Travers. Australian Society of World War I Aero Historians, 2001.

Sheremetevsky, Georgii Leonidovich: *"Memoirs of a Russian Aviator, The Recollections of Georgii Leonidovich Sheremetevsky."* Edited and translated by Margaret Travers. Cross & Cockade International 36/1, 37-53; Cross & Cockade International, Vol. 36, No. 2, (2005) 97-122.

Spatarel, Ivan Konstantinovich: *Protiv Chernogo Barona*. [Against the Black Baron] Soviet Union, 1967

Supf, Peter. *Das Buch der deutschen Fluggeschichte*. Band II. Vorkriegszeit - Kriegszeit - Nachkriegszeit. Berlin, GE, 1935.

Tkachev, Vyatcheslav Mikhailovich. *Kryl'ya Rossii. Vospominaniya o proshlom russkoy voyennoy aviatsii 1910-1917*. gg. [Wings of Russia. Memories of the Past of Russian Military Aviation 1910-1917]. SPb: GITS Novoye kul'turnoye prostranstvo, 2007.

Wulffen, Karl: *Die Schlacht bei Lodz*. Der große Krieg in Einzeldarstellungen. Heft 19. Oldenburg: Bernard & Graefe, 1918.

Yankevich, P. Poruchik. *Aerofotografiya Rukovodstvo Po Vozdushnoy Fotografii* [Aerial Photography Guide To Aerial Photography], Petrograd, RU, A.N. Lavrov and Co. 1917.

Voyennaya Entsiklopediya. Tom 1. [Military Encyclopedia. Vol.1.] St. Petersburg, RU, 1912.

SECONDARY SOURCES

Alexandrov, Andrei O. *An Illustrated History of the Imperial Russian Naval Air Service, 1894-1917*. St. Petersburg, RUS: Ed. Harry Woodman, 1991.

Bowden, Matt. *The Great War's Finest, An Operational History of the German Air Service, Volume 1: Western Front 1914*. Aeronaut Books, 2017.

Blume, August G. *The Russian Military Air Fleet in World War I, Volume 1 - A Chronology 1910-1917*. Atglen, PA: Schiffer Military History, 2010.

Buttar, Prit. *Collision of Empires - The War on the Eastern Front in 1914*. Oxford, UK: Osprey Publishing, 2014.

Buttar, Prit. *Germany Ascendant*. Oxford, UK: Osprey Publishing, 2014.

Chant, Christopher. *Austro Hungarian Aces of World War 1*. Oxford, UK: Osprey Publishing, 2002.

Cockfield, Jamie H. *With Snow on Their Boots, The Tragic Odyssey of the Russian Expeditionary Force in France During World War I*. New York: St Martin's Press, 1998.

Cuneo, John R. *Winged Mars. Volume 1: The German Air Weapon 1870-1914*. Harrisburg, PA: The Military Service Publishing Company, 1942.

Cuneo, John R. *The Air Weapon 1914-1916. Volume 2 of Winged Mars*. Harrisburg, PA: Military Service Publishing Company, 1947.

Davilla, James J., Arthur M. Soltan. *French Aircraft of the First World War*. Stratford, CT: Flying Machine Press, 1997.

DiNardo, Richard L. *Breakthrough, The Gorlice-Tarnow Campaign 1915*. Santa Barbara, CA: Praeger, 2010.

Domanski, Jacek and Wlodzimierz Wilczek. *Tannenberg 1914*. Warszawa Wydawnictwo Militaria 238, 2005.

Dowling, Timothy C. *The Brusilov Offensive*. Bloomington, IN: Indiana University Press, 2008.

Durkota, Alan, Thomas Darcey, Victor Kulikov. *The Imperial Russian Air Service, Famous Pilots & Aircraft of World War One*. Mountain View, CA: Flying Machine Press, 1995.

Duz', P. D. *Istoriya vozdukhoplavaniya i aviatsii v SSSR, do 1914-1918* [A History of Aeronautics and Aviation in The USSR for the Period 1914-1918]. Moscow: Oborongiz, 1960, and 1979.

Duz', P. D. *Istoriya vozdukhoplavaniya i aviatsii v SSSR, do 1914 g.* [A History of Aeronautics and Aviation in the USSR. pre-1914 Period]. Moscow: Mashinostroyeniye, 1981.

Echevarria II, Antulio J. *After Clausewitz, German Military Thinkers Before the Great War*. Lawrence, KS: University Press of Kansas, 2000.

Evans, Geoffrey C. *Tannenberg 1410 and 1914*. Harrisburg, PA: Stackpole Books, 1971.

Finnegan, Terrence J. *Shooting the Front, Allied Aerial Reconnaissance in the First World War*. Stroud, Gloucestershire, UK: The History Press, 2011 and 2014.

Forczyk, Robert. *Combat – 1914-15 German Infantryman Versus Russian Infantryman*. Oxford, UK: Osprey Publishing, 2015.

Gröschel, Dieter *H. M. Russische und Balkan-Kriegsschauplätze* [Russian and Balkan Fronts]. n.l, n.d.

Grosz, Peter M., George Haddow, and Peter Schiemer. *Austro-Hungarian Army Aircraft of World War One*. Mountain View, CA: Flying Machine Press. 1993.

Hardesty, Von. *Early Flight in Russia. In Russian Aviation and Air Power in the Twentieth Century*, edited by Robin Higham et al., 18-36. Portland, OR: Frank Cass, 1998.

Haruk, Andriy; Kondratiev, Vyacheslav; Khairulin, Marat. *«Anatra»: Litaky Odes'koho Aviabudivnoho Pidpryyemstva, 1910–1924* [Anatra: Aircraft of the Odessa Aircraft Building Enterprise, 1910–1924]. Tempora 2008.

Higham, Robin D. *Air power: A Concise History*. New York: St. Martin's, 1972.

Higham, Robin and Jacob W. Kipp, eds. *Soviet Aviation and Airpower: A Historical View*. Boulder, CO: Westview, 1977.

Higham, Robin, John T. Greenwood, and Von Hardesty, eds., *Russian Aviation and Air Power in the Twentieth Century*. Portland, OR: Frank Cass, 1998.

Haythornthwaite, Philip J. *The World War One Source Book*. London: Arms and Amour Pres, 1992.

Hoffmann, Max, *The War of Lost Opportunities*. London: Imperial War Museum. Facsimile edition, 1999.

Jäger, Helmut. *Erkundung mit der Kamera*. München, GE: Venorion, 2007.

Kahn, David. *The Codebreakers, The Story of Secret Writing*. New York: The Macmillan Company, 1967.

Keegan, John. *The First World War*. New York: Alfred A. Knopf, 1999.

Khairulin, Marat. *7-y Korpusnoy Aviatsionnyy Otryad* [7th Corps Aviation Detachment]. Hunter 2008.

Khairulin, Marat. *16-y Korpusnoy Aviatsionnyy Otryad*. [16th Corps Aviation Detachment], Hunter 2008.

Khairulin, Marat; Kulikov, Viktor. *Boyevyye Aviatsionnyye Gruppy Pervoy Mirovoy Voyny*. [Air Combat Groups of World War I], Gangut 2015.

Khairulin, Marat; *Kondratiev, Vyacheslav.Voyenlety Pogibshey Imperii. Aviatsiya v Grazhdanskoy Voyne* [Warplanes of the Lost Empire. Aviation in the Civil War]. Eksmo 2008.

Khairulin, Marat. *"Il'ya Muromets". Gordost' Russkoy Aviatsii.*["Ilya Muromets". Pride of Russian Aviation], Eksmo 2010

Khairulin, Marat., *Kraski Russkoy Aviatsii. 1909-1922 gg. Kniga 1* [Colors of Russian Aviation. 1909-1922 Book 1] Russian Knights 2013.

Khairulin, Marat; Stepanov, Boris. *Kraski Russkoy Aviatsii. 1909-1922 gg. Kniga 2* [Colors of Russian Aviation. 1909-1922 Book 2]. Russian Knights 2014.

Khairulin, Marat; *Kraski Russkoy Aviatsii. 1909-1922 gg. Kniga 3* [Colors of Russian Aviation. 1909-1922 Book 3]. Russian Knights 2014.

Khairulin, Marat; *Kraski Russkoy Aviatsii. 1909-1922 gg. Kniga 4* [Colors of Russian Aviation. 1909-1922 Book 4] Russian Knights 2014.

Khairulin, Marat. *Legendarnyy «Il'ya Muromets». Pervyy Tyazhelyy Bombardirovshchik* [The Legendary "Ilya Muromets". First Heavy Bomber] Eksmo 2018.

Khairulin, Marat; Kulikov, Viktor. *Rossiyskiye Letchiki-Istrebiteli Pervoy Mirovoy Voyny* [Russian Fighter Pilots of World War I] Russian Knights 2014.

Krasnopevtsev, Boris Viktorovich. *Osnovnyye Sobytiya V Istorii Fotogrammetrii V 19 i 20 Vekakh* [Main Events in the History of Photogrammetry in the 19th and 20th Centuries]. Moscow, 2008.

Lincoln, W. Bruce. *Passage Through Armageddon, The Russians in War and Revolution*. New York: Oxford University Press, 1994.

Machavariani, M. S. *Glaza na yug* [Eyes to the South]. Tbilisi: Merani, 1969.

Menning, Bruce. *Bayonets Before Bullets*. Bloomington: Indiana University Press, 1986.

McMeekin, Sean. *The Russian Origins of the First World War*. Cambridge, MA: The Belknap Press of Harvard University Press, 2011.

Morrow, John Howard Jr. *Building German Airpower, 1909-1914*. Knoxville, TN: The University of Tennessee Press, 1976.

Morrow, John Howard Jr. *The Great War in the Air: Military Aviation from 1909 to 1921*. Tuskaloosa, AL: University of Alabama Press, 2009.

Mückler, Jörg. D*eutsche Flugzeuge im Ersten Weltkrieg*. Stuttgart, 2013.

Murphy, Justin D. *Military Aircraft, Origins to 1918*. Santa Barbara, CA: ABC CLIO, 2005.

Napp, Niklas. *Die deutschen Luftstreitkräfte im Ersten Weltkrieg*. Paderborn, GE: Schöning, 2017.

Neiberg, Michael S., and David Jordan. *The Eastern Front 1914-1920, From Tannenberg to the Russo-Polish War*. London: Amber Books, 2008.

O'Connor, Martin. *Air Aces of the Austro-Hungarian Empire, 1914-1918*. Mesa, AZ: Champlin Fighter Museum Press, 1986.

Pawly, Ronald and Patrice Courcelle. *The Kaiser's Warlords: German Commanders of World War I*. London: Osprey Publishing, 2003.

Petrov, Gennady. *Imperial Russian Air Force, 1898-1917*. St. Petersburg: Uniform Press, 2013.

Pöhlmann, Markus. *Kriegsgeschichte und Geschichtspolitik. Der Erste Weltkrieg*. Paderborn, 2002.

Porch, Douglas. *The French Secret Services, From the Dreyfus Affair to the Gulf War.* New York: Farrar, Straus and Giroux, 1995.

Pope, Stephen and Elizabeth-Anne Wheal. *The Dictionary of The First World War*. New York: St. Martin's Press, 1995.

Rimell, R. L. *Zeppelins at War! 1914-1915*. Berkhamsted, Hertfordshire, UK: Albatros Productions, Ltd, 2014.

Robinson, Paul. *Grand Duke, Nikolai Nikolaevich, Supreme Commander of the Russian Army*. DeKalb, IL: Northern Illinois University Press, 2014.

Rosenboom, Sebastian. *Im Einsatz über der vergessenen Front. Der Luftkrieg an der Ostfront im Ersten Weltkrieg*. Potsdam, GE: Zentrum für Militärgeschichte und Sozialwissenschaften der Bundeswehr, 2013.

Seifert, Karl-Dieter. *Mit Sammelgroschen zur deutschen Luftmacht*. Berlin, GE, 2014.

Sergienko, Anatoly Mikhailovich. E*skadra vozdushnykh korabley Il'ya Muromets* [Aircraft squadron "Ilya Muromets": The history of Creation and Combat use in Documents and Memoirs, 3 volumes]. Belgorod: Ostaschenko A.A.

Shavrov, V. B. *Istoriya konstruktsii samoletov v SSSR do 1938 g.* [A History of Aircraft Design in the USSR for the Period Before 1938]. Moscow: Mashinostroyeniye, 1969; 3. ed., 1985.

Showalter, Dennis E. *Tannenberg, Clash of Empires*. Hamden, CT: Archon Books, 1991.

Smoot, Andrew H. *A battle lost: re-examining the role of German radio intelligence in the Battle of Gumbinnen*. Intelligence and National Security, 32:3, 286-299, DOI: 10.1080/02684527.2016.1270990, 2017.

Sobelev, Dmitry A. *Istoriia sameletov: nauchnyi period* [Aircraft history: Scientific Period]. Moscow: Rosspen, 1995

Solzhenitsyn, Alexander. *August 1914*. Trans. Michael Glenny.New York: Farrar, Straus and Giroux, 1971.

Stamper, George L. *Sikorsky S-16 and Russian Aviation During the Great War*. Master's thesis, Athens: University of Georgia, 1995.

Stokesbury, James L. *A Short History of World War I*. New York: William Morrow and Company, Inc., 1981.

Stone, David R. *The Russian Army in the Great War, The Eastern Front, 1914-1917*. Lawrence, KS: University of Kansas Press, 2015.

Stone, Norman. *The Eastern Front, 1914–1917*. New York: MacMillan, 1976.

Sweetman, John. *Tannenberg 1914*. Cassell & Co., London, 2002.

Trunov, K. I. *Petr Nesterov*. Moscow: Sovetskaiia Rossiia, 1975.

Tuchman, Barbara. *The Guns of August*. New York: The Macmillan Company, 1962.

Tunstall, Graydon A. *Blood on the Snow, The Carpathian Winter War of 1915*. Lawrence, KS: University Press of Kansas, 2010.

Ul'yanin, Yuriy Alekseyevich, Pioner rossiyskoy aviatsii [Pioneer of Russian Aviation]. Moscow: Nezavismoye Izd., 2001.

Uhle-Wettler, Franz. Erich *Ludendorff in seiner Zeit*. Berg, GE, 1995.

Vitarbo, Gregory. *Army of the Sky, Russian Military Aviation before the Great War, 1904-1914*. New York: Peter Lang, 2012.

Watson, Alexander. *Ring of Steel, Germany and Austria-Hungary in World War I*. New York: Basic Books, 2014.

The West Point Atlas of War. Edited by Brigadier General Vincent J. Esposito. New York: Tess Press, 1997.

Zabecki, David T. *Steel Wind, Colonel Georg Bruchmuller and the Birth of Modern Artillery*. Westport, CT: Praeger, 1994.

Vogler, Philipp. *Die deutsche militärische Luftbildaufklärung. Von den Anfängen bis 1945*. Karlsruhe, GE: Kit Scientific Publishing, 2020.

PERIODICALS AND NEWSPAPERS

Alexandrov, Andrei O. and Ron Kintner, "The Battle of Angernsee," Over the Front, Vol. 18, No. 4 (2003).

Armstrong, Richard N., "Tactical Triumph at Tannenberg," Military History, Vol. 14, No. 3, (August 1997), 58-64.

Flight Aviation Journal. 18 September 1914.

Kulikov, Viktor P., Polskii Orel v Russkom Nebe – Letchik-Istrebitel Donat Makienok [Polish Eagle in the Russian Sky - Fighter Pilot Donat Makienok]. AviaMaster 6/2000.

Marshall, Alex, "Russian Military Intelligence, 1905-1917: The Untold Story behind Tsarist Russian in the First World War," *War in History, Vol. 11, No. 4,* (November 2004), 393-423.

Mahncke, Jochen OEO, "German Aircraft Operations in East Prussia During August and September 1914," *The South African Military History Society, Military History Journal,* Vol. 8, No. 3 (June 1990).

Norwitz, Jeffrey. H., "Leveraging Operational Intelligence – The Battle of Tannenberg and Masurian Lakes (1914)", *Naval War College,* Newport, RI (May 2001).

Noskoff, A.A., General. "Der Russische Nachrichtendienst vor dem Gorlice Durchbruch," *Militär-Wochenblatt Nr. 42.* (Berlin: 1935), 1670 f. [The Russian Intelligence Service before the Gorlice Breakthrough].

Pöhlmann, Markus, "German Intelligence at War, 1914-1918" *Journal of Intelligence History* (Winter 2005), 25-54.

Rosenboom, Sebastian, "Leopold Anslinger, An Aviation Pioneer on the Eastern Front," *Over the Front,* Vol 28, No. 3 (Autumn 2013).

Schindler, John. "Steamrollered in Galicia: The Austro-Hungarian Army and the Brusilov Offensive, 1916," *War in History,* No. 10 (2003), 27-59.

Serviss, Garrett P. "A Biplane That Will Carry Ten Passengers," *The Bee, Omaha,* (November 1913).

Wilson, Tim. "Eskadra Vozdushnikh Korablei (EVK) and the Brusilov Offensive: June-August 1916," *Over the Front* (2003).

IMPERIAL RUSSIAN ERA PERIODICALS

Automobil' i vozdukhoplavaniye, 1911-12.

Automobil'naya zhizn'i aviatsiya, 1913-14.

Aero, 1909-10.

Aero-i-automobil'naya zhizn', 1910-14.

Aeromobil', 1912.

Armiya i flot Niva, 1897-17.

Biblioteka vozdukhoplavaniya, 1909-10.

Byulleten' Moskovskogo obshchestva vozdukhoplavaniya, 1910.

K'sport!, 1911-17.

Letaniye, 1910.

Morskoi sbornik, 1908-16.

Sevastopol'skiy aviatsionnyy illustrirovannyy zhurnal, 1910-.

Vestnik letchikov i aviatsionnykh motoristov svobodnoi Rossii, 1917.

Vestnik vozdukhoplavaniya, 1910-13.

Vestnik vozdukhoplavaniya i sporta, 1913.

Vestnik vozdushnogo flota, 1918-38.

Voyenno-vozdushnyy flot, 1914.

Voennyi Letchik, zhurnal Sevastopol'skoi aviatsionnoi shkoly, 1917.

Voyennyy mir, 1912-14.

Voyennyy sbornik, 1908-17.

Vozdukhoplavaniye i sport, 1910.

Vozdukhoplavaniye, nauka i sport, 1910.

Vozdukhoplavatel', 1903-16.

Vozdushnyy put', 1911.

PERSONAL ARCHIVES

Eliseev, Sergey Pavlovich, *Strategiya Organizovannogo Stroitel'stva Otechestvennykh Voyenno-Vozdushnykh Sil* [The Strategy of The Organized Construction of The Domestic Air Force]. Moscow. 2015.

Gröschel, Dieter H. M. *Russische und Balkan-Kriegsschauplätze* [Russian and

Balkan Fronts]". n.l, n.d.

Karinskiy, Nikolai S. *«Istoriya aerovokzala i aviatsii v Rossii»* [The Story of Aerostation and Aviation in Russia]. Manuscript from New York Public Library.

Kulikov, Viktor, *Iz Istorii Vozdushnykh Voyn Za Otechestvo! Khronika boyevykh deystviy 15-go aviaotryada Russkoy armi na frontakh Pervoy mirovoy voyny* [From the History of Air Wars for the Fatherland! Chronicle of the Fighting of the 15th Squadron of the Russian Army on the Fronts of the First World War]. Avia-khobbi, 2007.

Nicolsky, S.N., *Posledniy bogatyr'* [The Last Knight]. Manuscript - circa 1920's

Nicolsky, M. N. *Organizatsiya boyevykh chastey vozdushnykh korabley «Il'ya Muromets»* [Organization of the Combat Units of the Aerial Ships 'Ilya Muromets']. Translated by Alex Nicolsky. Manuscript.

WEBSITES

Der Erste Weltkrieg; Die Schlacht bei Tannenberg. http://www.faz.net/aktuell/politik/der-erste-weltkrieg/die-schlacht-bei-tannenberg-13114302.html

Mobilmachung, Aufmarsch und Einsatz der deutschen Luftstreitkräfte im August 1914: Mit Anlagen, Taf. & Karten. http://flugplatzgeschichte-grossenhain.de.tl/1913-bis-1919.htm

Österreichische Nationalbibliothek Anno Historische Zeitungen und Zeitschriften. https://anno.onb.ac.at

The Prussian Machine. http://prussianmachine.com/

Die Schlacht bei Tannenberg. https://wikivividly.com/lang-de/wiki/Schlacht_bei_Tannenberg_(1914)

Tannenberg 1914. http://www.tannenberg1914.de/h_armeen/a_dt8.htm

INDEX

[English translation in brackets] Page references in italics refer to captions. When a term is used frequently, the page references given refer to the most germane definition or information and other entries are indicated as passim.

PERSONNEL INDEX

[US Army equivalent rank in brackets]

Helmut Jäger, Carl Bobrow and Terry Finnegan.

About the Authors

Terry Finnegan is a retired USAF Colonel and senior Defense Department civilian with 50 years of service. He served many missions, to include European Command, Pacific Command, Central Command during Desert Storm, NSA and Space Command. In his last years he provided joint training expertise to the US National Guard for the entire western United States. His military intelligence expertise has been internationally praised for support to operations in all facets of warfare, to include Information Operations. He has applied his depth of understanding military with two books, *Shooting the Front* (praised as the definitive work on aerial reconnaissance in World War I) and *A Delicate Affair on the Western Front* (cited as one of the most detailed and incisive works on World War I since the master piece *All Quiet on the Western Front*). Terry is retired, living in New Hampshire and racing triathlons.

Carl J. Bobrow is a former Verville Fellow at the Smithsonian Institution's National Air and Space Museum. His succeeding 20-year career at NASM brought him direct contact with many historic objects in the National Collection. He holds an interdisciplinary degree from New York State University in the History of Science and Technology. His long-term research expertise is in technological innovations and early flight, specifically the advent and development of Russian Aviation. He has written and lectured extensively on the subject, including the 1987 publication of *Igor Sikorsky, The Russian Years* with Dr. Von Hardesty.

Helmut Jäger is a retired German Airforce Lieutenant Colonel. He served in the 1960 - 1980 cold war period as fighter bomber and reconnaissance pilot, squadron commander and wing operations officer. Having finished the German General Staff course, he was in the Military Attaché service, followed by an assignment as lecturer on air war at the German Command and Staff College, Hamburg. After leaving the German Airforce, he worked for German industry firms and retired in 2002. Then he turned to study the World War I air service history, and authored two books (in German language) on German photographic aerial reconnaissance: *Erkundung mit der Kamera*, and *Luftbilder auf der Karte finden*.

Elaine Mintzer is a freelance proofreader and copy editor and has worked in various disciplines—publishing, psychology, and finance. She published research in the Journal of Clinical Psychology, while working on a masters degree. She subsequently transitioned into finance, where she worked for Sanford C. Bernstein & Co., as a senior fixed income trader for nearly 20 years. She was a proofreader at Dell Publishing early in her career, received an M.A. in Psychology at State University of New York at New Paltz, then went on to do doctoral work in psychology, at the University of Iowa.

Bridget Finnegan is an award-winning designer, illustrator, animator, and writer. She is the author of four illustrated books and one novel, *Odette's: A Quality Men's Club*. She has a BFA from Washington University and an MBA from Plymouth State University.
Learn more about her work at bridgetfinnegan.com.

Made in the USA
Las Vegas, NV
03 November 2026

57423143R00219